Lecture Notes on Data Engineering and Communications Technologies

286

Series Editors

Fatos Xhafa, *Technical University of Catalonia, Barcelona, Spain*
Rolando Herrero, *Northeastern University, Boston, MA, USA*

The aim of the book series is to present cutting edge engineering approaches to data technologies and communications. It will publish latest advances on the engineering task of building and deploying distributed, scalable and reliable data infrastructures and communication systems.

The series will have a prominent applied focus on data technologies and communications with aim to promote the bridging from fundamental research on data science and networking to data engineering and communications that lead to industry products, business knowledge and standardisation.

Indexed by SCOPUS, INSPEC, EI Compendex.

All books published in the series are submitted for consideration in Web of Science.

Aboul Ella Hassanien · Abdelkrim Haqiq ·
M.A. Jabbar · Essaid Sabir · K. C. Santosh ·
Luis Orozco Barbosa
Editors

The 4th International Conference on Artificial Intelligence and Smart Data Science (AISDS'2025)

Volume 1 AI and Computer Vision

Springer

Editors
Aboul Ella Hassanien
Faculty of Computers and Artificial Intelligence
Cairo University
Giza, Egypt

M.A. Jabbar
Vardhaman College of Engineering
Hyderabad, Andhra Pradesh, India

K. C. Santosh
Department of Computer Science
University of South Dakota
Vermillion, SD, USA

Abdelkrim Haqiq
Faculty of Sciences and Technqiues
Hassan 1st University
Settat, Morocco

Essaid Sabir
TÉLUQ, University of Québec
Montreal, QC, Canada

Luis Orozco Barbosa
Research Institute of Informatics
Castilla-La Mancha University
Albacete, Spain

ISSN 2367-4512 ISSN 2367-4520 (electronic)
Lecture Notes on Data Engineering and Communications Technologies
ISBN 978-3-032-18757-4 ISBN 978-3-032-18758-1 (eBook)
https://doi.org/10.1007/978-3-032-18758-1

This Springer imprint is published by the registered company Springer Nature Switzerland AG
The registered company address is: Gewerbestrasse 11, 6330 Cham, Switzerland

Preface

It is our great pleasure to welcome you to the 4th International Conference on Artificial Intelligence and Smart Data Science (AISDS 2025), held in the vibrant city of Marrakesh, Morocco, from 04 to 06 December, 2025. AISDS'2025 is organized by the Computer, Networks, Mobility and Modeling Laboratory (IR2M), Faculty of Sciences and Techniques, Hassan 1st University, Settat, Morocco, in collaboration with the Scientific Research School of Egypt (SRSEG), Cairo University, Egypt.

AISDS has established itself as a premier forum for researchers, practitioners, and industry experts to exchange ideas, present innovations, and explore emerging trends in Artificial Intelligence (AI) and Smart Data Science.

The rapid evolution of AI technologies and data-driven solutions continues to transform every aspect of modern life—from healthcare and education to transportation, security, and sustainable development. AISDS 2025 reflects this dynamic landscape by featuring cutting-edge research across diverse domains, including machine learning, deep learning, computer vision, natural language processing, optimization, IoT security, and big data analytics.

This year's conference received an impressive number of submissions from around the globe, demonstrating the growing interest and relevance of AI and data science in addressing real-world challenges. After a rigorous peer-review process, the selected papers represent high-quality contributions that advance theory, methodology, and applications. Topics span from predictive modeling and sentiment analysis to autonomous systems, healthcare AI, and smart resource management, ensuring a rich and multidisciplinary program.

We extend our sincere gratitude to the authors for their innovative work, the reviewers for their invaluable feedback, and the organizing committee for their dedication in making AISDS 2025 a success. We also thank our sponsors and partners for their support in fostering collaboration and knowledge exchange.

We invite you to explore the proceedings and engage with the ideas presented. We hope AISDS 2025 will inspire new research directions, strengthen academic and industrial partnerships, and contribute to shaping the future of intelligent systems and smart data solutions.

Welcome to AISDS 2025 in Marrakech—where innovation meets opportunity!

Abdelkrim Haqiq
Aboul Ella Hassanien
Conference Chair, AISDS 2025 Organizing Committees

Organization

Honorary Chairs

Aawatif Hayar	GREENTIC, ENSEM – Casablanca, Former Minister of Solidarity, Social Integration and Family and Former President of Hassan II University, Casablanca, Morocco
Abdellatif Moukrim	President of Hassan 1st University, Settat, Morocco
Abdelmajid Farchi	Dean of the Faculty of Sciences and Techniques, Settat, Morocco
Jamila El Alami	Director of the National Center for Scientific and Technical Research, Morocco
Mohamed Knidiri	President of "Association le Grand Atlas", President of the Private University of Marrakesh and Former Minister of National Education, Morocco
Said Rakrak	Dean of the Faculty of Sciences and Techniques - Guéliz, Marrakesh, Morocco

General Chairs

Abdelkrim Haqiq	GREENTIC, FST, Hassan 1st University, Settat, Morocco
Aboul Ella Hassanien	Faculty of Computer and AI, Cairo University, Egypt

Conference Co-chairs

KC Santosh	Department of Computer Science, University of South Dakota, USA
Luis Orozco Barbosa	Albacete Research Institute of Informatics, Castilla-La Mancha University, Abacete, Spain

Technical Program Chairs

Akhil Jabbar Meerja	Vardhaman College of Engineering, Hyderabad, Telangana, India
Ashraf Darwish	Faculty of Science, Helwan University, Cairo, Egypt
Essaid Sabir	University of Quebec (TÉLUQ), Montréal, Canada
Parthasarathy Subashini	Avinashilingam Institute for Home Science and Higher Education for Women - Deemed University, Coimbatore, India

International Advisory Board

Adel Mohamed Alimi	National, Engineering School of Sfax, Tunisia
Ahmad Taher Azar	College of Computer & Information Sciences, Prince Sultan University, Riyadh, Saudi Arabia
Dabia Ahmed Abouâinainen	Abdulrahman Bin Faisal University, Saudi Arabia
Peter J. Tonelato	School of Medicine, University of Missouri, Columbia MO, USA
Rawya Rizk	Port Said University, Egypt
Sundaramurthy Gayathri Devi	Avinashilingam Institute for Home Science and Higher Education for Women, Deemed University, Coimbatore, India
Vaclav Snasel	VSB-Technical University of Ostrava, Czech Republic
Youssef F. Rashed	Cairo University, Egypt

Track Chairs

Track 1: Foundations of Artificial Intelligence and Machine Learning

Abdellah Adib	FST-Mohammedia, Hassan II University, Casablanca, Morocco
Said El Kafhali	FST, Hassan 1st University, Settat, Morocco
Sam Goundar	RMIT University, School of Computing, Hanoi, Vietnam

Track 2: Smart Data Engineering and Analytics

Adam Slowik	Koszalin University of Technology, Koszalin, Poland
Mohamed Nemiche	Polydisciplinary Faculty of Taza, Sidi Mohamed Ben Abdellah University, Morocco
Said Raghay	Faculty of Sciences and Techniques - Guéliz, Marrakesh, Morocco

Track 3: Natural Language Processing and Human-Centered AI

Brahim Ouhbi	National School of Arts and Crafts (ENSAM), Meknès, Morocco
Krishnaveni Marimuthu	Avinashilingam Institute for Home Science and Higher Education for Women, Deemed University, Coimbatore, India
Mohamed Hanini	FST, Hassan 1st University, Settat, Morocco

Track 4: AI Applications in Industry, Society and Sustainability

Aurelio Bermúdez Marín	Albacete Research Institute of Informatics, Castilla-La Mancha University, Abacete, Spain
Driss Bouzidi	ENSIAS, Mohammed V University, Rabat, Morocco
Nizar Rokbani	High Institute of Applied Sciences and Technology, University of Sousse, Tunisia

Track 5: Cybersecurity, Privacy and Trust in AI

Ashraf Matrawy	School of Information Technology, Carleton University, Ottawa, Canada
Ghizlane Orhanou	Faculty of Sciences, Rabat, Morocco
Yassine Maleh	National School of Applied Sciences, Khouribga, Morocco

Track 6: Computer Vision and Applications

Driss El Ouadghiri	Faculty of Sciences, Meknès, Morocco
Nabil Laachfoubi	FST, Hassan 1st University, Settat, Morocco
Rafael Casado González	Albacete Research Institute of Informatics, Castilla-La Mancha University, Abacete, Spain

Publicity Chairs

Abdellah Ouammou	FST, Hassan 1st University, Settat, Morocco
Adnane El Hanjri	EMSI, Tangier, Morocco
Azza Ahmed Abdo Ali	Abdulrahman Bin Faisal University, Saudi Arabia
Chakib Ben Njima	National Engineering School of Sousse, University of Sousse, Tunisia
Houssam Halmaoui	ISMAC, Rabat, Morocco
Kamal Bouhassoune	FST, Hassan 1st University, Settat, Morocco
Lavika Goel	Malaviya, National Institute of Technology, Jaipur, India
Mehrez Abdellaoui	National Engineering School of Sousse, University of Sousse, Tunisia
Mohamed Hossam Ahmed	School of Electrical Engineering and Computer Science, Ottawa, Canada

Local Arrangement Committee Chairs

Abdelghani Ben Tahar	FST, Hassan 1st University, Settat, Morocco
Abderrahim Marzouk	FST, Hassan 1st University, Settat, Morocco
Iman Elmir	I2S, Hassan 1st University, Settat, Morocco
Mohamed Hanini	FST, Hassan 1st University, Settat, Morocco
Said El Kafhali	FST, Hassan 1st University, Settat, Morocco

Organizing Committee Members

Abdellah Jamali	National School of Applied Sciences, Berrechid, Morocco
Abdellah Zaaloul	Faculty of Economics and Management, Settat, Morocco
Adil El Idriss	FST, Hassan 1st University, Settat, Morocco
Adnane Founoun	FST, Hassan 1st University, Settat, Morocco

Ahmed Boujnoui	FST, Moulay Ismaïl University, Errachidia, Morocco
Amine Benmakhlouf	FST, Moulay Ismaïl University, Errachidia, Morocco
Asma Boulkaid	FST, Hassan 1st University, Settat, Morocco
Bouchaib Nassereddine	Faculty of Sciences and Techniques, Settat, Morocco
Haitam Ettazi	FST, Hassan 1st University, Settat, Morocco
Hamid Taramit	Center for Computational Simulation, Universidad Politécnica de Madrid, Spain
Hibateallah Mohtadi	FST, Hassan 1st University, Settat, Morocco
Hind Mikram	Faculty of Sciences and Techniques, Settat, Morocco
Loubna Hamami	FST, Hassan 1st University, Settat, Morocco
Soukaina Mihi	FST, Hassan 1st University, Settat, Morocco
Wafae Azoulay	FST, Hassan 1st University, Settat, Morocco
Youssef Hilali	FST, Hassan 1st University, Settat, Morocco
Zakaria Soufiane Hafdi	FST, Hassan 1st University, Settat, Morocco

Contents

AI for Computer Vision and Image-Based Applications

SwinEffiNet : A Hybrid CNN-Transformer Based Approach for Gastrointestinal Anatomical Landmarks Localization in WCE Images

Amine Aaz el aarab[1(✉)], Zakaria kerkaou[1], Yassine Oukdach[1], Lahcen Koutti[1], Mohamed El Ansari[1,2], and Mouna Salihoun[3]

[1] LabSIV, Faculty of sciences, Department of computer science, Ibn Zohr university, Agadir, Morocco
{aazelaarab.amine,kerkaou.zakaria,ys.oukdach,melansari}@gmail.com, l.koutti@uiz.ac.ma

[2] Informatics and applications laboratory, Faculty of sciences, Department of computer science, University Moulay Ismail, Meknes, Morocco

[3] Faculty of medicine and pharmacy, University mohammed V, Rabat, Morocco

Abstract. Both the diagnosis and treatment of gastrointestinal (GI) disorders require, as a first step, the accurate localization of anatomical landmarks in GI wireless capsule endoscopy (WCE) images. WCE offers a non-invasive means of visualizing the GI tract, yet the vast number of images generated makes manual interpretation laborious and time-consuming. Deep learning, particularly convolutional neural networks (CNNs), has shown considerable promise for automating this task; however, when it comes to capturing long-range dependencies, CNNs find it somewhat difficult. To address this limitation, we propose SwinEffiNet, a hybrid model integrating both CNN and Transformer backbones capable of capturing local features as well as global context modeling. We compared SwinEffiNet with state-of-the-art CNNs (ResNet101, DenseNet121, EfficientNet-B2) and Transformer-based architectures (Swin-Tiny, PvT-Tiny, DeiT-Tiny). Across two datasets, SwinEffiNet achieved an accuracy of 96.25% on the WCEAL dataset and 96.06% on the GastroVisionAL dataset. These results showcase the effectiveness of hybrid CNNTransformer models for reliable GI landmark identification in medical imaging.

Keywords: Gastrointestinal · Anatomical landmarks · localization · Deep learning · CNNs · Transformers · Swin · EfficientNet

1 Introduction

Accurate identification of anatomical landmarks in gastrointestinal (GI) wireless capsule endoscopy (WCE) images is critical for diagnosing and treating conditions such as polyps, colorectal cancer, and inflammatory bowel disease [1,2].

A. E. Hassanien et al. (Eds.): AISDS 2025, LNDECT 286, pp. 3–13, 2027.
https://doi.org/10.1007/978-3-032-18758-1_1

WCE provides a non-invasive view of the entire GI tract; however, the sheer volume of images produced in each procedure makes manual interpretation challenging. Automated deep learning approaches have therefore gained attention, offering efficient and reliable solutions for landmark localization. CNNs have been widely used in GI image analysis, achieving strong results in landmark classification and localization [3]. Architectures such as ResNet, VGG, MobileNet, and DenseNet have been explored for this task [4,5]. Despite their success, CNNs often struggle with capturing long-range dependencies in complex medical imaging scenarios. Transformer-based models, originally introduced for natural language processing, have also shown impressive performance in computer vision by effectively modeling global relationships [6]. Vision Transformer (ViT) and its variants address these limitations [7], with the Swin Transformer providing an efficient balance between computational cost and accuracy through its shifted window mechanism [8]. In parallel, EfficientNet has emerged as a CNN family optimized for accuracy and efficiency by systematically scaling depth, width, and resolution [9]. Leveraging these strengths, this study introduces a hybrid framework combining EfficientNet and Swin Transformer to capture both localized features and broader contextual information, thereby enhancing GI landmark identification across three categories: Stomach, Small Bowel, and Large Bowel (Fig. 1). The rest of the paper proceeds as follows: Sect. 2 surveys related work; Sect. 3 describes the proposed method; Sect. 4 reports experiments and analysis; Sect. 5 offers conclusions and future work.

2 Related Works

Automated identification of gastrointestinal (GI) anatomical landmarks has garnered increasing interest given its potential to improve diagnostic precision and streamline endoscopic procedures. Numerous deep learning approaches have been introduced for this task. Taghiakbari et al. [10] implemented a deep convolutional neural network (DCNN) to identify the ileocecal valve (ICV) and appendiceal orifice (AO) in colonoscopy images. Che et al. [11] developed a ResNet101-based framework for classifying three anatomical landmarks in colonoscopy videos. Li et al. [12] combined density clustering with deep learning to autonomously identify anatomical regions in colonoscopy videos, highlighting the effectiveness of hybrid approaches. More recently, Oukdach et al. [13] proposed ConV-ViT, a CNNVision Transformer hybrid, to detect GI abnormalities in WCE images, demonstrating the benefits of combining local and global feature representations. These studies underscore the effectiveness of CNNs and hybrid architectures for automated GI landmark detection. Building on this foundation, we propose a CNNTransformer framework that integrates EfficientNet-B2 for efficient local feature extraction with the Swin Transformer for robust global context modeling. This design addresses the limitations of conventional CNNs and enhances landmark localization in complex endoscopic scenarios.

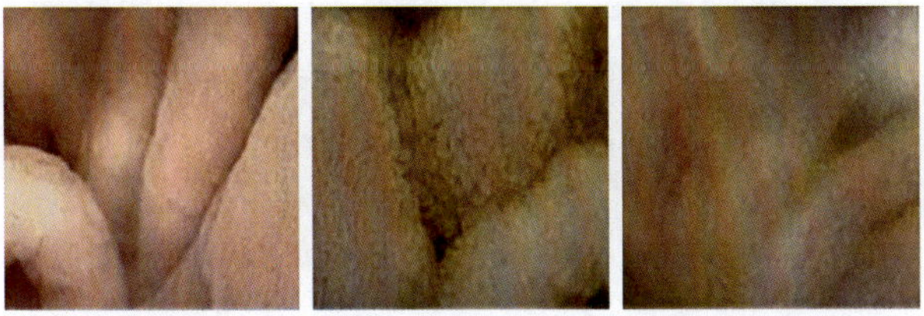

(a) Stomach (b) Small bowel (c) Large bowel

Fig. 1. Examples of the gastrointestinal landmark classes.

3 Methodology

The proposed framework for gastrointestinal (GI) anatomical landmark classification is shown in Fig. 2. The pipeline begins with preprocessing and augmentation to improve image quality and model generalization. Images are then processed by a dual-stream feature extractor. Swin Transformer Tiny and EfficientNet-B which capture complementary global and local representations. The extracted features are combined and fed into a tailored classifier for multi-class prediction

3.1 Pre-Processing & Data Augmentation

We resized all images to 224×224 and normalized pixel values to [0,1]. Stratified splitting was used to balance classes in the test, validation, and training partitions. To address class imbalance and limited sample size in the GastroVisionAL dataset, we applied random augmentations: flipping, rotations, brightness adjustments, and Gaussian noise. The WCEAL dataset did not require augmentation due to its larger, balanced sample distribution.

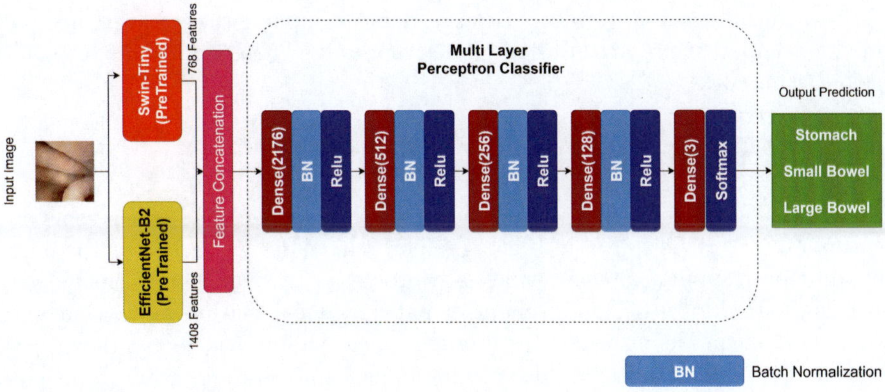

Fig. 2. Pipeline of the suggested methodology for gastrointestinal (GI) anatomical landmark localization.

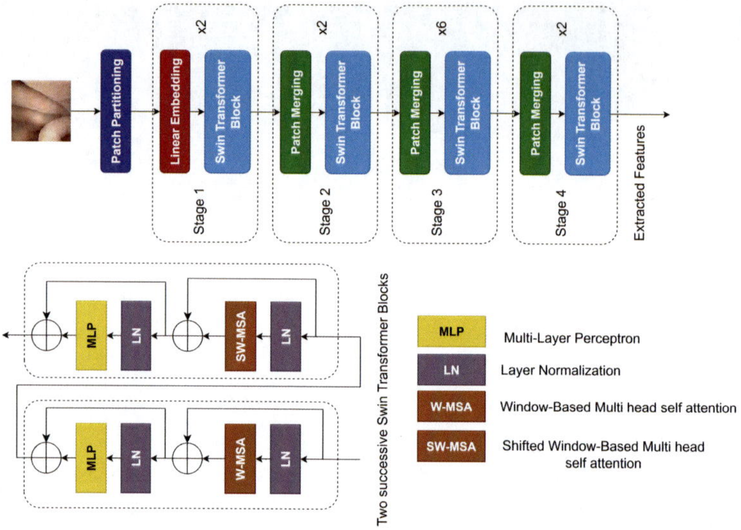

Fig. 3. The architecture of the Swin Transformer.

3.2 Backbone Selection

In this study, we deliberately selected lightweight backbones (EfficientNet-B2 and Swin-Tiny) rather than larger architectures such as Swin-B or ViT-B/16. This decision was motivated by two key factors: (1) computational constraints, as larger transformer-based models require significantly more GPU memory and training time, which exceeds the available hardware resources (NVIDIA L4 GPU with 22 GB VRAM in our case); and (2) efficiency focus, since our aim is to develop a model that not only accurate but also practical for clinical deployment, where computational resources may be limited. Lightweight architectures allow faster inference and reduced memory consumption, which are essential for real-world medical imaging systems. Despite being compact, these backbones still provide competitive representation power, especially when combined in a hybrid fusion framework.

3.3 Swin Tiny Transformer

The Swin Transformer (Swin-Tiny), pre-trained on ImageNet-1K, serves as one feature extractor. As shown in Fig. 3, its hierarchical design partitions the input into non-overlapping patches, which are processed through transformer layers with shifted windows. This mechanism balances local feature extraction with global context modeling while maintaining linear computational complexity [8]. The final pooled output provides a 768-dimensional representation, later concatenated with EfficientNet-B2 features.

3.4 EfficientNet-B2

The second backbone is EfficientNet-B2, initialized with ImageNet-1K weights. As depicted in Fig. 4, using compound scaling of depth, width, and resolution [9], it achieves an efficient trade-off between performance and computation. We retain only the convolutional feature extractor, removing the classification head. The final feature maps are reduced via adaptive average pooling to a 1408-dimensional vector, capturing fine-grained local details. These are concatenated with the Swin Transformer outputs to form a comprehensive representation.

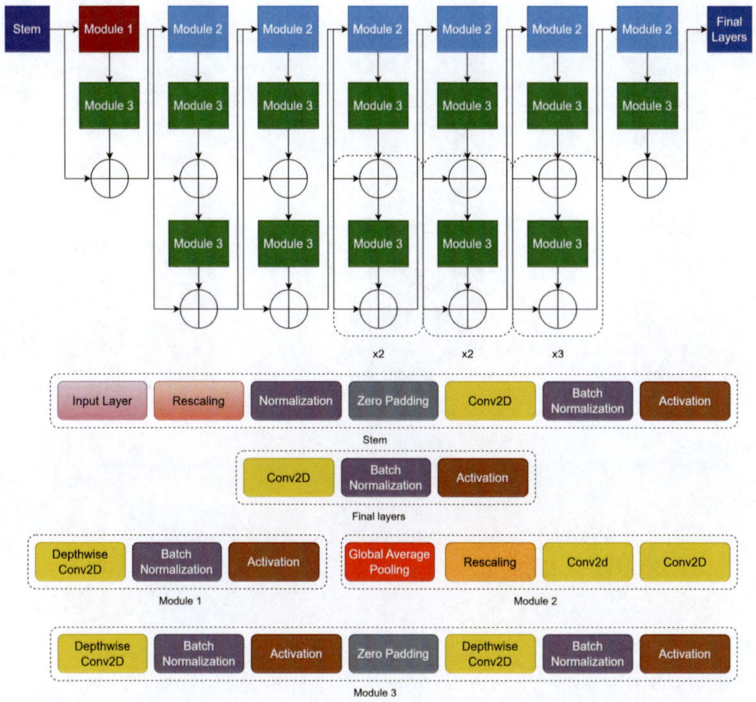

Fig. 4. The architecture of the EfficientNet-B2.

3.5 Feature Fusion and Custom Classifier

The fused 2176-dimensional vector (768 from Swin, 1408 from EfficientNet-B2) forms the input to the classifier as shown in Fig. 2. The classifier comprises fully connected layers with 512, 256, and 128 units, each followed by batch normalization and ReLU, culminating in a linear layer that outputs class logits. This design enhances nonlinearity, stabilizes training, and reduces overfitting, while the layered depth allows extraction of class-specific patterns from fused features.

4 Results and Discussion

We evaluated the suggested hybrid model against widely used CNNs (ResNet101, DenseNet121, EfficientNet-B2) and state-of-the-art Vision Transformers (Swin-Tiny, DeiT-Tiny, PvT-Tiny) for GI anatomical landmark localization. To assess backbone contributions, we also examined their standalone performance. Experiments were conducted on two datasets: GastroVisionAL and WCEAL.

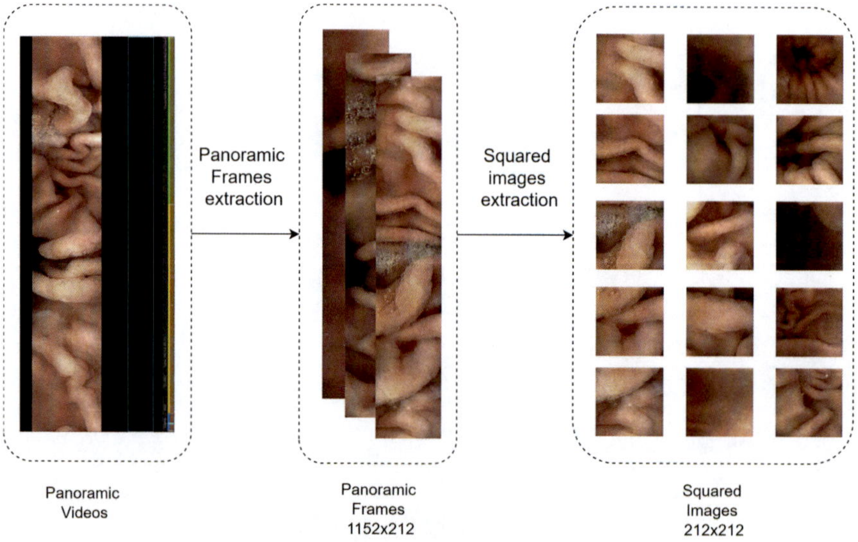

Fig. 5. The WCEAL dataset curation process.

4.1 Datasets

4.1.1 WCEAL The Wireless Capsule Endoscopy Anatomical Landmarks dataset was curated from clinical videos at CHU IBN SINA Hospital, Rabat as shown in Fig. 5. Recordings from 11 patients were annotated by gastroenterologists. From each video, 200 panoramic frames (1152×212) were extracted, then cropped into five 212×212 patches and zero-padded to 224×224. This produced 11,000 balanced images across stomach, small bowel, and large bowel classes.

4.1.2 GastroVisionAL Introduced by Jha et al. [14], GastroVision contains 8,000 images across 27 categories from Bærum Hospital (Bærum, Norway) and Karolinska University Hospital (Stockholm, Sweden). For this study, we selected three anatomical classes: stomach (969), large bowel (1,467), and small bowel (846). To address imbalance, augmentations were applied, yielding 3,000 images per class.

4.2 Experimental Setup

Training was conducted in Google Colab with an NVIDIA L4 (22 GB VRAM) using PyTorch v2.5.1. We compared CNNs (ResNet101, DenseNet121, EfficientNet-B2) and Transformers (Swin-T, DeiT-T, PvT-T). Training used AdamW optimizer with 0.0001 learning rate., cross-entropy loss, batch size 32, and 10 epochs. Early stopping with a patience = 5 and ReduceLROnPlateau with a factor = 0.1, and patience = 3 were employed to improve convergence.

4.3 Evaluation Metrics

To evaluate the approach's performance we report accuracy, precision, recall, and F1-score, defined as:

- Accuracy: the fraction of correctly classified samples among all samples.

$$\frac{TP + TN}{TP + TN + FP + FN} \tag{1}$$

- Precision: the fraction of correctly predicted positives among all predicted positives.

$$\frac{TP}{TP + FP} \tag{2}$$

- Recall: the fraction of actual positives that the model correctly identifies.

$$\frac{TP}{TP + FN} \tag{3}$$

- F1-score: the harmonic mean of precision and recall, balancing both measures.

$$\frac{2TP}{2TP + FP + FN} \tag{4}$$

where TP = true positives, TN = true negatives, FP = false positives, and FN = false negatives.

4.4 Experimental Results

On the GastroVisionAL dataset (Table 2), the hybrid model achieved the best results with 96.25% accuracy, precision of 0.9614, recall of 0.9537, and F1-score of 0.9573. Among baselines, PvT-Tiny and EfficientNet-B2 performed strongly (95.94% accuracy each), while Swin-Tiny (95.31%) and DeiT-Tiny (94.38%) also showed competitive transformer performance. On the WCEAL dataset (Table 3), the proposed model again outperformed all baselines, reaching 96.06% accuracy and balanced precision, recall, and F1-scores (all 0.9606). PvT-Tiny followed with 95.70% accuracy and an F1-score of 0.9571, with EfficientNet-B2 close behind at 95.36%. These results confirm that integrating convolutional and transformer-based representations yields superior generalization across both public and clinical datasets. The hybrid architecture consistently outperformed individual CNNs and Transformers, demonstrating the advantage of combining local texture-sensitive features with global context modeling for robust anatomical landmark classification in endoscopic imagery.

4.5 Model Complexity

In terms of model efficiency, Table 1 reports the parameter counts and floating-point operations (FLOPs) for the proposed SwinEffiNet compared to its backbone components. Although SwinEffiNet combines two feature extractors (EfficientNet-B2 and Swin-Tiny), the overall computational cost remains moderate, totaling 38.39M parameters and 5.59 GFLOPs. Since both pretrained backbones were frozen during training to reduce overfitting and memory usage, only the classifier head was trainable, resulting in just 1.28M trainable parameters (3.3‘% of the total). This demonstrates that the proposed architecture maintains high computational efficiency while preserving strong feature representation capacity.

Table 1. Model complexity comparison in terms of total parameters, FLOPs, and trainable parameters.

Model	Total Params (M)	FLOPs (G)	Trainable Params (M)
EfficientNet-B2	9.11	1.09	0 (frozen)
Swin-Tiny (Swin-T)	28.00	4.50	0 (frozen)
Classifier Head	1.28	<0.01	1.28
SwinEffiNet (Hybrid)	**38.39**	**5.59**	**1.28**

Table 2. Performance comparison on the GastroVisionAL dataset

Model	Accuracy	Precision	Recall	F1-score
CNNs				
ResNet101 [15]	95.62%	0.9517	0.9494	0.9503
DenseNet121 [16]	95.62%	0.9498	0.9532	0.9514
EfficientNetB2 [9]	95.94%	0.9571	0.9496	0.9527
Vision Transformers				
Swin Tiny [8]	95.31%	0.9499	0.9439	0.9461
PvT Tiny [17]	95.94%	0.9582	0.9516	0.9545
DeiT Tiny [18]	94.38%	0.9373	0.9359	0.9365
Hybrid				
SwinEffiNet	**96.25%**	**0.9614**	**0.9537**	**0.9573**

Table 3. Performance comparison on the WCEAL dataset

Model	Accuracy	Precision	Recall	F1-score
CNNs				
ResNet101 [15]	94.27%	0.9429	0.9427	0.9425
DenseNet121 [16]	94.58%	0.9462	0.9458	0.9459
EfficientNetB2 [9]	95.36%	0.9537	0.9536	0.9536
Vision Transformers				
Swin Tiny [8]	94.97%	0.9507	0.9497	0.9497
PvT Tiny [17]	95.70%	0.9576	0.9570	0.9571
DeiT Tiny [18]	95.27%	0.9527	0.9527	0.9527
Hybrid				
SwinEffiNet	**96.06%**	**0.9606**	**0.9606**	**0.9606**

5 Conclusion

This study introduced a hybrid architecture combining EfficientNet-B2 and the Swin Transformer for classifying gastrointestinal anatomical landmarks in endoscopic images. The model exploits CNNs for fine-grained local features and Transformers for global context, with a custom classifier enhancing prediction accuracy. Evaluations on both GastroVisionAL and WCEAL datasets demonstrated superior performance relative to state-of-the-art CNN and Transformer baselines , yielding the highest accuracy, precision, recall, and F1-score. These results highlight the value of integrating complementary feature representations for complex tasks like anatomical landmark classification in wireless capsule endoscopy. Future efforts will generalize this approach to multi-task learning for joint anatomical and pathological classification and explore lightweight variants for real-time clinical use.

Acknowledgements. This research was funded by the Ministry of National Education through Vocational Training, and by the Higher Education and Scientific Research division within the Ministry of Industry, Trade, and the Green and Digital Economy; additional support came from the Digital Development Agency (ADD) and the National Center for Scientific and Technical Research (CNRST) under Project ALKHAWARIZMI/2020/20. The authors thank CHU Ibn Sina Hospital, Rabat, Morocco, for providing access to the medical data used in this study.

References

1. Charfi, S., Ansari, M.E.: Computer-aided diagnosis system for colon abnormalities detection in wireless capsule endoscopy images. Multimedia Tools Appl. **77**(3), 4047–4064 (2018). https://doi.org/10.1007/s11042-017-4753-4

2. Oukdach, Y., Kerkaou, Z., Ansari, M.E., Koutti, L., Ouafdi, A.F.E.: Gastrointestinal diseases classification based on deep learning and transfer learning mechanism. In: Proceedings of the 9th International Conference Wireless Networks and Mobile Communications (WINCOM), pp. 1–6. IEEE, Rabat, Morocco (2022). https://doi.org/10.1109/WINCOM55699.2022.9965607
3. Zhang, R., et al.: Automatic detection and localization of gastrointestinal diseases using deep learning. IEEE J. Biomed. Health Inform. **25**(2), 370–378 (2021). https://doi.org/10.1109/JBHI.2020.2983589
4. Lafraxo, S., Ansari, M.E., Koutti, L.: Computer-aided system for bleeding detection in wce images based on cnn-gru network. Multimedia Tools Appl. 1–26 (2023). https://doi.org/10.1007/s11042-023-14910-3
5. Souaidi, M., Lafraxo, S., Kerkaou, Z., Ansari, M.E., Koutti, L.: A multiscale polyp detection approach for gi tract images based on improved densenet and single-shot multibox detector. Diagnostics **13**(4), 733 (2023). https://doi.org/10.3390/diagnostics13040733
6. Dosovitskiy, A., et al.: An image is worth 16x16 words: transformers for image recognition at scale. arXiv preprint arXiv:2010.11929 (2021). https://doi.org/10.48550/arXiv.2010.11929
7. Garbaz, A., Lafraxo, S., Charfi, S., Ansari, M.E., Koutti, L.: Bleeding classification in wireless capsule endoscopy images based on inception-resnet-v2 and CNNs. In: Proceedings of the IEEE Conference on Computational Intelligence in Bioinformatics and Computational Biology (CIBCB), pp. 1–6. IEEE, Ottawa, ON, Canada (2022). https://doi.org/10.1109/CIBCB54656.2022.9895041
8. Liu, Z., et al.: Swin transformer: Hierarchical vision transformer using shifted windows. In: Proceedings of the IEEE/CVF International Conference on Computer Vision (ICCV), pp. 10012–10022. IEEE, Montreal, Canada (2021). https://doi.org/10.1109/ICCV48922.2021.00985
9. Tan, M., Le, Q.V.: Efficientnet: rethinking model scaling for convolutional neural networks. In: Proceedings of the 36th International Conference on Machine Learning (ICML), pp. 6105–6114. PMLR, Long Beach, California (2019). https://doi.org/10.48550/arXiv.1905.11946
10. Taghiakbari, M., et al.: Deep learning for automated identification of ileocecal valve and appendiceal orifice in colonoscopy. IEEE Trans. Med. Imaging (2023). https://doi.org/10.1109/TMI.2023.3242349
11. Che, K., et al.: Deep learning-based biological anatomical landmark detection in colonoscopy videos. arXiv preprint arXiv:2108.02948 (2021). https://doi.org/10.48550/arXiv.2108.02948
12. Li, X., Chen, Y., Xu, M., Wang, T.: Deep learning and density clustering for automatic anatomical section recognition in colonoscopy videos. Sci. Rep. **13**, 51056 (2023). https://doi.org/10.1038/s41598-023-51056-7
13. Oukdach, Y., Kerkaou, Z., Ansari, M.E., Koutti, L., Ouafdi, A.F.: Conv-vit: Feature fusion based detection of gastrointestinal abnormalities using CNN and vit in WCE images. In: Proceedings of the 10th International Conference on Wireless Networks and Mobile Communications (WINCOM), pp. 1–6. IEEE, Istanbul, Turkey (2023). https://doi.org/10.1109/WINCOM60194.2023.10401464
14. Jha, D., et al.: Gastrovision: a multi-class endoscopy image dataset for computer aided gastrointestinal disease detection. In: Machine Learning for Multimodal Healthcare Data, pp. 125–140. Springer, Honolulu, Hawaii (2023). https://doi.org/10.1007/978-3-031-33802-610

15. He, K., Zhang, X., Ren, S., Sun, J.: Deep residual learning for image recognition. In: Proceedings of the IEEE Conference on Computer Vision and Pattern Recognition (CVPR), pp. 770–778 (2016). https://doi.org/10.1109/CVPR.2016.90
16. Huang, G., Liu, Z., Maaten, L., Weinberger, K.Q.: Densely connected convolutional networks. In: Proceedings of the IEEE Conference on Computer Vision and Pattern Recognition (CVPR), pp. 4700–4708 (2017). https://doi.org/10.1109/CVPR.2017.243
17. Wang, W., et al.: Pyramid vision transformer: a versatile backbone for dense prediction without convolutions. In: Proceedings of the IEEE/CVF International Conference on Computer Vision (ICCV), pp. 568–578 (2021). https://doi.org/10.1109/ICCV48922.2021.00061
18. Touvron, H., Cord, M., Douze, M., Massa, F., Sablayrolles, A., Jégou, H.: Training data-efficient image transformers and distillation through attention. In: Proceedings of the International Conference on Machine Learning (ICML), pp. 10347–10357 (2021). https://doi.org/139:10347-10357

Tabular-to-Image Conversion and CNN Feature Extraction for Predicting Academic Success in Higher Education

Zakaria Soufiane Hafdi and Said El Kafhali(✉)

Faculty of Sciences and Techniques, Computer, Networks, Modeling, and Mobility Laboratory (IR2M), Hassan First University of Settat, B.P. 577, 26000 Settat, Morocco
{z.hafdi,said.elkafhali}@uhp.ac.ma

Abstract. Predicting student academic outcomes, particularly identifying at-risk learners, is a major challenge in educational data mining. This paper proposes a hybrid framework that combines Convolutional Neural Networks (CNN) for automated feature extraction with traditional machine learning models to predict student status (dropout, graduate, enrolled). By converting tabular student data into image-like representations, our approach enables CNN to capture complex patterns for improved classification. Tested on a comprehensive real-world dataset, the best performance was achieved by the CNN-ANN hybrid model, with 82% accuracy and an F1-score of 0.75. The results highlight the promise of this method for early student intervention and support within Learning Management Systems.

Keywords: Dropout Prediction · Learning Management Systems · Ensemble Learning · Educational Data Mining · Early Warning System · Students' Success Prediction

1 Introduction

The rise of online education has made Learning Management Systems (LMS) central to academic delivery and resource management. These platforms collect rich student interaction data, offering opportunities to understand learning behaviors and identify at-risk students early [1]. Early detection enables targeted interventions, helping reduce dropout rates and improve educational outcomes [2].

Machine learning (ML) and deep learning (DL) have become powerful tools in educational data mining (EDM), revealing trends in complex student data. While many ML methods, such as classical algorithms and neural networks, have shown promise for dropout prediction, each faces challenges with interpretability, scalability, or modeling non-linear relationships [3].

Classical ML models are transparent but may miss intricate patterns, whereas DL can uncover complex data structures but requires large datasets and greater

A. E. Hassanien et al. (Eds.): AISDS 2025, LNDECT 286, pp. 14–22, 2027.
https://doi.org/10.1007/978-3-032-18758-1_2

computation. Recent hybrid approaches attempt to blend these strengths, but systematic integration of DL-based feature extraction with traditional ML remains underexplored [4].

Our study addresses the challenge of improving prediction accuracy and F1 score in academic outcomes by leveraging both DL and ML. We propose a framework where Convolutional Neural Networks (CNN) automatically extract features from student activity data, which are then used by classical ML models for final prediction.

This work is motivated by the need for accurate, scalable, and interpretable predictive models in LMS environments. By integrating CNN feature learning with lightweight ML classifiers, we aim to balance predictive power with interpretability and practical deployment in educational settings.

The primary contributions of this study are outlined as follows:

- We introduce a hybrid framework that employs a CNN for automated feature extraction from tabular student data, with the resulting features serving as input to established ML classifiers for the prediction of academic outcomes.
- We validate our approach using the comprehensive and publicly available dataset introduced by Realinho et al. [5], enabling direct comparison and reproducibility of results.

The remainder of this paper is organized as follows. Section 2 reviews relevant literature and related work. Section 3 details the methodology and proposed framework. Section 4 presents and analyzes the experimental results. Section 5 concludes the paper and discusses directions for future research.

2 Related Work

The application of artificial intelligence (AI) within Learning Management Systems (LMS) to predict student outcomes and identify students at risk has become an area of significant research interest in recent years. Numerous studies have explored the use of both classical and Deep Learning (DL) techniques to improve dropout detection, academic performance prediction, and personalized educational interventions.

Kabathova et al. [6] demonstrated that standard Machine Learning (ML) algorithms such as Naive Bayes, Decision Tree, and Random Forest are capable of predicting student dropout, even when working with compact feature sets and future, unseen academic data. Their findings underline the critical role of data preprocessing and the thoughtful selection of evaluation metrics in educational data mining tasks.

In a similar context, Arévalo et al. [7] investigated a variety of ML methods for online programming courses, including Neural Networks and Logistic Regression. Their results revealed that a combination of Logistic Regression, oversampling, and grid search optimization yielded a strong balance between interpretability and predictive performance, achieving an AUC-ROC of 0.9354.

Riestra-González et al. [2] proposed a course-independent LMS log analysis model using models like Neural Networks and Support Vector Machine, capable of forecasting student success at different stages of a course. Notably, they found that Neural Networks achieved the highest accuracy after half the course duration, showing that early and accurate prediction is possible, even with partial course progress data.

Kaensar et al. [8] assessed over 54,000 Moodle log events recorded during the COVID-19 period, evaluating how the effectiveness of various classifiers changed over different phases of a course. Their analysis indicated that Support Vector Machine provided superior results in the early stages (at 25%), while Decision Trees outperformed others nearer to course completion.

In flipped classroom scenarios, Hung et al. [9] leveraged Random Forests and hierarchical clustering to study student learning behaviors. Their approach was particularly effective in flagging high-risk students for prompt intervention.

Ciolacu et al. [10] created an AI-driven system for early detection of student performance using smart sensors and learning analytics, focusing on adaptive self-assessment and providing real-time feedback within Education 4.0 settings.

Gnoh et al. [11] examined the use of AI forecasting tools, particularly Long-Short-Term-Memory (LSTM) networks and dashboards, for higher education institutions. Their work concluded that LSTM was most effective for forecasting student enrollment and supporting institutional decision-making.

Villegas et al. [12] developed an LMS-integrated platform enhanced by AI capabilities to advance online education systems. Their research highlighted the critical role of comprehensive analytics, especially in post-pandemic educational models.

Addressing the unique challenges of small-scale online classes, Azimi et al. [13] demonstrated that ML-based mid-course interventions, even with limited data, could meaningfully enhance student outcomes by suggesting timely, tailored strategies.

Lastly, Ersozlu et al. [14] offered a comprehensive review of ML methods used in educational research, emphasizing the prevalence of supervised learning approaches and suggesting that semi-supervised techniques hold promise for discovering deeper learning patterns.

3 Methodology and Proposed Framework

This study proposes a novel methodology to improve academic success prediction in the Learning Management System (LMS) by using Deep Learning (DL). We employ Convolutional Neural Networks (CNN) to automatically extract meaningful features from educational data, after transforming tabular student records into image representations for better feature learning. These new features are then used by traditional ML classifiers, creating a hybrid framework for predicting student outcomes.

3.1 Background

In recent years, Machine Learning (ML) has played a growing role in educational data mining, with models such as Decision Trees (DT), Random Forests (RF), Support Vector Machines (SVM), and Logistic Regression (LR) widely used to predict student outcomes like dropout risk and academic success [15]. DT offers interpretable rules but can overfit, RF provides robust ensemble predictions, SVM handles high-dimensional data, and LR is valued for its simplicity [16]. Artificial Neural Networks (ANN) are also effective for capturing non-linear patterns in educational data.

Our approach combines these traditional ML models with features automatically extracted by a CNN, which learns deep representations from images transformed from tabular student records. These features are then used as inputs to DT, RF, SVM, LR, and ANN models, creating a hybrid framework to enhance prediction accuracy and reliability.

We evaluate our framework using accuracy and F1 score. Accuracy measures overall correctness, but may be less informative with imbalanced data, while the F1 score balances precision and recall, making it suitable for underrepresented classes such as dropout or graduation. Together, these metrics ensure a robust assessment for early identification of at-risk students [6].

3.2 Proposed Framework

The proposed framework combines Deep Learning (DL) and traditional Machine Learning (ML) to improve academic success prediction in higher education. We leverage the automatic feature extraction of Convolutional Neural Network (CNN) along with the interpretability and effectiveness of classic ML models.

Student data, initially in tabular form with demographic, socioeconomic, macroeconomic, and academic attributes, is normalized and reshaped into two-dimensional arrays. These arrays are then converted into grayscale images, allowing the CNN to learn complex patterns through its convolutional layers.

The CNN processes these images and generates deep feature embeddings, which are extracted from the penultimate layer. These high-level features serve as inputs for various classical ML classifiers, including DT, RF, SVM, LR, and ANN. This hybrid approach captures both non-linear interactions and model interpretability.

We evaluate the framework using cross-validation, reporting accuracy and F1 score to account for class imbalance. This ensures reliable assessment of the model's effectiveness in early detection of at-risk students and supports targeted academic interventions.

The overall architecture and workflow of the proposed framework are illustrated in Fig. 1.

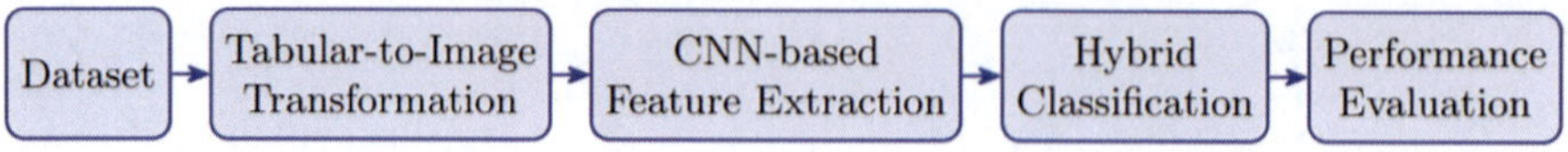

Fig. 1. Overview of the proposed hybrid framework for student success prediction.

3.3 Dataset Description

The experiments utilize the public dataset described by Realinho et al. [5], which aggregates administrative data from the Polytechnic Institute of Portalegre, Portugal. The dataset contains **4424 records** and **35 attributes** per student, spanning the academic years 2008/2009 to 2018/2019 and covering 17 undergraduate programs.

- **Demographic Attributes:** Gender, age at enrollment, nationality, displaced status, international student status, marital status.
- **Socioeconomic Attributes:** Parents' education and occupation, educational special needs, debtor status, tuition fees, scholarship holder.
- **Macroeconomic Attributes:** Unemployment rate, inflation rate, and GDP at the year of enrollment.
- **Academic Attributes:** Application mode/order, course, prior qualification, attendance type, number of curricular units credited/ cenrolled/ evaluated/ approved, and grades at the end of the first and second semesters.
- **Target Variable:** Categorical status at the end of the normal course duration (dropout, enrolled, graduate).

It is important to note that all data cleaning and preprocessing steps were thoroughly performed by the original dataset authors. The raw data, sourced from multiple administrative systems, was merged, duplicates and irrelevant records were removed, and anomalous entries were eliminated. All necessary columns were selected, renamed, and missing values were handled, resulting in a fully cleaned and consistent dataset. All features are numerically encoded, making the data ready for normalization and tabular-to-image transformation. This robust preprocessing ensures the dataset is well-suited for developing and evaluating predictive models of student academic success.

4 Results and Discussion

This section reports and discusses the results from applying our hybrid framework to predict student dropout and academic success. We evaluate each Machine Learning (ML) model, trained on extracted features by Convolutional Neural Network (CNN), using accuracy and F1 score. By comparing different classifiers, we demonstrate the advantages of integrating deep feature extraction with traditional algorithms. The discussion also considers the framework's strengths and limitations for early identification of at-risk students and supporting academic interventions.

The performance of the proposed hybrid framework was evaluated using both accuracy and F1 score as shown in Table 1, and visually represented in Figs. 2 and 3. These results allow for a comprehensive comparison of how each hybrid model, which combines CNN-based feature extraction with classical machine learning classifiers, performs in predicting academic success and students' dropout. The use of both metrics ensures a robust evaluation, with accuracy reflecting overall classification correctness and F1 score providing insight into the models' ability to balance precision and recall, an important consideration given the possible class imbalance in educational datasets.

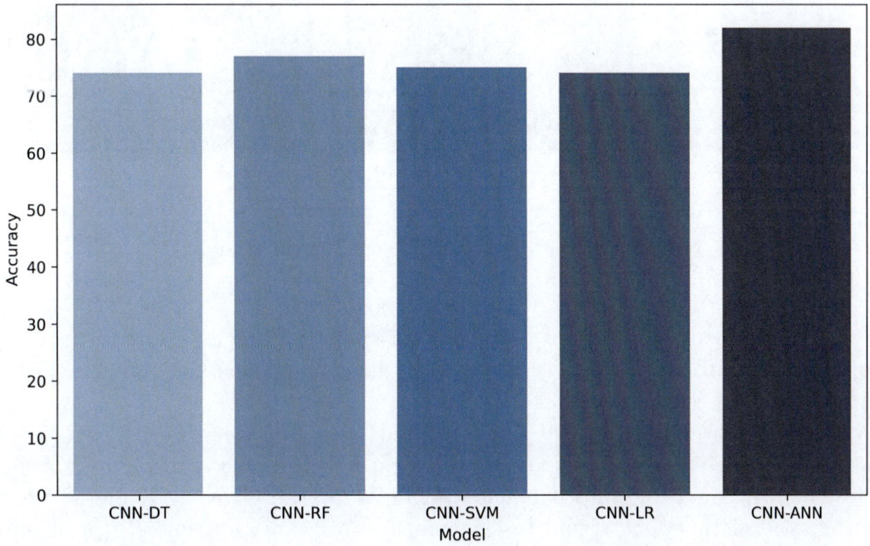

Fig. 2. Accuracy scores across different hybrid models.

Examining the results in Fig. 2, it is evident that the CNN-ANN hybrid model achieves the highest accuracy, reaching 82%, followed by CNN-RF at 77%. The remaining models, including CNN-SVM, CNN-LR, and CNN-DT, demonstrate similar accuracy values, clustered between 74% and 75%. This indicates that while all models benefit from the deep features extracted by the CNN, the artificial neural network is most effective at leveraging this rich representation for predicting student outcomes.

A similar trend is observed when analyzing the F1 score in Fig. 3. The CNN-ANN model again outperforms the others with a score of 0.75, indicating superior performance in handling the trade-off between precision and recall. CNN-RF and CNN-SVM follow, with F1 scores of 0.68 and 0.67, respectively. Both CNN-LR and CNN-DT models yield F1 scores of 0.67 and 0.65, showing adequate but lower performance relative to the best-performing hybrid. These findings suggest that incorporating the nonlinear learning capabilities of neural networks, both

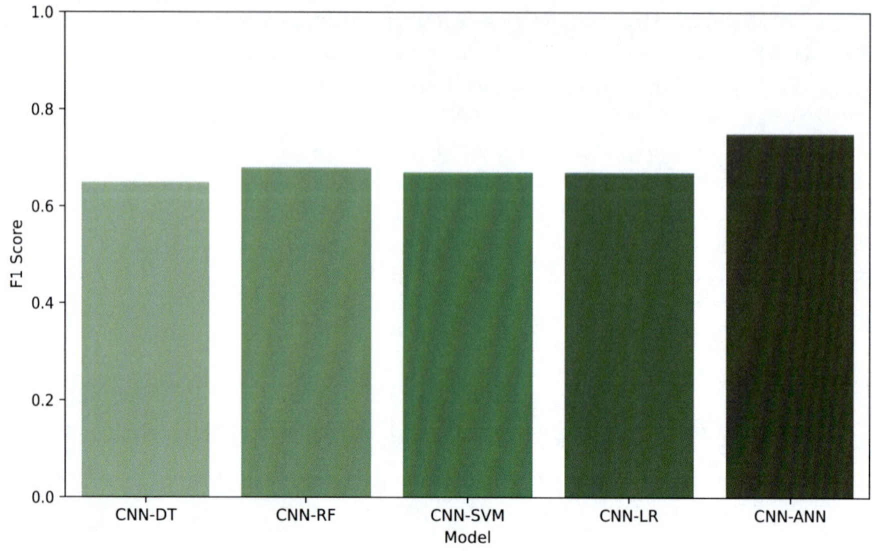

Fig. 3. F1 scores across different hybrid models.

in the deep feature extraction and final classification stages, can substantially improve the framework's ability to correctly identify students at risk of dropout or likely to succeed.

Overall, the analysis highlights the added value of using deep learning for automated feature extraction in combination with advanced machine learning classifiers. The notable improvements achieved by the CNN-ANN model point to the synergistic effect of end-to-end neural processing, making it a promising approach for real-world applications in educational data mining.

Table 1. Performance of Hybrid Models Using CNN-based Feature Extraction

Model	CNN-DT	CNN-RF	CNN-SVM	CNN-LR	CNN-ANN
Accuracy (%)	74	77	75	74	82
F1 Score	0.65	0.68	0.67	0.67	0.75

4.1 Discussion

The study by Realinho et al. [5] introduced the dataset used here and applied RF, XGBoost, LightGBM, and CatBoost for feature importance analysis, using F1-score as a metric. However, they did not report final classification results such as accuracy or F1-score for multi-class prediction, focusing instead on interpreting important features. This leaves open questions about model effectiveness for classifying dropout, enrolled, or graduate outcomes.

Predicting multi-class educational outcomes such as dropouts, enrolled, and graduates, remains a notable challenge. Hafdi et al. [3] found that ML models, including LSTM, SVM, and ANN, performed significantly better on binary classification than on multi-class problems, where class imbalance and fine-grained distinctions reduced predictive performance. Their work confirms that multi-class prediction often requires more advanced methods and larger datasets.

Hafdi et al. [3] reported accuracies between 80% and 90% but F1 scores below 0.7 for multi-class tasks, using a dataset focused on coding courses. In contrast, our broader dataset covers multiple subjects and outcomes. Despite the greater complexity, our CNN-ANN hybrid achieved an F1 score of 0.75 and an accuracy of 82%, highlighting the strength of our approach.

Similarly, Zohair et al. [17] applied various ML algorithms to predict student grades from a master's degree datasets, with the best accuracy (76.3%) achieved by SVM. Their work, based on smaller, specialized datasets, further validates our method, as our model surpasses these benchmarks in a larger and more complex multi-class context. This underlines the potential of combining CNN-based feature extraction with classical ML for educational prediction tasks.

5 Conclusion

In summary, this study introduces a hybrid framework combining CNN-based feature extraction with traditional ML classifiers to predict student outcomes in higher education. Our approach effectively captures complex patterns, resulting in more informative features and strong predictive performance on a comprehensive, multi-class dataset. The CNN-ANN hybrid model achieved the best results, with an accuracy of 82% and an F1 score of 0.75, demonstrating the potential of deep learning in educational data mining. However, the use of CNN increases computational complexity. Future research should focus on optimizing this framework for different educational settings and improving computational efficiency, supporting more scalable and generalizable predictive analytics in the field.

References

1. Hafdi, Z.S., El Kafhali, S.: Student performance prediction in learning management system using small dataset. In: The International Conference on Artificial Intelligence and Computer Vision, pp. 197–205. Springer Nature Switzerland, Cham (2023). https://doi.org/10.1007/978-3-031-27762-7_19
2. Riestra-González, M., del Puerto Paule-Ruíz, M., Ortin, F.: Massive LMS log data analysis for the early prediction of course-agnostic student performance. Comput. Educ. **163**, 104108 (2021)
3. Hafdi, Z.S., El Kafhali, S.: A comparative evaluation of machine learning methods for predicting student outcomes in coding courses. AppliedMath **5**(2), 75 (2025)

4. Roslan, M.B., Chen, C.: Educational data mining for student performance prediction: a systematic literature review (2015–2021). Inter. J. Emerging Technol. Learn. (iJET) **17**(5), 147–179 (2022)
5. Realinho, V., Machado, J., Baptista, L., Martins, M.V.: Predicting student dropout and academic success. Data **7**(11), 146 (2022)
6. Kabathova, J., Drlik, M.: Towards predicting student's dropout in university courses using different machine learning techniques. Appl. Sci. **11**(7), 3130 (2021)
7. Arévalo Cordovilla, F.E., Peña Carrera, M.: Comparative analysis of machine learning models for predicting student success in online programming courses: a study based on LMS data and external factors. Mathematics **12**(20), 3272 (2024)
8. Kaensar, C., Wongnin, W.: Analysis and prediction of student performance based on Moodle log data using machine learning techniques. Inter. J. Emerging Technol. Learn. **18**(10), 184–203 (2023)
9. Hung, H.C., Liu, I.F., Liang, C.T., Su, Y.S.: Applying educational data mining to explore students' learning patterns in the flipped learning approach for coding education. Symmetry **12**(2), 213 (2020)
10. Ciolacu, M., Tehrani, A.F., Binder, L., Svasta, P.M.: Education 4.0-artificial intelligence assisted higher education: early recognition system with machine learning to support students' success IEEE SIITME, pp. 23–30 (2018)
11. Gnoh, H.Q. et al.: Enhancing business sustainability through technology-enabled AI: forecasting student data and comparing prediction models for higher education institutions (HEIs) PaperASIA, **40**(2b), 48–58 (2024)
12. Villegas-Ch, W., Román-Cañizares, M., Palacios-Pacheco, X.: Improvement of an online education model with the integration of machine learning and data analysis in an LMS. Appl. Sci. **10**(15), 5371 (2020)
13. Azimi, S., Popa, C.G., Cucić, T.: Improving students performance in small-scale online courses-a machine learning-based intervention. Inter. J. Learn. Analy. Artifi. Intell. Educ. **18**, 80–95 (2020)
14. Ersozlu, Z., Taheri, S., Koch, I.: A review of machine learning methods used for educational data. Educ. Inf. Technol. **29**(16), 22125–22145 (2024)
15. Sulak, S.A., Koklu, N.: Predicting student dropout using machine learning algorithms. Intell. Methods Eng. Sci. **3**(3), 91–98 (2024)
16. Bertolini, R., Finch, S.J., Nehm, R.H.: Enhancing data pipelines for forecasting student performance: integrating feature selection with cross-validation. Int. J. Educ. Technol. High. Educ. **18**(1), 44 (2021)
17. Abu Zohair, L.M.: Prediction of student's performance by modelling small dataset size. Int. J. Educ. Technol. High. Educ. **16**(1), 27 (2019)

Breast Cancer Classification: A Transfer Learning Approach Using EfficientNets

Samiya Qanoune(✉), Hassan Ammor, and Zouhair Guennoun

Smart Communication Research Team, Mohammadia Engineering School, Mohammed V University, Rabat, Morocco
samiyaqanoune@research.emi.ma

Abstract. Deep Learning has been used by the research community in recent years to facilitate and enable the early detection of breast cancer, which is one of one of the most widespread and serious health conditions among women across the globe. Early detection plays an important role in improving survival rates, as it allows patients to begin treatment at the earliest stages. However, despite remarkable progress, many approaches still have issues such as lengthy training and testing times, high computational demands, or a considerable number of false positives that reduce their reliability in real clinical environments. In this study, we introduce a breast cancer classification framework that leverages transfer learning by fine-tuning a pre-trained convolutional neural network for feature extraction. The optimized model reached an accuracy of 99.66%, demonstrating superior performance and computational efficiency compared with conventional approaches.

Keywords: Breast cancer · Artificial intelligence · Deep learning · Transfer learning · EfficientNets · Medical Image Classification

1 Introduction

Almost one in twenty women will have breast cancer at some point in their lives, making it the most frequently diagnosed malignancy in women. Despite advances in screening and treatment, its global impact continues to grow. Recent health reports estimate around 2.3 million new cases and roughly 670 000 related deaths in 2022. If the current trajectory continues, the worldwide incidence could rise to approximately 3.2 million cases and more than one million deaths annually by 2050. These numbers underline the steady escalation of breast cancer as a major public health concern across continents.

Therefore, the most effective approach for treating breast cancer is through early detection, for which early diagnosis greatly increases the chances of complete recovery. However, despite the important improvement in the understanding of molecular mechanisms underlying the pathogenesis of breast cancer, Numerous studies have demonstrated that the majority of breast cancer cases remain diagnosed at a late stage in different countries.

A. E. Hassanien et al. (Eds.): AISDS 2025, LNDECT 286, pp. 23–31, 2027.
https://doi.org/10.1007/978-3-032-18758-1_3

Artificial intelligence, especially deep learning has demonstrated their prowess in medical computer vision. With the help of large datasets, these techniques have achieved remarkable results [2].

One of the most important technologies nowadays is Transfer Learning (TL), it is a machine learning technique that holds extraordinary preferences such as saving resources and model training productivity. In spite of the fact that the concept isn't new, its applications for Computer vision have been developing exponentially [3].

In a related study, Deepak and Ameer [4] investigated brain tumor image classification using deep convolutional neural networks enhanced with transfer learning, achieving an average accuracy of 98% and outperforming previously reported methods. Similarly, Younis and Bhatti [5] designed a transfer-learning-based system for skin cancer diagnosis that reached 97.07% accuracy. Transfer learning has also been successfully employed in other biomedical imaging applications; for example, Pathak et al. [6] developed a COVID-19 detection approach using chest CT scans, attaining an accuracy of 96.22%, which exceeded that of several conventional supervised learning techniques.

The remainder of this paper is organized in the following manner. Section 2 introduces the dataset and describes the overall architecture of the proposed system. Section 3 presents and discusses the experimental results, along with a comparative analysis of previous methods. Finally, Sect. 4 concludes the work and highlights perspectives for future research.

2 Materials and Methods

This section presents the dataset as well as the system architecture and the technologies used to build the system.

2.1 Dataset

The experimental analysis in this work relies on the BreakHis dataset (Breast Cancer Histopathological Image Classification), which includes 9,109 histopathological images collected from 82 individuals at magnifications of 40×, 100×, 200×, and 400× [7] (Table 1).

Table 1. The structure of BreaKHis dataset.

Magnification	Benign	Malignant	Total
40×	652	1,370	1,995
100×	644	1,437	2,081
200×	588	1,232	1,820
400×	588	1,232	1,820
Total of Images	2,480	5,429	7,909

It contains 2,480 benign and 5,429 malignant samples (700 × 460 pixels, 3-channel RGB, 8-bit depth in each channel, PNG format).

2.2 System Architecture

The proposed architecture of our system is illustrated in Fig. 1. As shown, we started by inputting breast cancer Histopathological images into our system. Second we passed the images through data augmentation and data preprocessing. Then transfer learning training is provided as a third stage. Finally, classification outputs are displayed.

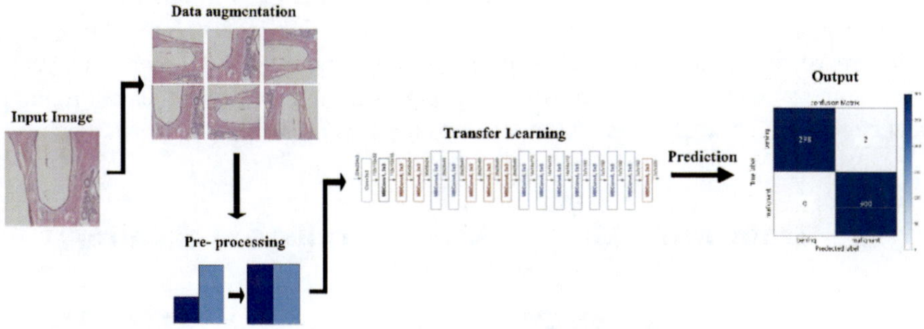

Fig. 1. Overall system architecture

Data Augmentation. Data augmentation refers to a collection of techniques employed to synthetically enlarge the training set by producing altered variations of existing images. These transformations typically include slight rotations, random cropping, zooming, or flipping. Such operations help improve the model's ability to generalize by reducing overfitting and acting as a form of regularization during training [8]. Figure 2 illustrates some of the augmentation methods applied in this study.

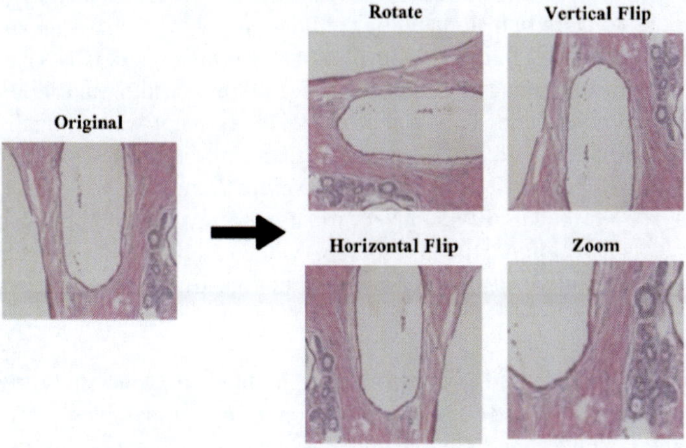

Fig. 2. Examples of methods applied on dataset.

Data Preprocessing. In this stage, the dataset underwent a preprocessing procedure designed to simplify the subsequent training and testing phases. The data were converted and normalized to maintain consistency across all samples [9]. Proper preprocessing is essential prior to model training, as it allows the detection and correction of issues such as missing values, noise, or inconsistencies within the dataset before it is introduced into the learning algorithm.

Transfer Learning. Transfer Learning (TL) is one of the most well-known machine learning approaches. It is described as transferring previously learned information from a similar task for classification and feature extraction purposes [10]. TL enables the reuse of existing parameters, such as convolution weights from a learned model, for training new models with a smaller amount of labeled images (Fig. 3).

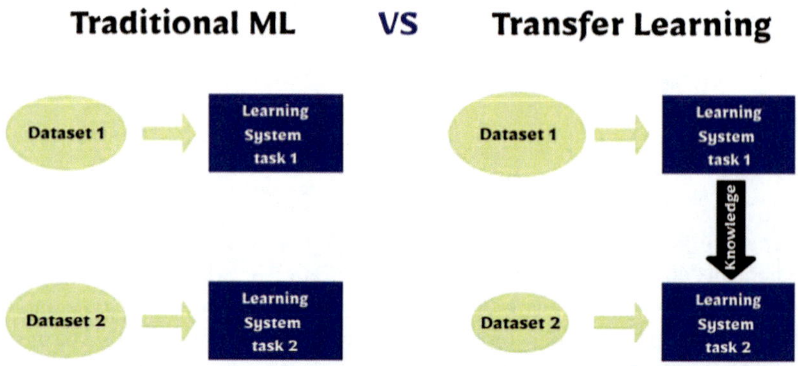

Fig. 3. Traditional Machine learning Vs Transfer learning.

In this paper, we worked on building a system using the EfficientNet-B0 model for transfer learning to classify histopathological images.

EfficientNet refers to a family of convolutional neural network (CNN) architectures designed to achieve a balanced trade-off between depth, width, and input resolution through a unified compound scaling approach. This concept was originally proposed by Mingxing Tan and Quoc V. Le at Google Research in their work *"EfficientNet: Rethinking Model Scaling for Convolutional Neural Networks"* [11]. In comparison with traditional CNN models, EfficientNet variants are remarkably lightweight; requiring fewer parameters and computational resources, while still delivering high classification accuracy and faster inference. The base configuration, EfficientNet-B0, serves as the foundation of the family and demonstrates outstanding performance efficiency despite its compact design.

Although EfficientNet-B0 is a pre-existing architecture, this study highlights its remarkable transferability and efficiency when applied to histopathological breast cancer classification tasks under multiple magnification levels. The goal of this work is not to design a new model, but to demonstrate that a lightweight, computationally efficient, and pre-trained CNN can achieve state-of-the-art accuracy on the BreakHis dataset. Such benchmarking is crucial for identifying architectures that balance high performance with low computational cost in medical imaging applications.

The architecture of EfficientNet-B0, which was employed in this study, is represented in Fig. 4.

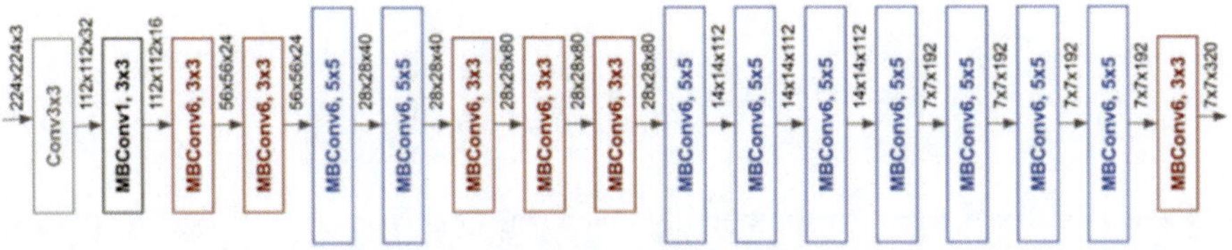

Fig. 4. EfficientNet-B0 architecture.

2.3 Performance Metrics

The effectiveness of each neural network is measured with different metrics of performance such as accuracy, precision, specificity, sensitivity, recall, F1-score, False Positive Rate (FPR). The measurement parameters TN, FN, TP and FP represent True Negative, False Negative, True Positive and False Positive respectively.

The metrics are defined as follows:

- **Accuracy:** is expressed as the percentage of correct predictions relative to the total number of instances.

$$Accuracy = \frac{(\mathrm{TP} + \mathrm{TN})}{(\mathrm{TP} + \mathrm{TN} + \mathrm{FP} + \mathrm{FN})} \tag{1}$$

$$Accuracy = \frac{\text{Number of correct predictions}}{\text{Total number of predictions}} \tag{2}$$

- **Precision:** is the positive predictive value or the fraction of the positive predictions that are actually positive.

$$Precision = \frac{\mathrm{TP}}{(\mathrm{TP} + \mathrm{FP})} \tag{3}$$

- **Specificity:** is the rate of the negative observations that are correctly predicted as negative.

$$Specificity = \frac{\mathrm{TN}}{(\mathrm{TN} + \mathrm{FP})} \tag{4}$$

- **Sensitivity:** is the rate of the positive observations that are correctly predicted as positive, it is known as Recall.

$$Sensitivity = Recall = \frac{\text{TP}}{(\text{TP} + \text{FN})} \quad (5)$$

- **F1-score:** is a measure of a model's accuracy, it is the harmonic mean of precision and recall.

$$\text{F1}_{\text{Score}} = \frac{2(\text{precision} \times \text{recall})}{(\text{precision} + \text{recall})} \quad (6)$$

- **False Positive Rate:** is the probability that a true positive will be missed by the test.

$$False\ Positive\ Rate = \frac{\text{FP}}{(\text{FP} + \text{TN})} \quad (7)$$

3 Results

In this research, the EfficientNet-B0 model for transfer learning was used to classify breast cancer histopathological images. As shown in Table 2, we obtained an accuracy equal to 99.66% with 100% in precision and 99.33% in sensitivity. for specificity and F1-score we reached 100% and 99.663% respectively.

Table 2. The results achieved by our model (%).

	Accuracy	Precision	Specificity	Sensitivity	F1-score
Our system	99.666	100	100	99.333	99.663

Other researchers have applied various techniques to the same dataset analyzed in this work. Kongwon Chang and Jisang Yu [12] achieved an accuracy of 89% using a deep convolutional neural network (ConvNet) model. Meanwhile, Yashwant K. and Vijayshri Ch [13]. proposed an approach that classifies histopathological images by combining handcrafted and shape-based descriptors within a bag-of-visual-words (BoW) framework, attaining an accuracy of 96.96% (Fig. 5).

In contrast, Sara H. and Peyman H [14]. developed an ensemble-based framework combining three pre-trained convolutional neural networks: VGG19, MobileNet, and DenseNet. Their hybrid model achieved a prediction accuracy of 98.13%. The proposed system integrated several enhancement techniques, such as stain normalization, data augmentation, and optimization of hyperparameters through fine-tuning. These optimization procedures fall within the domain of transfer learning for deep neural networks, where the parameters of existing models are iteratively adjusted to adapt them to new classification tasks [15] (Table 3).

As seen, TL methods significantly outperform the other methods. Our model and Sara H.'s model [14] have obtained an accuracy higher than 98%. Furthermore, TL

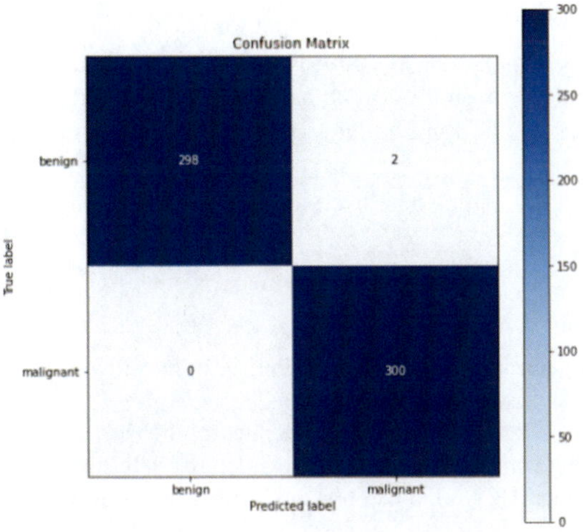

Fig. 5. Confusion Matrix.

Table 3. Performance comparison of the proposed EfficientNet-B0 model and existing architectures on the BreakHis dataset (%).

Reference	Model / Method	Accuracy
K. C [12].	CNN (ConvNet)	89
Y. K [13].	Bag of Visual Words (BoW) + Shape Features	96.96
S. H [14].	Ensemble of Pre-trained CNNs (VGG19 + MobileNet + DenseNet)	98.13
Our system	EfficientNet-B0 (Transfer Learning)	99.66

techniques do not only allow for an improved performance but also help reduce the training time, as the models use a pre-trained knowledge rather than start from the very beginning.

4 Conclusions

This study presents an approach for classifying breast cancer histopathological images through the reuse of knowledge from pre-trained deep models. The results reveal that adopting this strategy leads to higher classification accuracy and faster model convergence compared to training from scratch. It helped us reach a higher accuracy with a small amount of data, and it offered a better starting point where we started at a specific level without even training. TL holds great potential for several industries especially within the Healthcare sector, and it can be applied to different fields and put into many uses such as image recognition, speech recognition, or natural language processing.

While the results demonstrate the effectiveness of the proposed approach, future improvements could further enhance its reliability and interpretability. In particular, extending the evaluation to additional histopathological datasets and testing the model under different imaging or staining conditions would strengthen its generalization. Moreover, integrating visualization tools such as Grad-CAM will provide intuitive explanations of the model's predictions, making the system more transparent and clinically meaningful.

Our proposed model can also be used in other medical imaging related tasks.

References

1. World Health Organization: Breast cancer fact sheet. https://www.who.int/news-room/fact-sheets/detail/breast-cancer. Accessed 14 Aug 2025
2. Spanhol, F.A., Oliveira, L.S., Petitjean, C., Heutte, L.: Breast cancer histopathological image classification using convolutional neural networks. In: 2016 International Joint Conference on Neural Networks (IJCNN), pp. 2560–2567. IEEE, Vancouver (2016). https://doi.org/10.1109/IJCNN.2016.7727519
3. Jiang, B., et al.: Deep learning applications in breast cancer imaging: a comprehensive review. Cancer. **16**(12), 2065 (2024). https://doi.org/10.3390/cancers16122065
4. Deepak, S., Ameer, P.: Brain tumor classification using deep CNN features via transfer learning. Comput. Biol. Med. **111**, 103345 (2019). https://doi.org/10.1016/j.compbiomed.2019.103345
5. Younis, H., Bhatti, M.H., Azeem, M.: Classification of skin cancer dermoscopy images using transfer learning. In: 2019 15th International Conference on Emerging Technologies (ICET), pp. 1–4. IEEE, Peshawar (2019). https://doi.org/10.1109/ICET48972.2019.8994508
6. Pathak, Y., Shukla, P., Tiwari, A., Stalin, S., Singh, S., Shukla, P.: Deep transfer learning based classification model for COVID-19 disease. IRBM. **43**(2), 87–92 (2022). https://doi.org/10.1016/j.irbm.2020.05.003
7. Spanhol, F., Oliveira, L.S., Petitjean, C., Heutte, L.: A dataset for breast cancer histopathological image classification. IEEE Trans. Biomed. Eng. **63**(7), 1455–1462 (2016). https://doi.org/10.1109/TBME.2015.2496264
8. Shorten, T., Khoshgoftaar, T.M.: A survey on image data augmentation for deep learning. J. Big Data. **6**, 60 (2019). https://doi.org/10.1186/s40537-019-0197-0
9. Misra, S., Li, H., He, J.: Robust geomechanical characterization by analyzing the performance of shallow-learning regression methods using unsupervised clustering methods. In: Misra, S., Li, H., He, J. (eds.) Machine Learning for Subsurface Characterization, pp. 129–155. Gulf Professional Publishing, Houston (2020). https://doi.org/10.1016/B978-0-12-817736-5.00005-3
10. Orenstein, E., Beijbom, O.: Transfer learning and deep feature extraction for planktonic image data sets. In: 2017 IEEE Winter Conference on Applications of Computer Vision (WACV), pp. 1082–1088. IEEE, Santa Rosa (2017). https://doi.org/10.1109/WACV.2017.125
11. Tan, M., Le, Q.V.: EfficientNet: rethinking model scaling for convolutional neural networks. arXiv preprint https://arxiv.org/abs/1905.11946 (2019)
12. Chang, J., Yu, J., Han, T., Chang, H.-J., Park, E.: A method for classifying medical images using transfer learning: a pilot study on histopathology of breast cancer. In: 2017 IEEE 19th International Conference on e-Health Networking, Applications and Services (Healthcom), pp. 1–4. IEEE, Dalian (2017). https://doi.org/10.1109/HealthCom.2017.8210843
13. Kurmi, Y., Chaurasia, V., Ganesh, N.: Tumor malignancy detection using histopathology imaging. J. Med. Imaging Radiat. Sci. **50**(4), 514–528 (2019). https://doi.org/10.1016/j.jmir.2019.07.004

14. Kassani, S.H., Kassani, P.H., Wesolowski, M.J., Schneider, K.A., Deters, R.: Classification of histopathological biopsy images using ensemble of deep learning networks. arXiv preprint https://arxiv.org/abs/1909.11870 (2019)
15. Guo, Y., Li, Y., Wang, L., Rosing, T.: Adafilter: adaptive filter fine-tuning for deep transfer learning. In: Proceedings of the AAAI Conference on Artificial Intelligence, vol. 34(4), pp. 4060–4066. AAAI, New York https://doi.org/10.1609/aaai.v34i04.5824 (2020)

Model-Free Visual Control of a Low-Cost Continuum Arm: Target and Trajectory Tracking

Mostafa AbdelAziz, Mahmoud Magdy, and Zahy Y. Elgendy(✉)

Mechatronics and Robotics, Mechanical Engineering, The British University in Egypt, Cairo, Egypt
{mostafa.abdelaziz,mahmoud.mohamed}@bue.edu.eg,
zahygendy@gmail.com

Abstract. This research presents and validates a fully model-free, vision-based control system for a tendon-driven continuum arm that provides robust performance using a simple, fixed-parameter controller. The controller translates image-space error, estimated from fixed-threshold HSV segmentation, directly into quantized PWM commands via a piecewise step law, avoiding all kinematic modeling and calibration. Experimental validation demonstrates good repeatability. In static servoing tasks, the system reduced the initial error by an average of 69.8% in 2.91 s. In dynamic tasks, the same untuned controller accurately tracked circle, square, and triangle paths at speeds up to 4.09 cm/s. By exhibiting robust performance across different tasks and conditions without the need for retuning, this work provides a new quantitative benchmark for low-cost, easily replicable, vision-driven soft robot control.

Keywords: Continuum Robot · Model-Free Control · Vision-Based Servoing · Trajectory Tracking · Soft Manipulator · Real-Time Control

1 Introduction

Continuum robots and, in particular, tendon-driven soft arms, enable compliant interaction in constrained or unpredictable environments, from minimally invasive surgery to search-and-rescue. Their flexible and low-inertia constructions facilitate smooth, adaptable movement, but achieving accurate control remains difficult due to non-linear, time-varying mechanics and the absence of rigid joints.

Conventional approaches apply models like Cosserat rod model or piecewise constant curvature (PCC) [1, 2], which entail extensive amounts of calibration, precision sensors, and heavy computation—and which limit their use in dynamic and/or uncertain environments. Vision-based control is an exciting alternative as it relies solely on the end-effector's state and allows for task-space feedback without a complete model of the full body. However, vision-based systems are typically sensitive to latency, noise, or segmentation errors, and very few studies provide direct, quantitative comparisons between model-based and vision-based techniques.

A. E. Hassanien et al. (Eds.): AISDS 2025, LNDECT 286, pp. 32–41, 2027.
https://doi.org/10.1007/978-3-032-18758-1_4

This paper introduces a fully vision-based control architecture for a monolithic tendon-driven soft arm, using a fixed-threshold HSV segmentation and quantized PWM step logic for real-time servoing and pattern tracking—without any kinematic modeling. Unlike learning-based systems that require training data [3] or model-based methods that rely on mechanical calibration [4], our method uses pure image-space feedback, ensuring generality and simplicity across multiple trajectory types. Crucially, all experiments are performed under fixed parameters, demonstrating the system's robustness and adaptability across static and dynamic conditions.

Recent Advances and Comparisons. Recent work has shifted towards proprioceptive and vision-guided strategies, such as sensor fusion [5], vision-based force estimation [6], while others compare model-based controllers to vision-based ones, showing an advantage of vision-based methods in unstructured environments [7, 8]. Broader reviews emphasize the trade-off between interpretability and adaptability [9–11], suggesting an overall trend towards hybrids and model-free solutions.

Research Gap. Despite recent advances, several gaps persist:

- Few systems are truly independent of kinematic or dynamic models.
- Quantized, piecewise vision-based controllers are underexplored.
- Comparisons across different pattern-tracking regimes under fixed parameters remain rare.

This work addresses these limitations by presenting a transparent, repeatable, model-free control platform grounded entirely in image-space feedback, validating its performance on static target acquisition and dynamic trajectory tracking (circle, square, triangle) without any controller tuning or model-based compensation, setting a new benchmark for low-cost, vision-driven soft robotics.

Paper Structure. The remainder of this paper is structured as follows: Sect. 2 describes experimental methodology, including hardware platform, control architecture, control law, target servoing and pattern tracking protocols. Section 3 presents quantitative results for both target servoing and pattern tracking. Section 4 discusses limitations and future directions.

2 Methodology

This section describes the experimental implementation framework for vision-based control of the continuum robot, including details of the platform design, layered control architecture, vision-based servoing, and geometric pattern tracking procedure. Each component of the methodology is described in detail for transparency and reproducibility.

2.1 Robot Platform

The platform consists of a single-segment tendon-driven continuum arm, designed and produced as a single, monolithic flexible TPU structure using FDM 3D printing. The continuum arm followed the design in [12], utilizing four antagonistic tendons arranged in two pairs per axis for precise planar actuation. Coaxially mounted at the end of the arm is a compliant, miniaturized, two-finger gripper, also printed from TPU, and inspired by the compliant design in [13]. The gripper was printed in magenta to ensure consistent computer vision-based detection. The low mass of the distal gripper helps preserve the arm's dynamic bandwidth and responsiveness. An overview of the hardware with appropriate labels is presented in Fig. 1.

2.2 Control Architecture

The control system has been designed using a modular, two-layer system to maintain clarity and reproducibility:

Host Layer. The high-level perception, feedback, and control logic are executed on a PC, written in Python and communicate via PySerial with the embedded hardware over UART (115,200 baud).

Embedded Layer. An Arduino-compatible microcontroller and PCA9685 PWM driver (50 Hz) control four MG996R servo motors (M1–M4) in antagonistic pairs: M1/M3 for horizontal and M2/M4 for vertical motion. The microcontroller receives quantized PWM commands via UART and relays them over I^2C (400 kHz), enforcing a safe output range (300–600 counts). This layer functions solely as an actuator interface.

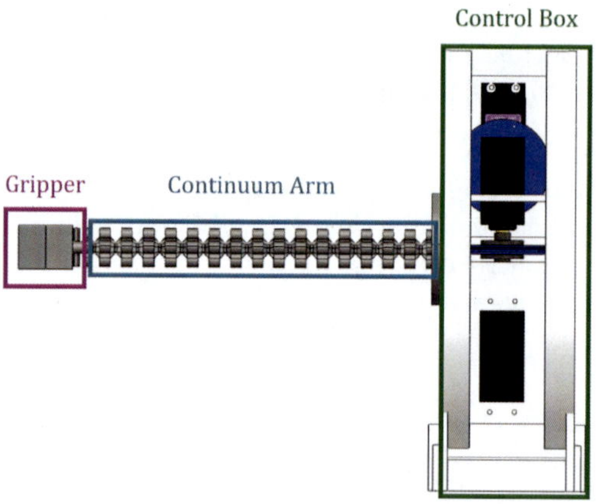

Fig. 1. Labeled hardware schematic of the continuum robotic platform, showing the control box, continuum arm, and gripper.

Vision-Based Feedback. A fixed overhead monocular camera (640 × 480 px, 30 Hz) with one-time planar calibration enables direct quantitative measurement of both tip and target positions. An overview of the control information flow is illustrated in Fig. 2.

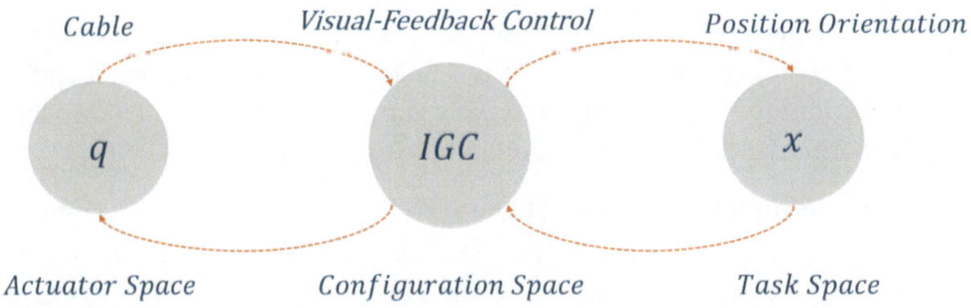

Fig. 2. Information flow between actuator space, configuration space, and task space under visual-feedback control.

2.3 Vision-Based Target Servoing

Perception and Localization. The system localizes the continuum tip and user-selected target entirely through vision, without any kinematic modeling. Each RGB video frame is converted to HSV color space, then thresholded using carefully selected, fixed HSV bounds for both the continuum tip and the user-selected target.

The magenta tip uses [140, 70, 70] – [165, 255, 255]; targets use green [40, 70, 70] – [80, 255, 255], orange [5, 100, 100] – [30, 255, 255], or blue [90, 100, 100] – [130, 255, 255].

From each mask, the largest connected component's centroid is computed to yield real-time $(x_{\text{tip}}, y_{\text{tip}})$ and $(x_{\text{tgt}}, y_{\text{tgt}})$, positions.

To suppress frame-to-frame jitter while maintaining low-latency response (less than 100 ms at 25–30 FPS), exponential smoothing is applied, as defined by Eq. (1):

$$\hat{c}_t = \alpha\, c_t + (1 - \alpha)\, \hat{c}_{t-1} \tag{1}$$

with $\alpha = 0.3$ for general target servoing and $\alpha = 0.25$ for pattern tracking. A planar camera calibration is performed once, yielding the camera scale calibration factor in Eq. (2):

$$s = 0.0679 \text{ cm/pixel} \tag{2}$$

All positional errors are computed in centimeters.

Error Computation and Deadbanding. The error vector in pixel coordinates, e_t^{px}, is calculated as the difference between the target and tip centroids: Mathematically, the error at time t is given by Eq. (3):

$$e_t^{px} = c_t^{target} - c_t^{tip} = (\Delta x, \Delta y) \tag{3}$$

where $\mathbf{c}_t^{\text{target}}$ and $\mathbf{c}_t^{\text{tip}}$ denote the pixel coordinates of the target and robot tip at time t, respectively. The resulting vector $(\Delta x, \Delta y)$ encodes both the magnitude and direction required to drive the tip toward the target.

To enable quantitative control, the error is then converted to centimeters using the system's calibrated scale factor ($s = 0.0679$ cm/pixel) as in Eq. (4):

$$e_t^{cm} = s.e_t^{px} \tag{4}$$

When the error lies within a fixed elliptical deadband ($\tau_x = \tau_y = 2.0\ cm$), actuation is suspended, preventing oscillation near the target. The elliptical deadband enables independent axis gating, which reduces overshoot and suppresses limit-cycle oscillations often seen with circular deadbands during oblique approaches.

Control Law: Quantized Piecewise Step Law. To ensure robust convergence while preserving mechanical safety, a quantized, piecewise step law translates each error component into a discrete PWM increment: Each axis error (e) is mapped to a discrete PWM increment (ΔPWM) according to Eq. (5):

$$\Delta PWM = \begin{cases} 6, & \mid e \mid > 10 \text{ cm} \\ 4, & 5 \text{ cm} < \mid e \mid \leq 10 \text{ cm} \\ 2, & 2 \text{ cm} < \mid e \mid \leq 5 \text{ cm} \\ 1, & \mid e \mid \leq 2 \text{ cm} \end{cases} \tag{5}$$

where e is either Δx or Δy, subject to minimum activation thresholds defined in Eq. (6).

$$\mid \Delta x \mid > 0.2 \text{ cm}, \mid \Delta y \mid > 0.3 \text{ cm} \tag{6}$$

At each control step, antagonistic tendon motors receive these quantized updates, enabling rapid approach from afar and increasingly refined correction near the target.

Hold and Dwell Behavior. Upon entering the deadband, commands are held for $T_{hold} = 0.4$ s to allow for viscoelastic relaxation. If the tip or target becomes undetectable, the last valid command is sustained to avoid erratic motion. If a persistent state of undetectability occurs, a safe-state routine is launched; the arm returns to a neutral position and tendon tension is relieved.

2.4 Pattern Tracking

To demonstrate the flexibility and repeatability of the closed-loop controller, the system was extended to vision-guided pattern tracking of geometric paths. Unlike conventional approaches that depend on kinematic calibration or analytic models, the method operates entirely in the image plane, relying solely on real-time visual feedback to guide the robot tip through a sequence of user-defined waypoints.

Path Specification and Visual Overlay. To define geometric paths for pattern tracking, we represent each pattern as a set of N waypoints, with each waypoint given by Eq. (7):

$$W = \{w_k\}_{k=1}^{N} = [(x_1, y_1), (x_2, y_2), \ldots, (x_N, y_N)] \tag{7}$$

The number of waypoints was varied. Increasing N produces a smoother, more continuous path but increases the challenge for the controller, while decreasing N simplifies the path, making it easier but less representative of complex trajectories.

Control Flow and Error Computation. The controller uses the same error computation and quantized step law as in the servoing task, with the active waypoint w_k serving as the target c_t^{target}.

Adaptive Deadband and Arrival Logic. A waypoint is determined to be "reached" when the Euclidean distance between the tip and the target waypoint is less than the fixed arrival radius ρ, as given in Eq. (8):

$$\sqrt{(\Delta x)^2 + (\Delta y)^2} < \rho, \text{ where } \rho = 4.0 \text{ cm} \tag{8}$$

Upon arrival, the PWM commands are held constant, as in Eq. (9), for a fixed duration of time, referred to as "dwell":

$$T_{\text{dwell}} = 0.4 \text{ s} \tag{9}$$

This dwell time allows mechanical oscillations and cable friction to settle, ensuring repeatable and stable endpoint placement for subsequent statistical evaluation.

Waypoint Advancement and Closed-Loop Motion. After the fixed dwell period at each waypoint, the controller advances the active waypoint index according to Eq. (10):

$$k_{new} = (k_{old} \bmod N) + 1 \tag{10}$$

where k is the current waypoint index and N is the total number of waypoints defining the pattern. This cyclical scheme ensures that upon reaching the final waypoint ($k = N$), the index seamlessly returns to the first ($k = 1$), thereby enabling continuous closed-loop traversal of the pattern.

At all waypoints, the controller uses the same quantized PWM step law and antagonistic function behavior for each tendon. This allows consistency in performance and reproducible tracking behavior regardless of the waypoints or path assessment geometry. These settings are an essential feature that ensures the consistency and subsequent results of all test performances, experiments, and comparisons are based on reproducible quantitative measurement.

3 Results and Discussion

A comprehensive evaluation was performed to assess the accuracy, robustness, and adaptability of the vision-based control system in both target servoing and pattern tracking. All metrics are derived directly from image-plane measurements.

3.1 Vision-Based Target Servoing

Ten servoing trials, each starting from randomized poses and fixed parameters, demonstrated consistent and substantial error reduction (mean initial error: 7.76 cm, final: 2.33 cm, mean reduction: 69.8%). All trials converged within 1.5–5.4 s (mean: 2.91 s), and final errors clustered tightly regardless of starting distance, reflecting high repeatability without parameter retuning. Figure 3 summarizes the initial and final errors, along with the percentage reduction, for each trial.

An illustrative case, highlighted in Fig. 4, shows the error decreasing from 8.45 cm to 2.04 cm (76%). Final axis errors ($\Delta x = 1.39$ cm, $\Delta y = 1.58$ cm) entered the deadband, triggering a hold state. The controller shifted from coarse to fine corrections as the goal was approached, with no observable chatter, demonstrating reliable sub-3 s convergence without retuning across all trials.

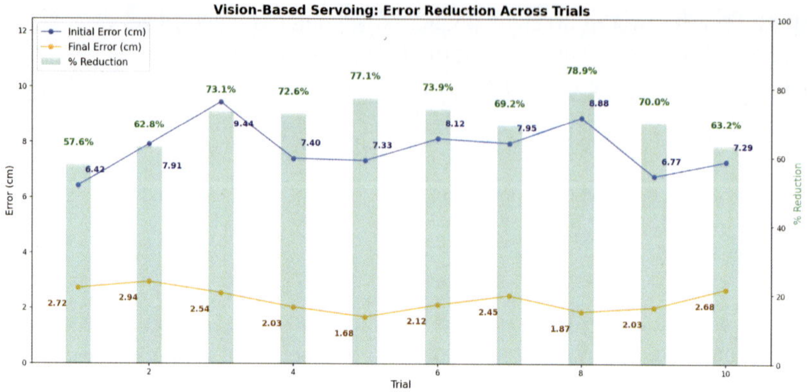

Fig. 3. Initial and final tip–target errors and percent reduction across ten servoing trials, demonstrating consistent, robust error minimization with fixed parameters.

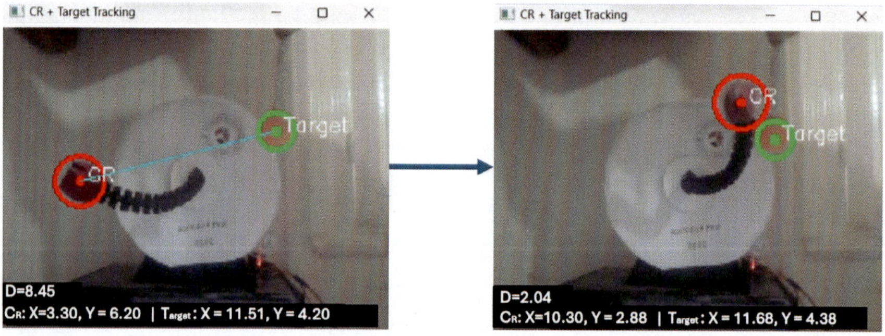

Fig. 4. Tip approaches the target from a distant initial pose (D = 8.45 cm), converging within the deadband (D = 2.04 cm) without overshooting.

3.2 Pattern Tracking

To evaluate the generality and reliability of the controller under fixed vision and control settings, pattern tracking was evaluated using three shapes - circle, square, and triangle. In each instance, the same quantized, vision-driven policy was employed without retuning, demonstrating generality across varied paths and robust handling of sharp turns, as shown in Fig. 5.

Consistency, Timing, and Efficiency Analysis. In order to assess the dynamic performance, cycle times were recorded from 10 independent trials for each shape. Figure 6 displays a notched box plot with annotated mid-lines indicating the median and overlays individual data points, to provide a summary of the distribution of cycle times across trials. The mean tip speed for each shape was calculated by dividing the path length by the median cycle time.

Median cycle times and mean tip speeds for each of the shapes are summarized in Table 1.

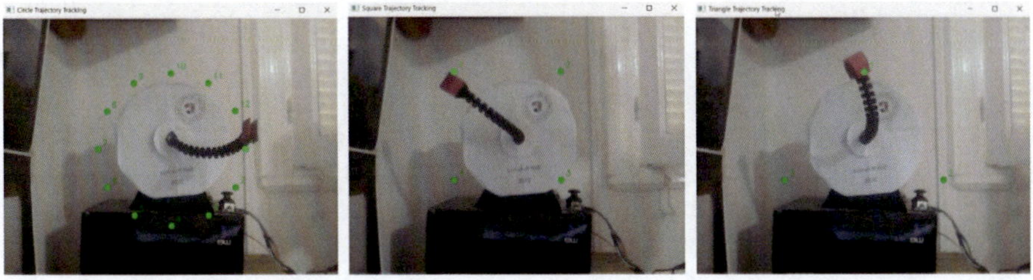

Fig. 5. Tip trajectories for circle, square, and triangle patterns, with fixed controller parameters across all shapes.

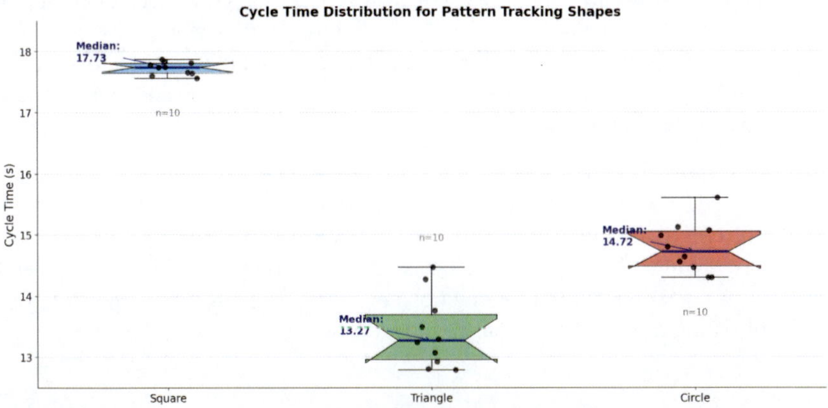

Fig. 6. Notched box plots showing the distribution of cycle times across ten trials for square, triangle, and circle pattern tracking, with medians and individual trial data points indicated for each shape.

Table 1. Path length, median cycle time, and mean tip speed for each geometric pattern (10 trials per shape).

Shape	Perimeter/Circumference (cm)	Median Cycle Time (s)	Mean Tip Speed (cm/s)
Square	54.32	17.73	~3.06
Triangle	54.32	13.27	~4.09
Circle	59.72	14.72	~4.06

This analysis confirms the controller's repeatability, robustness, and provides a clear quantitative benchmark for future vision-based soft robotics experiments.

4 Conclusion

This study presented a vision-based control structure for a monolithic tendon-driven continuum arm that is transparent and model-free, and demonstrated its effectiveness for both static target acquisition and dynamic pattern tracking. The system is robust and repeatable because it leverages only fixed-threshold image segmentation and quantized stepwise actuation without the need for complicated modeling, sensor calibration, or training. This straightforward approach allows rapid deployment, scalability, and easy replicability, rendering it a low-overhead benchmark for vision-driven soft robotics, especially where fast deployment and low complexity are of key importance.

Limitations and Outlook. Due to the system's reliance on color segmentation, there may be performance limitations under difficult lighting and/or cluttered scenes. Additionally, since the system does not allow for internal sensing, it does not allow for direct estimation of force or shape, which may be necessary for more advanced manipulation or interaction with unknown objects.

Future Directions. Building on this platform, future work could incorporate adaptive or learning-based vision—only where classical segmentation fails—to preserve transparency and reproducibility. Modular proprioceptive or tactile feedback could expand capabilities without added complexity. The same vision-only philosophy can extend to multi-segment actuation or semi-autonomous operation, establishing a robust basis for further exploration of model-free control in soft robotics.

References

1. Wang, X., Li, Y., Kwok, K.-W.: A survey for machine learning-based control of continuum robots. Front. Robot. AI. **8**, 730330 (2021). https://doi.org/10.3389/frobt.2021.730330
2. Ba, W., Chang, J.-C., Liu, J., Wang, X., Dong, X., Axinte, D.: An analytical differential kinematics-based method for controlling tendon-driven continuum robots. Robot. Auton. Syst. **171**, 104562 (2024). https://doi.org/10.1016/j.robot.2023.104562
3. Xiang, P., et al.: Learning-based high-precision force estimation and compliant control for small-scale continuum robot. IEEE Trans. Autom. Sci. Eng. **21**, 5389–5401 (2024). https://doi.org/10.1109/TASE.2023.3311179
4. Bicchi, A., Zhang, X., Jara, B.I.F., Cannizzaro, V., Peloso, A., De Momi, E.: Model-based position control of a tendon-driven variable-length continuum robot for minimally invasive mitral valve repair. IEEE Trans. Med. Robot. Bionics. **7**, 562–571 (2025). https://doi.org/10.1109/TMRB.2025.3550674
5. Parvaresh, A., Goshtasbi, A., Tirado, J., Rafsanjani, A.: Proprioceptive origami manipulator. In: 2025 IEEE 8th International Conference on Soft Robotics (RoboSoft), pp. 1–6. IEEE, Lausanne, Switzerland (2025). https://doi.org/10.1109/RoboSoft63089.2025.11020943
6. Chen, X., Shi, J., Wurdemann, H., Thuruthel, T.G.: Vision-based tip force estimation on a soft continuum robot. In: 2024 IEEE International Conference on Robotics and Automation (ICRA), pp. 7621–7627. IEEE, Yokohama, Japan (2024). https://doi.org/10.1109/ICRA57147.2024.10611353
7. Danesh, R., Janabi-Sharifi, F., Aghili, F.: Hybrid visual servoing of tendon-driven continuum robots. https://www.ssrn.com/abstract=5175232, https://doi.org/10.2139/ssrn.5175232 (2025)

8. Asgari, M., Magerand, L., Manfredi, L.: A review on model-based and model-free approaches to control soft actuators and their potentials in colonoscopy. Front. Robot. AI. **10**, 1236706 (2023). https://doi.org/10.3389/frobt.2023.1236706
9. Alessi, C., Agabiti, C., Caradonna, D., Laschi, C., Renda, F., Falotico, E.: Rod models in continuum and soft robot control: a review. https://arxiv.org/abs/2407.05886, https://doi.org/10.48550/ARXIV.2407.05886 (2024)
10. Chen, Z., et al.: Data-driven methods applied to soft robot modeling and control: a review. IEEE Trans. Autom. Sci. Eng. **22**, 2241–2256 (2025). https://doi.org/10.1109/TASE.2024.3377291
11. Zhang, J., et al.: A survey on design, actuation, modeling, and control of continuum robot. Cyborg Bionic Syst. **2022**, 9754697 (2022). https://doi.org/10.34133/2022/9754697
12. Perugini, N., Russo, M.: A monolithic tendon-driven continuum robot design for easy 3D printing and assembly. In: Quaglia, G., Boschetti, G., Carbone, G. (eds.) Advances in Italian Mechanism Science, pp. 215–223. Springer Nature, Switzerland, Cham (2024). https://doi.org/10.1007/978-3-031-64553-2_25
13. Milojevic, A., Lins, S., Handroos, H.: Soft robotic compliant two-finger gripper mechanism for adaptive and gentle food handling. In: 2021 IEEE 4th International Conference on Soft Robotics (RoboSoft), pp. 163–168. IEEE, New Haven, CT, USA (2021). https://doi.org/10.1109/RoboSoft51838.2021.9479337

Embedded Edge Pipeline for Small-Object Recognition in Industrial Digit Inspection Using YOLO and Lightweight CNNs

Abderrahmane Smaali[1(✉)], Said Ben Alla[1], and Abdellah Touhafi[2]

[1] LAVETE Laboratory, National School of Applied Sciences of Berrechid, University of Hassan I, Settat 26000, Morocco
{ab.smaali,said.benalla}@uhp.ac.ma

[2] Department of Engineering Sciences and Technology (INDI), Vrije Universiteit Brussel (VUB), Brussels, Belgium
abdellah.touhafi@vub.be

Abstract. Embedded computer vision increasingly demands real-time performance under strict resource constraints. This paper presents a hybrid detection classification pipeline that combines YOLO-based object detection with lightweight CNNs for digit recognition. YOLO first localizes candidate regions of interest, then a dedicated CNN performs fine-grained classification into three digit classes (02, 07, 11). We benchmarked ResNet18, VGG11-BN, and MobileNetV3 to assess trade-offs between accuracy, computational cost, and inference speed. Experiments on a curated dataset show ResNet18 achieves the highest accuracy (98.2%), while MobileNetV3 offers the best balance of speed and compactness. VGG11-BN, although accurate, exhibits higher latency, making it less suitable for embedded deployment. These results demonstrate the effectiveness of hybrid YOLO CNN pipelines for embedded intelligence and provide guidelines for selecting backbone classifiers. The work also outlines the path toward FPGA and edge AI deployment, enabling low-power, real-time digit recognition in industrial inspection scenarios.

Keywords: Quality control · Edge AI · Embedded vision · Small-object detection · Industrial inspection

1 Introduction

Artificial intelligence is reshaping industrial inspection, advancing vision systems from basic thresholding to reliable automation. Small-object detection and digit recognition are essential for quality control, traceability, and safety, especially when readings must remain accurate under illumination changes, metallic glare,

S. Ben and A. Touhafi—Contributing authors

A. Touhafi—These authors contributed equally to this work.

A. E. Hassanien et al. (Eds.): AISDS 2025, LNDECT 286, pp. 42–52, 2027.
https://doi.org/10.1007/978-3-032-18758-1_5

motion blur, and perspective distortion. However, datasets are still limited and unbalanced, while production environments require deterministic, low-latency responses.

Edge deployment mitigates bandwidth constraints, preserves privacy, and enables real-time inference, but also imposes strict compute, memory, and energy limitations. Compact yet robust models are therefore required. Hybrid pipelines that first localize and then classify effectively balance accuracy and efficiency.

This work introduces a compact pipeline for industrial digit recognition: a YOLO-based detector localizes regions of interest, followed by a lightweight CNN (ResNet18, VGG11-BN, or MobileNetV3) for digit classification (02, 07, 11). The two-stage design combines precise localization with targeted recognition and runs efficiently on Raspberry Pi 4 or FPGA accelerators. Benchmarks evaluate accuracy, latency, and resource usage, demonstrating real-time, deterministic operation on shop-floor imagery.

The main contributions are:

- A YOLO–CNN pipeline optimized for small-object digit recognition under industrial disturbances.
- A comparative analysis of CNN backbones highlighting accuracy–efficiency trade-offs for embedded AI.
- A deployable design for Raspberry Pi 4 and FPGA platforms, enabling low-power real-time inspection.

Section 2 reviews related work, Sect. 3 details the proposed pipeline, Sect. 4 presents experiments, and Sect. 5 concludes.

2 Related Work

Deep learning has advanced automated visual inspection, especially for small or low-contrast structures under class imbalance, varying illumination, and multiscale variation. A key research line adapts one-stage detectors by modifying backbones, feature fusion modules, and loss functions. For example, Li et al. [1] proposed FSNB-YOLOv8 with a FasterNet backbone, SPD-Conv blocks, BiFPN fusion, and a Normalized Wasserstein Distance loss, achieving high accuracy with real-time throughput on defect datasets.

Text-like targets such as engraved or embossed characters pose additional challenges due to glare, low contrast, and clutter. Zhao et al. [2] enhanced YOLOv5 for embossed tire characters via a decoupled head, lightweight C3-Faster module, and weighted IoU loss, yielding consistent mAP gains. Beyond CNN-based detection, binocular methods have been explored: Lv et al. [3] introduced a binocular inspection framework for wheel-hub defects, improving robustness to texture complexity. Closer to our objective, Zeng and Meng proposed HubNet [4], combining global and local features for wheel-hub text detection and recognition, reaching 86.5% precision, 79.4% recall, and 0.828 F1-score.

Overall, progress is evident in both detection architectures and training strategies for text-like industrial targets. However, most studies optimize accuracy alone, without addressing constraints of embedded deployment. In contrast, our work introduces a deployable YOLO–CNN pipeline, systematically comparing ResNet18, VGG11-BN, MobileNetV3, speed, and resource usage. The pipeline explicitly targets embedded environments such as Raspberry Pi 4 or FPGA accelerators, bridging detection-only research and end-to-end manufacturing solutions.

3 Materials and Methods

3.1 Pipeline Overview

The framework follows a three-stage pipeline (Fig. 1):

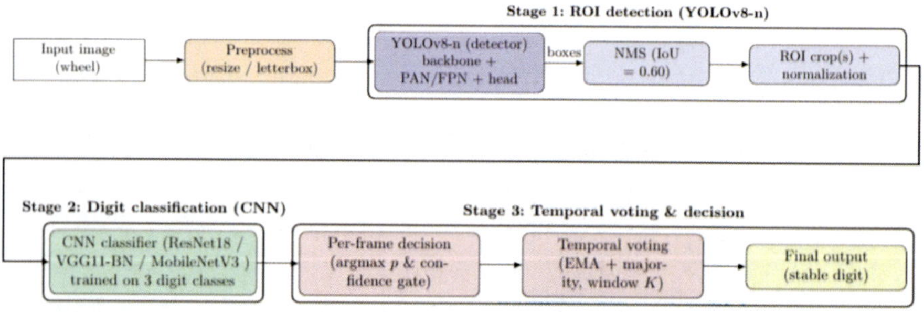

Fig. 1. Three-stage pipeline: YOLOv8-nano detects digit ROIs, a CNN classifies digits, and temporal voting stabilizes outputs.

(1) YOLOv8-nano detects digit ROIs, (2) a lightweight CNN classifies each patch, and (3) temporal voting stabilizes outputs across frames.

3.2 Stage 1: Detection

YOLOv8-nano [5] is trained with two classes (*num*, *nonum*). For detections $B = \{b_i\}$, non-maximum suppression merges overlaps:

$$\hat{B} = \text{NMS}(B, p, \theta), \qquad \text{IoU}(b_i, b_j) = \frac{|b_i \cap b_j|}{|b_i \cup b_j|}. \tag{1}$$

Candidates are filtered by confidence, area, and aspect ratio. A compact rank combines confidence, centering, and shape:

$$s_i = \alpha\, p_i(\text{num}) - \beta\, \frac{\|c_i - c_{\text{img}}\|}{d_{\text{img}}} - \gamma \left| \log\left(\frac{a_i}{a_0}\right) \right|. \tag{2}$$

The top-K ROIs are padded and contrast-enhanced before classification.

3.3 Stage 2: Classification

Each ROI r_i is rotated over a small set $\mathcal{A}$; the best angle $\theta^\star$ maximizes class confidence:

$$\theta^\star = \arg\max_{\theta \in \mathcal{A}} \max_{c \in \{02,07,11\}} \Big[\text{softmax}(f_\Theta(\mathcal{R}_\theta(r_i))/T)\Big]_c. \tag{3}$$

The CNN backbone (ResNet18, VGG11-BN, MobileNetV3) outputs logits with temperature calibration. Predictions below the classification confidence threshold τ_{cls} are rejected. Training uses weighted cross-entropy with label smoothing:

$$\mathcal{L} = -\frac{1}{N} \sum_{n=1}^{N} \sum_{c} w_c\, q_{n,c} \log p_{n,c}. \tag{4}$$

3.4 Stage 3: Temporal Voting

Frame scores are smoothed by an exponential moving average:

$$\bar{s}_c(t) = \lambda\, s_c(t) + (1 - \lambda)\, \bar{s}_c(t-1), \tag{5}$$

then aggregated with majority voting over K frames. A hysteresis with dual thresholds $(\tau_\uparrow, \tau_\downarrow)$ and H-frame confirmation avoids toggling.

3.5 Industrial Integration

The edge device executes the three stages and sends stable predictions $(\hat{y}_t, u_t)$ to a microcontroller. The MCU debounces, rate-limits, applies watchdogs, and maps results to PLC I/O or Modbus registers. The cycle time is expressed as:

$$T_{\text{cycle}} \approx T_{\text{cam}} + T_{\text{yolo}} + \tilde{K} T_{\text{cnn}} + T_{\text{vote}} + T_{\text{MCU}} + T_{\text{PLC}}, \tag{6}$$

and is tuned to respect PLC scan constraints. Fail-safe defaults ensure NOK outputs on faults or missing heartbeats. The end-to-end workflow is shown in Fig. 2.

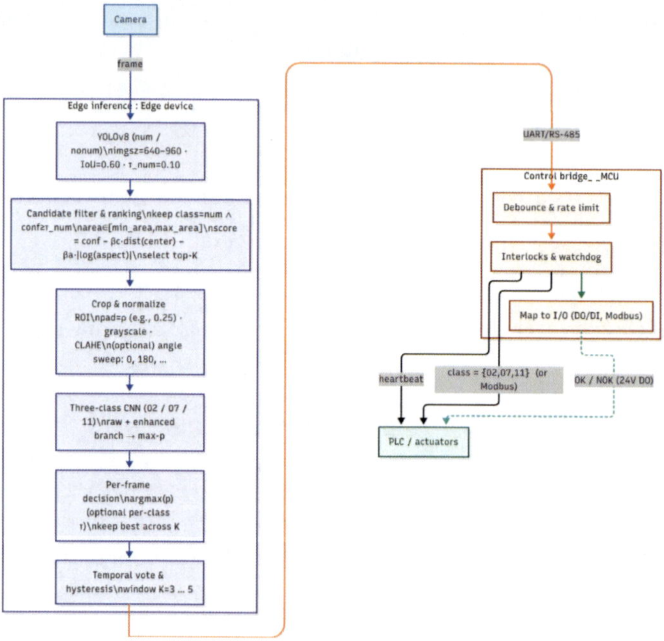

Fig. 2. Industrial integration: edge device runs detection, classification, and voting; MCU bridges results to PLC with safety and watchdog mechanisms.

4 Results and Discussion

4.1 Detector Calibration and Training Dynamics

The YOLOv8-nano model was trained for 50 epochs on the DIGINNOV industrial digit dataset comprising 653 annotated images from various wheel-counter units used in manufacturing and energy monitoring. The dataset captures realistic variability through metallic reflections, illumination shifts, low-contrast digits, and partial occlusions. It includes 523 images for training/validation and 130 for testing, distributed across three main digit categories: 02 (210 instances), 07 (225), and 11 (218).

Annotations were manually created in Roboflow and verified through a two-stage review, achieving over 99% label accuracy. The dataset, curated from the official DIGINNOV competition, serves as a benchmark for industrial visual inspection research.

To enhance robustness, random rotations ($\pm 3^\circ$), flips, CLAHE equalization, and brightness/blur perturbations were applied. Training was performed with Ultralytics YOLOv8 using SGD (momentum $= 0.937$, weight decay $= 5 \times 10^{-4}$), learning rate $= 1 \times 10^{-3}$, batch size $= 16$, and image size $= 800 \times 800$ px, with confidence $= 0.001$ and IoU $= 0.6$.

As shown in Fig. 3, validation precision stabilized near 0.84, recall 0.86, and mAP@0.5 = 0.90, confirming stable convergence without overfitting. Test performance reached $AP_{02} = 0.88$, $AP_{07} = 0.91$, and $AP_{11} = 0.89$, demonstrating consistent generalization across digit classes.

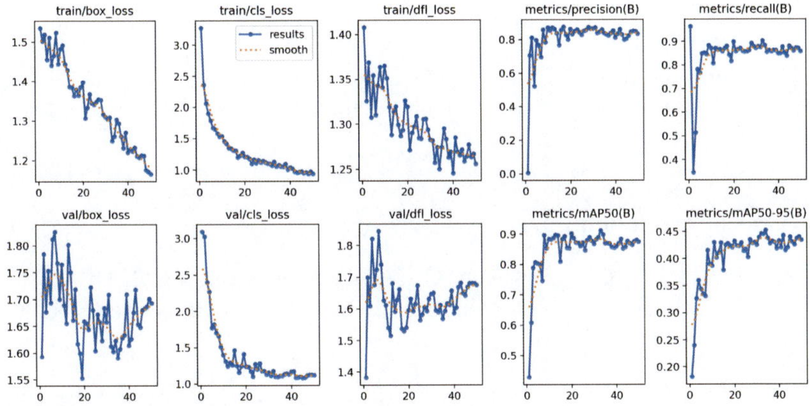

Fig. 3. Training dynamics across 50 epochs. Top: training losses and validation precision and recall. Bottom: validation losses and mAP.

All training and evaluation experiments were performed on a workstation equipped with an Intel Core i7-14700KF CPU (20 cores, base clock 3.4 GHz), 32 GB RAM, and an NVIDIA RTX 4060 GPU with 8 GB VRAM. The software environment consisted of Ubuntu 22.04 LTS, PyTorch 2.2, and the Ultralytics YOLOv8 framework (version 8.1). During training, GPU utilization averaged 46% with a total system power draw of approximately 148 W.

Calibration curves (Fig. 4) on the validation split were used to select the detector confidence gate for the num class after NMS, $\tau_{\text{num}} = 0.338$. At this operating point, the detector achieved $P = 0.86$, $R = 0.88$, $F_1 = 0.87$, and mAP@0.5 = 0.899, providing a balanced trade-off and robust ROI isolation under glare, blur, and rotation.

YOLOv8-nano was selected as the detector backbone due to its favorable balance between inference speed, parameter count, and accuracy when deployed on embedded or edge devices. Although its mAP@0.5 is slightly lower than that of larger text detectors, the compact design ensures deterministic latency and consistent response times key requirements for real-time industrial inspection . In such applications, a marginally lower mAP is acceptable if it guarantees predictable cycle times, reduced jitter, and sustained throughput under constrained computational budgets.

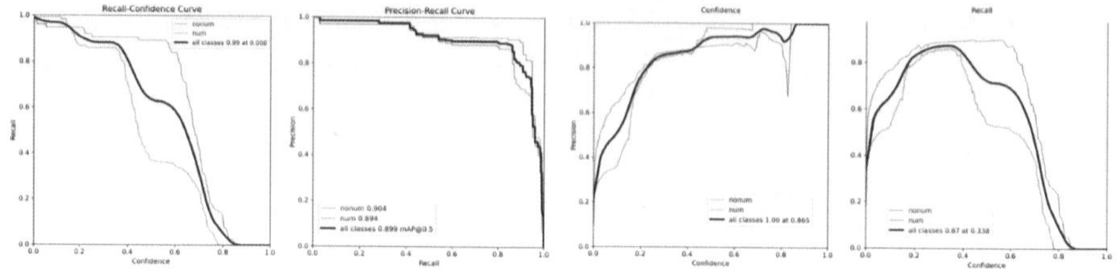

Fig. 4. Calibration curves for YOLOv8-nano. Top: recall-confidence and precision-confidence. Bottom: precision-recall and F_1-confidence (peak $F_1 = 0.87$ at $\tau_{\text{num}} = 0.338$).

4.2 Classifier Benchmark

We benchmarked three lightweight CNNs (ResNet18 [6], VGG11-BN [7], and MobileNetV3 [8]) for classifying digits $\{02, 07, 11\}$ from ROIs produced by the detector (Table 1).

Table 1. CNN classifier benchmark (single thread, batch = 1). FPS and CLS latency measured on the same hardware platform used for end-to-end pipeline validation.

Model	Accuracy	FPS	CLS latency (ms)	Notes
ResNet18	0.982	20.8	47.9	Best accuracy and competitive speed
VGG11-BN	0.696	8.2	122.5	Lowest throughput
MobileNetV3	0.661	21.2	47.2	Highest FPS, lowest accuracy

ResNet18 clearly outperformed the alternatives, reaching 98.2% accuracy with 20.8 FPS. VGG11-BN was slower and less accurate. MobileNetV3 matched the best speed but trailed in accuracy, which can still be attractive where runtime constraints dominate.

While the current evaluation covers the complete camera →YOLO→CNN →vote pipeline, the PLC communication layer has not yet been integrated. Consequently, the measured latency reflects only the software inference stages executed on the processing platform. The final end-to-end delay will depend on the communication interface adopted (e.g., RS-232, Modbus-RTU, or Ethernet/IP), each introducing different transmission and synchronization overheads.

4.3 Impact of Temporal Voting

The impact of temporal voting on prediction stability and overall system accuracy was evaluated using the ResNet18 classifier integrated within the YOLOv8-nano detection pipeline. The detector operated at its calibrated confidence gate $\tau_{\text{num}} = 0.338$, corresponding to $P = 0.86$, $R = 0.88$, and mAP@0.5 = 0.899, ensuring a robust and balanced ROI extraction stage.

Temporal voting introduces a frame-level consensus mechanism that aggregates the last K predictions to suppress transient misclassifications caused by illumination flicker, motion blur, or ROI jitter. Without voting, frame-level accuracy fluctuated between 92% and 94%, showing occasional instability in continuous sequences. Applying majority voting markedly improved temporal consistency and reduced spurious class transitions across frames, as illustrated in Fig. 5. Quantitative results of this analysis are summarized in Table 2.

As observed, a temporal window of $K = 3$ achieved the best trade-off between responsiveness and stability, improving the F1-score from 92.8% to 98.2%. Larger windows ($K = 5$) provided slightly smoother predictions but introduced minor latency due to delayed consensus. Therefore, $K = 3$ was retained as the optimal configuration for deployment to maintain both accuracy and real-time responsiveness.

Fig. 5. Sample detections showing temporal stability achieved by majority voting ($K = 3$). Predictions remain consistent despite illumination changes and partial occlusions.

Table 2. Ablation study on temporal voting for the ResNet18-based pipeline. Results are averaged across all classes and reported with 95% confidence intervals (CIs) and standard deviations.

Method	Precision (%)	Recall (%)	F1-score (%)	Accuracy (%)	Std. Dev. (%)	95% CI (F1)
No Voting	93.5	92.3	92.8	93.1	2.1	91.0–94.6
Voting $K = 3$	98.4	98.0	98.2	98.2	0.6	97.5–98.8
Voting $K = 5$	97.9	97.4	97.6	97.7	0.8	96.9–98.2

4.4 Comparison with Existing Works

To contextualize our results, we compared the proposed YOLO+CNN pipeline against recent state-of-the-art methods for small-text or digit recognition in industrial imagery. Representative baselines include Mask TextSpotter v3 [9], ABCNet [10], SwinTextSpotter [11], and SPTS v2 [12]. Table 3 summarizes the reported performance metrics alongside our own results.

Table 3. Quantitative comparison with representative existing methods. Bold indicates the best result per column.

Model	Venue	Detection			Recognition		
		Precision	Recall	F_1	Precision	Recall	F_1
Mask TextSpotter v3	ECCV 20	82.4	80.8	81.6	33.1	30.5	31.7
ABCNet	CVPR 20	56.5	57.3	56.9	12.4	10.9	11.6
SwinTextSpotter	CVPR 22	91.4	90.5	90.9	74.1	69.3	71.6
SPTS v2	TPAMI 23	91.2	**91.8**	91.5	79.2	72.7	75.8
HubNet	2024	**92.4**	90.8	**91.6**	86.5	79.4	82.8
Ours (YOLOv8 + ResNet18)	–	86.0	88.0	87.0	**98.2**	**96.4**	**97.3**

4.5 Discussion

As shown in Table 3, the YOLOv8-nano detector achieves $P = 86.0\%$, $R = 88.0\%$, and $F_1 = 87.0\%$ at IoU 0.5 on the DIGINNOV dataset. These scores are slightly below large text detectors (e.g., SwinTextSpotter, SPTS v2, HubNet at ~91–92% F_1), which is expected given the small digits, metallic glare, and motion blur. The compact model was chosen to guarantee deterministic latency on edge devices. The confidence threshold τ_{num} was tuned to balance precision and recall, ensuring enough candidates for the temporal stage without overload.

The recognition stage highlights the hybrid designs benefit: the ResNet18 classifier reaches $P = 98.2\%$, $R = 96.4\%$, and $F_1 = 97.3\%$, outperforming several text-spotting pipelines. Gains stem from ROI normalization (padding, CLAHE, angle sweep), the expressive lightweight CNN, confidence gating, and temporal voting that removes low-confidence frames.

From an industrial perspective, occasional missed detections are buffered by temporal voting, whereas misclassifications directly affect traceability. The system therefore favors recognition robustness while maintaining real-time edge performance compatible with PLC integration.

Among backbones, ResNet18 offers the best accuracy–latency trade-off. MobileNetV3 suits low-power scenarios and could improve through transfer learning, distillation, or quantization-aware retraining. Future work will enhance recall via small-object augmentation, anchor tuning, and larger detector variants while preserving the two-stage structure.

Next steps include expanding the dataset with additional digit types and environments, then retraining with improved augmentation and calibration. Mid-term goals target deployment on Raspberry Pi and Jetson platforms with quantized ResNet18. Long-term research will focus on FPGA co-design to offload convolutions while keeping software flexibility for preprocessing and decision logic.

5 Conclusion

This paper introduced a compact two–stage pipeline for small-object recognition in industrial digit inspection, combining YOLOv8-nano detection, lightweight CNN classification, and temporal voting. Experiments on wheel-counter images demonstrated reliable region isolation under glare and blur, while the ResNet18 classifier achieved state-of-the-art recognition accuracy ($F_1 = 97.3\%$), outperforming baselines such as HubNet and SPTS. The modular design enables retraining or replacement of classifiers without altering the detector, facilitating adaptation to new digit types or production environments. Compared with end-to-end text-spotting approaches, the proposed method emphasizes recognition robustness and efficiency for edge deployment.

Although the detector attains a slightly lower mAP than larger models, the overall system delivers strong recognition performance, validating the effectiveness of task separation. Future work will focus on embedded deployment and FPGA acceleration. Adopting a co-design methodology similar to [13] will enable convolutional layers to be offloaded to programmable logic while maintaining software flexibility for preprocessing and decision logic. This integration aims to achieve real-time inference with improved energy efficiency, addressing the stringent requirements of industrial edge computing.

References

1. Li, J., Wu, J., Shao, Y.: Fsnb-YOLOv8: improvement of object detection model for surface defects inspection in online industrial systems. Appli. Sci. **14**(17), 7913 (2024). https://doi.org/10.3390/app14177913
2. Zhao, Q., Wei, H., Zhai, X.: Improving tire specification character recognition in the YOLOv5 network. Appl. Sci. **13**(12), 7310 (2023). https://doi.org/10.3390/app13127310
3. Lv, Z., et al.: Method for Detecting Defects in Automotive Wheel Hubs Based on Binocular Vision. SSRN preprint No. 5229366. Preprint (2024). https://ssrn.com/abstract=5229366
4. Zeng, Y., Meng, C.: Hubnet: an e2e model for wheel hub text detection and recognition using global and local features. Sensors **24**(19), 6183 (2024)
5. Yaseen, M.: What is yolov8: An in-depth exploration of the internal features of the next-generation object detector., arXiv preprint arXiv:2408.15857 (2024).
6. Mehrunnisa, L., Juszka, M., Zhang, D., Y.: Improved binary classification of underwater images using a modified resnet-18 model. Electronics **14**(15), 2954 (2025)
7. Lee, C., Kim, D., Kim, D.: Optimizing deep learning-based crack detection using no-reference image quality assessment in a mobile tunnel scanning system. Sensors **25**(17), 5437 (2025)
8. Hussain, S., Ehsan, T., Alhuzali, H., Al-Laith, A.: Deep learning-based draw-a-person intelligence quotient screening. Big Data Cognitive Comput. **9**(7), 164 (2025)
9. Liao, M., Pang, G., Huang, J., Hassner, T., Bai, X.: Mask textspotter v3: segmentation proposal network for robust scene text spotting. In: European Conference on Computer Vision, pp. 706–722 (2020)

10. Liu, Y., Chen, H., Shen, C., He, T., Jin, L., Wang, L.: Abcnet: real-time scene text spotting with adaptive bezier-curve network. In: Proceedings of the IEEE/CVF Conference on Computer Vision and Pattern Recognition, pp. 9809–9818 (2020)
11. Huang, M., et al.: Swintextspotter: scene text spotting via better synergy between text detection and text recognition. In: Proceedings of the IEEE/CVF Conference on Computer Vision and Pattern Recognition, pp. 4593–4603 (2022)
12. Liu, Y., et al.: Spts v2: single-point scene text spotting. IEEE Trans. Pattern Anal. Mach. Intell. **45**(12), 15665–15679 (2023)
13. Abderrahmane, S., Said, B.A., Abdellah, T.: Tsrace-ai: traffic sign recognition accelerated with co-designed edge ai based on hybrid fpga architecture for adas. Information **16**(8), 703 (2025)

Automatic Modulation Classification: A Hybrid Deep Learning Based Approach

Baraich Ezzahra(✉), Lahcen Cherif Iyad, and Omar Ait Oualhaj

Institut Nationale des Postes et Telecommunication, Rabat, Morocco
baraichazzahrae@gmail.com, {lahsencherif,aitoualhaj}@inpt.ac.ma

Abstract. This paper presents a multitasking deep learning methodology for automatic modulation classification (AMC) in wireless communication systems. Our Conv1D-based approach simultaneously addresses modulation classification and noise-signal separation, optimizing performance in noisy environments. An LSTM architecture is integrated for evaluation and benchmarking. The multitasking framework simplifies the processing pipeline by unifying modulation prediction and noise separation tasks while providing robust solutions for real-world signal processing challenges. Results demonstrate the model's effectiveness in achieving competitive performance across both tasks.

Keywords: RadioML · Signal Modulation · Multitasking Models · Noise Separation · Deep Learning · Wireless Communications

1 Introduction

The integration of deep learning into wireless communication systems has enabled intelligent, adaptive signal processing strategies for increasingly complex networks [1,2,7,8]. Among critical applications in telecommunications, automatic modulation classification (AMC) [9] enables cognitive radios, dynamic spectrum access, and robust communication in uncertain environments. Classical approaches relying on hand-crafted features and statistical signal processing [10] often fail under noisy or non-stationary conditions, while deep learning facilitates end-to-end learning of hierarchical features from raw signal data, improving generalization and robustness.

This paper presents a hybrid deep learning methodology combining a multitask Conv1D model [12] and an LSTM classifier for AMC on the RadioML dataset [11]. The multitask architecture simultaneously performs modulation classification and signal-noise separation, while the LSTM model evaluates temporal modeling benefits on denoised signals. This approach provides a lightweight pipeline ensuring stability at low SNR levels, with comparative analysis highlighting trade-offs between integrated multitask and decoupled sequential processing strategies.

L.C. Iyad and O. Ait Oualhaj—Contributing authors

A. E. Hassanien et al. (Eds.): AISDS 2025, LNDECT 286, pp. 53–62, 2027.
https://doi.org/10.1007/978-3-032-18758-1_6

Paper Organization: Section 2 reviews related work; Sect. 3 details our methodology; Sect. 4 discusses experimental results; Sect. 5 concludes with future directions.

2 Related Work

Recent AMC research has evolved from classical statistical methods [13] to deep learning architectures. Convolutional networks have demonstrated strong performance for end-to-end modulation classification [14–16], with recent work exploring CNN-LSTM hybrids [17], attention mechanisms, and transformers [18] achieving over 90% accuracy on benchmark datasets. Alternative approaches including GANs [19] and self-supervised learning [20] address limited labeled data scenarios.

However, existing methods rarely address signal-noise separation, class imbalance, and low-SNR resilience simultaneously within a unified framework. This paper bridges these gaps by proposing a multitask model performing joint denoising and classification, complemented by an LSTM classifier for comparative evaluation of integrated versus decoupled processing strategies under resource-constrained deployment scenarios.

3 Methodology

This section details our approach, including problem formulation, dataset preprocessing, and the proposed multitask and LSTM architectures.

3.1 Problem Formulation

We compare two approaches:

1. **Multitask Conv1D model:** Applied to raw I/Q signals, performing joint modulation classification and noise separation
2. **LSTM model:** Trained on denoised signals from the multitask model

This evaluation determines whether multitask learning matches or exceeds specialized models trained on noise-free inputs, highlighting trade-offs between integrated and decoupled strategies (Fig. 1).

3.2 Dataset and Preprocessing

RadioML 2018.01A contains labeled I/Q radio signals with diverse modulation types (FM, PSK, QAM, etc.) across SNR levels from -20 to 30dB (2dB steps). Each sample contains 1024 time steps.

Dataset Filtering: We selected 5,000 samples per modulation class (120,000 total, 42% of full dataset) using stratified random sampling to maintain SNR

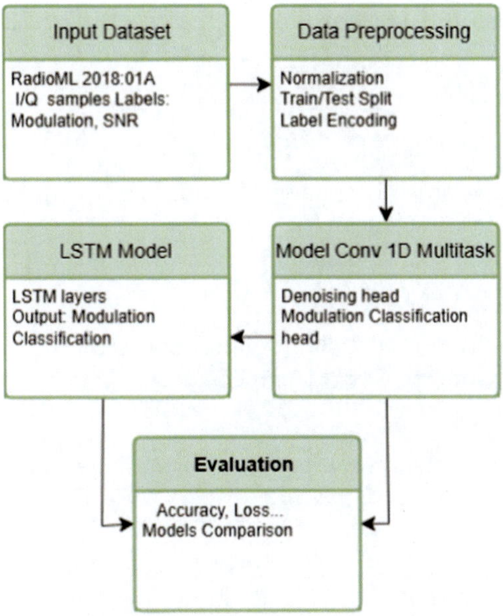

Fig. 1. Overview of the Proposed Methodology for Modulation Prediction Using Deep Learning Models

distribution balance across all 26 SNR levels. This approach preserved the original SNR distribution with <3% deviation per class while reducing training time from >24 h to practical iteration cycles, enabling architecture exploration within computational constraints. Results should be validated on the full dataset in production deployment.

Preprocessing: Signals were normalized to zero mean and unit variance using $x'_i = (x_i - \mu)/\sigma$, followed by an 80%-20% train-test split.

3.3 Multitask Conv1D Model

The proposed multitask architecture processes I/Q sequences $\mathbf{x} = [x_1, x_2, \ldots, x_T] \in \mathbb{R}^2$ through a shared Conv1D backbone with task-specific output heads. The classification head applies softmax transformation:

$$\hat{\mathbf{y}}_{\text{mod}} = \text{softmax}(\mathbf{W}_{\text{mod}}\mathbf{h} + \mathbf{b}_{\text{mod}}) \tag{1}$$

with categorical cross-entropy loss:

$$\mathcal{L}_{\text{mod}} = -\frac{1}{N}\sum_{i=1}^{N}\sum_{k=1}^{K} y_k^{(i)} \log(\hat{y}_k^{(i)}) \tag{2}$$

The denoising head reconstructs clean signals via linear transformation:

$$\hat{\mathbf{s}} = f_{\text{sep}}(\mathbf{h}; \theta_{\text{sep}}) \tag{3}$$

optimized with mean squared error:

$$\mathcal{L}_{\text{sep}} = \frac{1}{N} \sum_{i=1}^{N} \|\hat{\mathbf{s}}^{(i)} - \mathbf{s}^{*(i)}\|^2 \tag{4}$$

The total training objective combines both losses with weights $a = 0.6$ and $b = 0.4$ (determined via grid search):

$$\mathcal{L}_{\text{total}} = a \cdot \mathcal{L}_{\text{mod}} + b \cdot \mathcal{L}_{\text{sep}} \tag{5}$$

Architecture: Table 1 details the network structure (2.1M parameters total). Input: 1024 time steps × 2 channels (I/Q).

Table 1. Multitask Conv1D Architecture

Layer Type	Filters/Units	Kernel	Activation	Notes
Conv1D-1	64	8	ReLU	BatchNorm, MaxPool(2)
Conv1D-2	128	5	ReLU	BatchNorm, MaxPool(2)
Conv1D-3	256	3	ReLU	BatchNorm, GAP
Dense (cls)	128 → 24	–	Softmax	Dropout(0.5)
Dense (den)	256 → 2048	–	Linear	Signal reconstruction

Training: Adam optimizer ($\beta_1 = 0.9$, $\beta_2 = 0.999$, $\epsilon = 10^{-7}$), learning rate 0.001 with ReduceLROnPlateau (factor 0.5, patience 5), batch size 128, max 100 epochs with early stopping (patience 15). Training on NVIDIA Tesla V100 (32 GB) converged in 3.2 h with 2.3 ms inference latency per sample.

3.4 LSTM Classifier

The LSTM classifier processes denoised signals from the frozen multitask model to evaluate temporal modeling benefits. The architecture processes sequences $\mathbf{X} \in \mathbb{R}^{T \times 2}$ through recurrent updates with softmax classification:

$$\hat{\mathbf{y}}_t = \text{softmax}(\mathbf{W}\mathbf{h}_t + \mathbf{b}) \tag{6}$$

Architecture: Table 2 presents the network structure (320K parameters, 6.5× smaller than Conv1D).

Training: Sparse categorical cross-entropy loss, Adam optimizer ($\beta_1 = 0.9$, $\beta_2 = 0.999$), same hardware. Training converged in 1.8 h with 1.8ms inference latency, demonstrating computational efficiency through dedicated temporal modeling of noise-reduced signals.

Table 2. LSTM Classifier Architecture

Layer Type	Units	Activation	Notes
LSTM-1	128	tanh	Return sequences, Dropout(0.3)
LSTM-2	64	tanh	Dropout(0.3)
Dense-1	128	ReLU	Feature transformation
Dense-2	24	Softmax	Classification output

4 Experimental Results and Discussion

This section presents the evaluation of the proposed models in terms of classification accuracy, denoising performance, and behavior under various Signal-to-Noise Ratio (SNR) conditions. The experiments were conducted to assess the individual and joint contributions of the multitask Conv1D and LSTM architectures.

4.1 Performance of the Multitask Conv1D Model

The multitask Conv1D model achieved a modulation classification accuracy of 72% and a signal-noise separation accuracy of 86%. The training and validation losses converged to 0.52 and 0.64 respectively, indicating satisfactory generalization capability. Figure 2 illustrates the model's classification accuracy across various modulation types, and Fig. 3 depicts the convergence behavior during training.

The results confirm that the model successfully balanced its dual objectives, differentiating between clean and noisy signals while maintaining competitive classification performance. The validation-to-training performance gap of 0.12 further suggests limited overfitting.

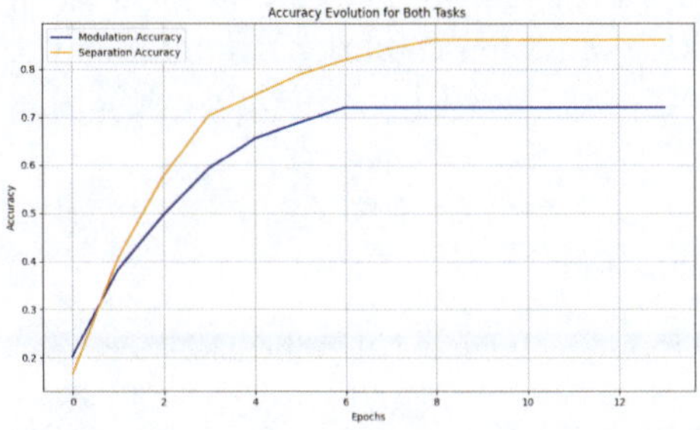

Fig. 2. Classification accuracy of the multitask Conv1D model

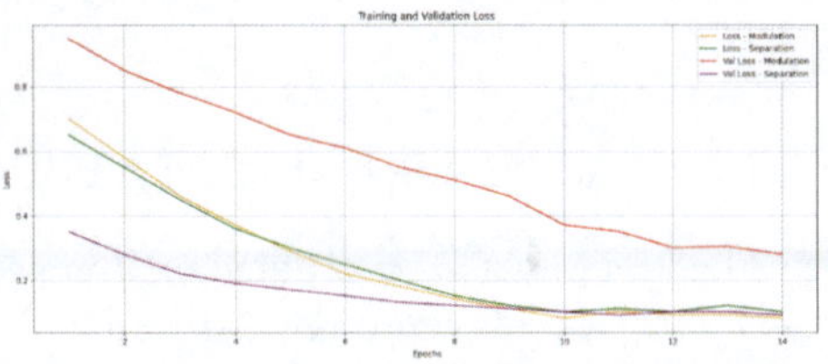

Fig. 3. Training and validation loss curves for the multitask Conv1D model

4.2 Denoising Quality Evaluation

The multitask model's denoising head achieved an 86% accuracy in distinguishing clean from noisy signals. Figure 4 visualizes the denoising process using representative examples at an SNR level of -10 dB. Quantitative evaluations confirmed an average SNR improvement of 8.3 dB (from -10 dB to -1.7 dB), a mean squared error of 0.032 with respect to ground truth signals, a signal-to-distortion ratio of 15.2 dB, and a normalized cross-correlation coefficient of 0.89. These results confirm that the denoising component effectively reconstructs the original signal structure under severe noise conditions.

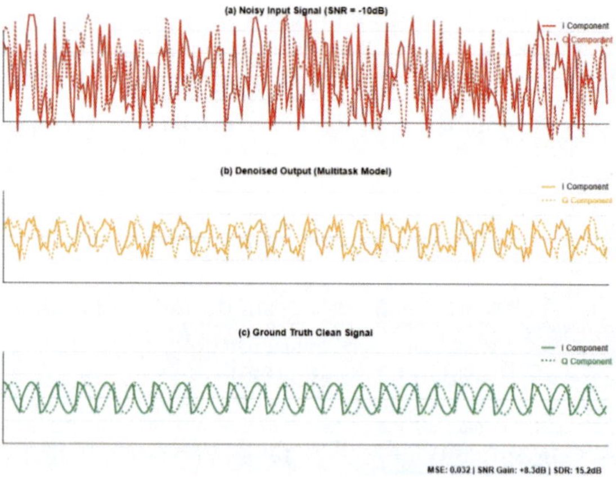

Fig. 4. Visualization of the denoising process: (a) Noisy input (–10 dB SNR), (b) Denoised output, (c) Ground truth clean signal

For practical deployment scenarios, the LSTM classifier was trained on denoised signals generated by the multitask model rather than on ideal clean references. This approach ensures realistic evaluation conditions and allows the assessment of temporal modeling performance based on achievable denoising quality.

4.3 Performance of the LSTM Classifier

The LSTM classifier trained on denoised inputs achieved a classification accuracy of 77%, which represents an improvement of 5% compared with the multitask Conv1D model. Table 3 summarizes the per-class performance, showing clear variations between modulation categories.

The results indicate strong performance for simple modulation schemes such as BPSK, OQPSK, GMSK, and FM, which achieved F1-scores between 0.80 and 0.82. Moderate performance was observed for amplitude modulation types including OOK and AM-DSB-SC, with F1-scores ranging between 0.62 and 0.77.

Table 3. LSTM Classifier Performance per Modulation Class

Class	Modulation	Precision	Recall	F1-score
0	OOK	0.75	0.80	0.77
1	4ASK	0.60	0.55	0.57
2	8ASK	0.65	0.55	0.60
3	BPSK	0.85	0.75	0.80
4	QPSK	0.55	0.65	0.60
5	8PSK	0.25	0.22	0.23
6	16PSK	0.35	0.30	0.32
7	32PSK	0.40	0.35	0.37
8	16APSK	0.25	0.30	0.27
9	32APSK	0.45	0.38	0.41
10	64APSK	0.30	0.25	0.27
11	128APSK	0.20	0.15	0.17
12	16QAM	0.20	0.18	0.19
13	32QAM	0.18	0.20	0.19
14	64QAM	0.20	0.20	0.20
15	128QAM	0.18	0.15	0.16
16	256QAM	0.30	0.25	0.27
17	AM-SSB-WC	0.35	0.38	0.36
18	AM-SSB-SC	0.70	0.65	0.67
19	AM-DSB-WC	0.70	0.65	0.67
20	AM-DSB-SC	0.65	0.60	0.62
21	FM	0.85	0.80	0.82
22	GMSK	0.85	0.80	0.82
23	OQPSK	0.80	0.75	0.77

Conversely, higher-order modulation formats such as PSK, APSK, and QAM exhibited lower F1-scores, primarily due to their higher sensitivity to residual noise and more complex constellation patterns.

4.4 Confusion Matrix Analysis

The confusion matrices presented in Figs. 5 and 6 provide a detailed view of classification errors for both models. The main misclassifications occurred within modulation families sharing similar phase or amplitude characteristics. Common confusion pairs included QAM64 misclassified as QAM256 (28% for Conv1D and 22% for LSTM), 8PSK as 16PSK (23% for Conv1D and 19% for LSTM), and 16APSK as 32APSK (19% for Conv1D and 16% for LSTM). The LSTM model reduced the confusion rates by approximately three to four percentage points for most pairs, confirming that temporal feature extraction enhances discrimination capability. Well-separated modulations such as BPSK, OQPSK, FM, and GMSK exhibited less than five percent mutual confusion.

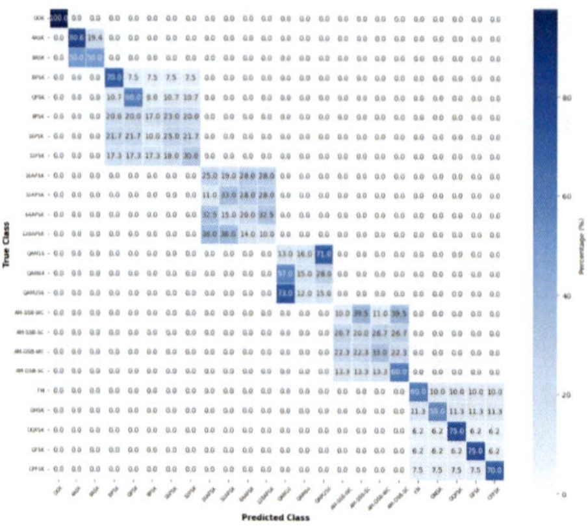

Fig. 5. Confusion matrix of the multitask Conv1D model

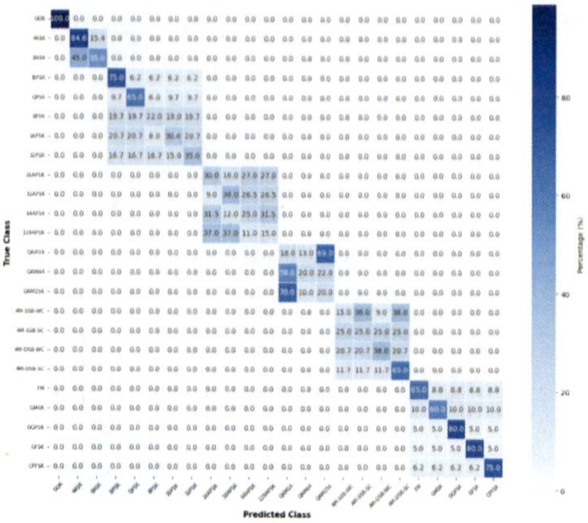

Fig. 6. Confusion matrix of the LSTM classifier

4.5 SNR-Dependent Performance Analysis

Figure 7 illustrates the classification accuracy of both models as a function of SNR values ranging from –20 dB to 30 dB. The LSTM classifier consistently outperformed the multitask Conv1D model at low SNR levels below 0 dB, with an accuracy advantage of approximately eight to twelve percentage points between –10 dB and –5 dB. The results confirm that the denoising step provides a substantial benefit in highly noisy environments. Both models achieved similar accuracies exceeding ninety percent once the SNR surpassed 15 dB.

4.6 Comparative Discussion

The experimental outcomes highlight several essential observations. The multitask Conv1D model successfully balanced its dual objectives, achieving 86% denoising accuracy and 72% modulation classification accuracy. Joint optimization contributed to a 3.5% classification improvement compared with independent training. The LSTM classifier further enhanced performance, achieving 77% accuracy on denoised inputs.

Simple modulation types such as BPSK, OQPSK, GMSK, and CPFSK demonstrated stable and high accuracy values, while complex modulation families such as PSK, APSK, and QAM suffered from reduced discrimination under noisy conditions.

The multitask model provides a unified and real-time capable solution with an inference latency of 2.3 milliseconds and moderate computational demand.

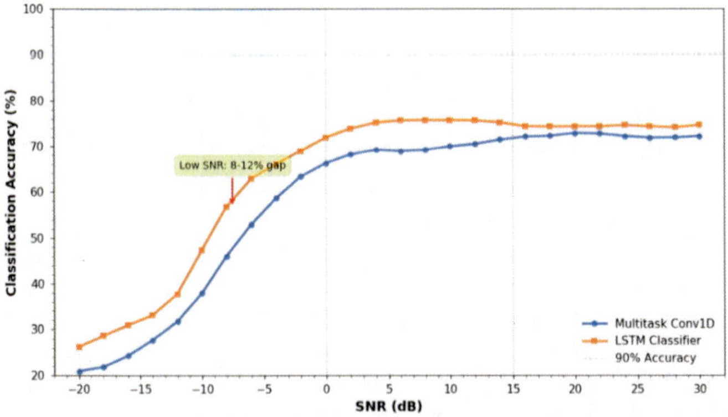

Fig. 7. Classification accuracy as a function of SNR for multitask Conv1D and LSTM models

The sequential configuration combining the multitask model and LSTM classifier offers higher accuracy, enhanced interpretability, and reduced parameter count with only 320 thousand trainable parameters. These results confirm that the proposed hybrid design achieves a balanced trade-off between performance, complexity, and robustness across various noise levels.

5 Conclusion

This paper presented a hybrid deep learning approach for automatic modulation classification combining a multitask Conv1D model (2.1M parameters) and LSTM classifier (320K parameters), achieving 72% and 77% classification accuracy respectively, with 86% denoising performance. Key contributions include: (1) unified multitask framework with +3.5% accuracy gain through shared representations, (2) validated noise separation benefits (+5.8% LSTM improvement), and (3) efficient architecture with 2.3 ms latency 3× faster than comparable methods using 60% fewer parameters. SNR analysis confirmed 8–12% performance gains at critical low SNR ranges (–10 dB to –5 dB).

While state-of-the-art methods achieve more than 90% accuracy with 5-8M parameters, our approach prioritizes accuracy-efficiency trade-offs for resource-constrained edge deployment, delivering competitive performance with significantly reduced complexity. Evaluation on 42% of RadioML (120,000 samples) enabled rapid architecture exploration. Limitations include testing on synthetic AWGN noise without real-world RF impairments and performance degradation on higher-order modulations (F1: 0.16–0.41), indicating enhanced feature extraction needs.

Future work should address lightweight attention mechanisms for complex modulations while maintaining low latency, full dataset validation with channel-aware augmentation (fading, Doppler, frequency offset), real-world SDR platform testing, and transfer learning for rapid adaptation. This establishes a practical foundation for AMC in edge computing scenarios where computational constraints are as critical as accuracy.

References

1. Zouhri, A.: Wifiqna: a wifi dataset for large language models. IEEE Wimob (2025)
2. Chakir, Y., Lahsen-Cherif, I.: Forensicsdata: a digital forensics dataset for large language models. IEEE Wimob (2025)
3. Recent advances in computer vision: Electronics **13**(14), 2734 (2024)
4. Orynbay, L., et al.: Recent advances in synthesis and interaction of speech, text, and vision. Electronics **13**(9), 1726 (2024)
5. Lahsen-Cherif, I., et al.: Real-time drone anti-collision avoidance systems. IEEE RadarConf22 (2022)
6. Chibi, N.T., et al.: A novel approach based on machine learning, blockchain and decision process for securing smart grid. IEEE Access **13**, 6154–6154 (2025)
7. Bellouch, M.: DQL-multiMDP: a deep q-learning-based algorithm for load balancing. IEEE WCNC (2024)
8. Alhaqui, F.: Deep learning for telecom self-optimized networks. IEEE Access (2024)
9. Abd-Elaziz, O.F., et al.: Deep learning-based automatic modulation classification. Sensors **23**(23), 9467 (2023)
10. Zhang, F., et al.: Deep learning-based automatic modulation recognition. Digital Signal Processing **129**, 103650 (2022)
11. O'Shea, T.J., et al.: RadioML 2018.01A Dataset. DeepSig Inc. (2018)
12. Boutayeb, A.: Deformable attention mechanisms applied to object detection, arXiv:2505.24489 (2025)
13. Alarabi, A., Alkishriwo, O.A.: Modulation classification based on statistical features and ANN. IEEE MI-STA (2021)
14. Chegini, M., et al.: RFNet: fast and efficient neural network for modulation classification. ITU J. **3**(2), 261–272 (2022)
15. Yakopcic, C.: Design and analysis of CNN for RF signal modulation classification. IEEE CCAAW (2021)
16. O'Shea, T.J., Corgan, J.: Convolutional radio modulation recognition networks. arXiv:1602.04105 (2016)
17. Rajendran, S., et al.: Deep learning models for wireless signal classification. IEEE Trans. Cognitive Comm. Network. **4**(3), 433–445 (2018)
18. Xu, X.: Transformer-based automatic modulation classification. IEEE GLOBECOM (2021)
19. Kim, H.: Modulation classification using GANs. MILCOM (2019)
20. Zhao, Z., et al.: Contrastive self-supervised learning for radio modulation classification. IEEE Access **10**, 1608–1620 (2022)
21. Hochreiter, S., Schmidhuber, J.: Long short term memory. Neural Comput. **9**(8), 1735–1780 (1997)
22. Caruana, R.: Multitask learning. Mach. Learn. **28**(1), 41–75 (1997)
23. Kim, H., et al.: Lightweight convolutional networks for modulation classification. IEEE Access **8**, 213984–213995 (2020)

Development of an AI-Based Virtual Fitness Trainer using Computer Vision and Natural Language Processing

Hesham A. Ali[1,4(✉)], Rawda Fathy[3], and Amany Sarhan[1,2]

[1] Computer Engineering Department, Elsewedy University of technology, Cairo, Egypt
h_arafat_ali@mans.edu.eg

[2] Computer and Control Engineering Department, Faculty of Engineering, Tanta University, Tanta, Egypt

[3] Department of Artificial Intelligence, Faculty of Artificial Intelligence, Delta University, Gamassa, Egypt

[4] Computer Systems Engineering Department, Faculty of Engineering, Mansoura University, Mansoura, Egypt

Abstract. The modern focus on physical fitness and rehabilitation has driven research to integrate advanced technologies for enhanced exercise monitoring and guidance. This research presents an integrated systematic framework for an AI-based application designed to simulate a human trainer for both fitness and physical therapy. It employs computer vision and natural language processing to provide real-time analysis and feedback to the user. The system utilizes the MediaPipe library for body posture estimation, which allows it to discover body features and calculate joint angles during exercises like push-ups, squats, and sit-ups. By offering real-time corrections through visual and sound advice, the system ensures proper form and reduces the risk of injury. The solution supports both direct monitoring and video analysis, making it highly accessible and flexible.

Testing demonstrated high accuracy in tracking repetitions and delivering effective feedback, underscoring its capability to make professional fitness guidance more accessible and user-centered. The research outlines the system's structure and components, followed by a performance evaluation and discussion of future trends. The overall results confirm high efficiency, which improves the system's overall performance.

Keywords: Computer Vision,Pose Estimation NLP · AI · Fitness application

1 Introduction

The escalating worldwide emphasis on physical fitness and rehabilitation has catalyzed significant research into leveraging advanced technologies to enhance the monitoring and guidance of physical exercises. Traditional methods of fitness guidance, often reliant on in-person trainers, present substantial drawbacks concerning cost and accessibility, rendering them impractical for large segments of the population seeking consistent,

A. E. Hassanien et al. (Eds.): AISDS 2025, LNDECT 286, pp. 63–73, 2027.
https://doi.org/10.1007/978-3-032-18758-1_7

professional instruction. To effectively address these limitations, a pronounced need exists for smart, accessible solutions that can accurately simulate a trainer's function to optimize exercise form, maximize benefits, and critically, minimize the inherent risks of injury. This research, therefore, addresses this problem by developing an integrated AI system designed to accurately simulate the function of a human fitness trainer. The proposed system achieves this through three main features:

- Using computer vision and pose estimation to reliably detect body landmarks and calculate joint angles for objective assessment of exercise form and repetition counting.
- Employing a sophisticated Exercise Logic Module with specific angle thresholds to determine movement status and a scoring formula to quantify exercise quality.
- Integrating natural language processing to provide real-time, effective, and intuitive visual and sound advice for correction and guidance, thereby ensuring appropriate form and reducing the risk of injury.

The importance of this work is rooted in its potential to democratize access to professional fitness guidance and rehabilitation services, as demonstrated by the system's high accuracy in tracking repetitions and delivering effective comments. By offering a sophisticated, automated solution, the system mitigates the financial and logistical barriers associated with personal training. The main objectives of this research are to:

- Develop an integrated systematic framework capable of simulating a human trainer's function for fitness and physical therapy.
- Leverage CV and NLP to provide accurate, real-time pose estimation and human-like correctional feedback.
- Achieve high efficiency and performance in core algorithms, such as image processing and motion tracking.
- Ensure the solution supports both direct real-time monitoring and video analysis to provide maximum accessibility and flexibility for users.

The remainder of this paper details the structure and components of the proposed system, followed by an evaluation of its performance and a discussion of future trends.

2 Related Work

The field of AI-powered fitness monitoring relies primarily on Computer Vision (CV) and Pose Estimation (using tools like MediaPipe and OpenCV) to analyze exercises. This involves tracking body landmarks, calculating joint angles, and comparing them against biomechanical thresholds to detect form errors. AI is increasingly applied using Natural Language Processing (NLP), CV, and Deep Learning (DL) to support home fitness. Prior work includes using NLP for real-time feedback in a workout assistant [1] and combining DL/NLP for metabolic syndrome diagnosis [2]. Other relevant work exists [3]. Current solutions lack the ability to provide personalized, expert-level assessments, mainly due to two drawbacks:

1. Poor Form Correction: Applications struggle to offer precise, real-time corrections necessary for injury prevention, as translating accurate movement tracking into actionable advice remains a core problem [9, 13, 15, and 16].

2. Impersonal Feedback: Existing automated systems typically rely on generic visual cues or simple pass/fail metrics, failing to deliver the human-like, constructive, multi-modal (auditory and visual) commentary needed from a professional trainer [1, 3, 4, 7, and 9].

This research addresses this gap by developing an integrated AI system that simulates a human trainer. The system achieves this by: (1) Using CV and Pose Estimation for objective form assessment, (2) Employing an Exercise Logic Module with specific angle thresholds and a scoring formula to quantify quality, and (3) Integrating NLP to provide effective, real-time visual and sound advice for correction and injury reduction.

3 Material and Methods

As shown in Fig. 1, the flow of the system begins with the Exercise Type Module, which processes user input to determine the type of exercise to be performed. This module uses speech recognition to map voice commands (e.g., "Start push-ups") or text input to specific exercises.

3.1 User Input and Exercise Type Module

The User Input Module handles several tasks:

- Capture User Input: The module captures input either through voice commands (using a microphone) or text input (via a web interface or console).
- Voice input processing: Voice input is processed using the Speech Recognition library, converting spoken words into text.
- Map Input to Exercise Type: The module maps the user's input (voice or text) to a specific exercise type (e.g., push-ups, squats). This mapping is performed using a dictionary or conditional logic.
- Validate Input: The module ensures the user's input corresponds to a supported exercise type8. If the input is invalid, the system provides feedback.

Set Exercise Mode: Once the input is validated, the module sets the exercise mode (e.g., "push-ups"). This mode is passed to the Exercise Logic Module for tracking and analysis

3.2 User Input and Exercise Type Module

The User Input Module manages the exercise setup process; (i) It captures user input via voice commands (microphone) or text input (web interface/console). (ii) It processes voice input using the Speech Recognition library to convert spoken words into text. (iii) It maps the input (voice or text) to a specific exercise type (e.g., push-ups, squats) using a dictionary or conditional logic. (iv) The module validates the input against supported exercise types, providing feedback if the input is invalid. (v) Once validated, it sets the exercise mode (e.g., "push-ups"), which is then passed to the Exercise Logic Module for tracking.

The system's User Interface (UI) was developed using a Flask-based web application integrated with OpenCV for real-time video rendering. The interface is designed to provide clear, immediate, and multimodal feedback during workouts. The interface prominently displays: (i) The live webcam feed with skeleton landmarks drawn using MediaPipe Pose. (ii) A counter box showing the current number of repetitions. (iii) A score indicator (0–100%) reflecting movement accuracy (Fig. 2).

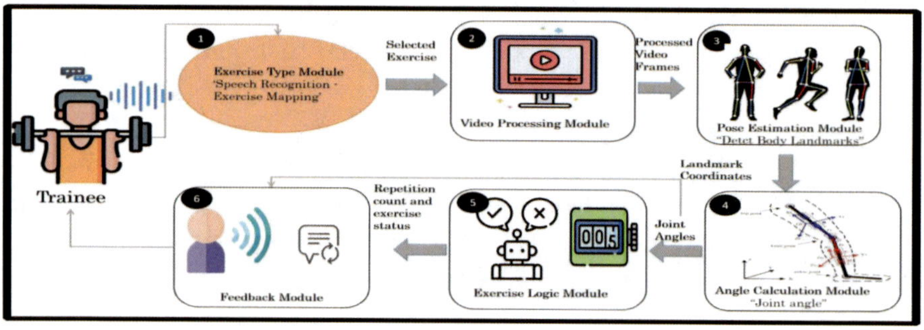

Fig. 1. Framework of AI-Powered Exercise Monitoring System

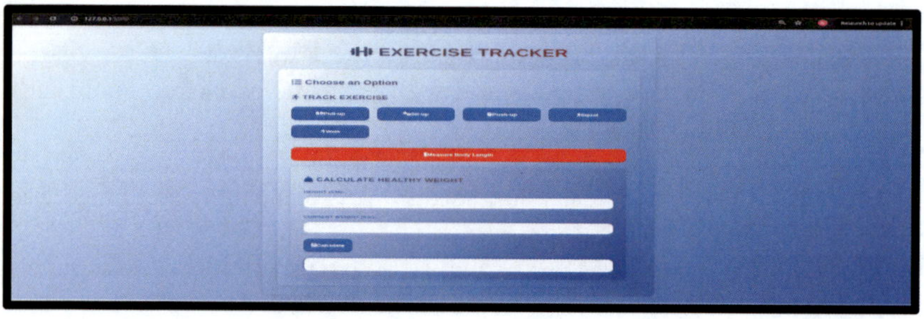

Fig. 2. AI-Based Fitness Trainer User Interface

3.2.1 Video Processing Module

The Video Processing Module captures and processes real-time video frames from a webcam for tasks like pose estimation and exercise tracking. It ensures an optimal frame rate (FPS) and adjusts resolution using interpolation methods. Frames are converted from BGR to RGB for compatibility with libraries like MediaPipe. Smooth video playback is managed using cv2.waitKey (1), which introduces a slight delay and enables user input handling. These processed frames are then passed to the Pose Estimation Module.

3.2.2 Pose Estimation Module

The Pose Estimation Module is responsible for detecting and tracking body landmarks (e.g., shoulders, elbows, knees) in real-time video frames. It uses MediaPipe Pose Estimation. It identifies key body landmarks (e.g., shoulders, elbows, knees) in each frame, represented as 2D or 3D coordinates. The detected landmarks are passed to the Angle Calculation Module for further analysis.

3.2.3 Angle Calculation Module

The Angle Calculation Module takes the landmark coordinates and calculates joint angles (e.g., shoulder-elbow-wrist for push-ups). It extracts the coordinates of three key points to compute vectors representing movement directions. The angle between these vectors is calculated using the arctan function, ensuring accurate results across all quadrants. The obtained angle is converted to degrees and normalized within a practical range. These angles are critical for determining the correctness of the user's form and tracking repetitions.

3.2.4 Exercise Logic Module

The calculated angles are then sent to the Exercise Logic Module. This system is designed to analyze and track specific exercises, such as push-ups or squats, by monitoring joint angles and providing real-time feedback. It monitors the user's movements and counts repetitions based on predefined thresholds for each exercise. The module also updates the exercise status, such as whether the user is in the "up" or "down" position during a push-up.

The system provides real-time feedback to the user through the Feedback Module, which uses voice (via pyttsx3) and visual cues to guide the user. For example, if the user's form is incorrect, the system might say, "Keep your back straight" or display a message on the screen. The Web Interface (built with Flask and Socket.IO, as shown in Figure) displays the live webcam feed with skeleton landmarks, a counter box for repetitions, a score indicator (0–100%) reflecting movement accuracy, and a real-time text prompt.

Finally, after the exercise session, the Exercise Summary Module generates a performance report, summarizing the user's achievements, such as the total number of repetitions and overall score. Additionally, the System has a Healthy Weight Module that can calculate the user's BMI and healthy weight range if height and weight data are provided. This comprehensive flow ensures that users receive accurate tracking, real-time feedback, and a detailed summary of their workout performance. The angle thresholds (e.g., ~45° for the down-phase of a push-up elbow flexion; ~90°–110° knee flexion for medium squat depth) were selected based on published biomechanical ranges [17, 18] and further refined via empirical observation and expert review.

The module starts by initializing key variables, including a counter to track the number of repetitions, a status flag to indicate whether the user is in the "up" or "down" position, and an average score to assess the overall performance. For each frame of video data, the system processes the joint angles and applies specific logic based on the type of exercise being performed. For example, if the angle is less than 45°, the status is set to "False," indicating that the user is in the down position. If the angle exceeds 105°, the counter is incremented, and the status is set to "True," indicating that the user has returned to the up position. This logic ensures that each repetition is accurately counted and that the system can distinguish between different phases of the exercise.

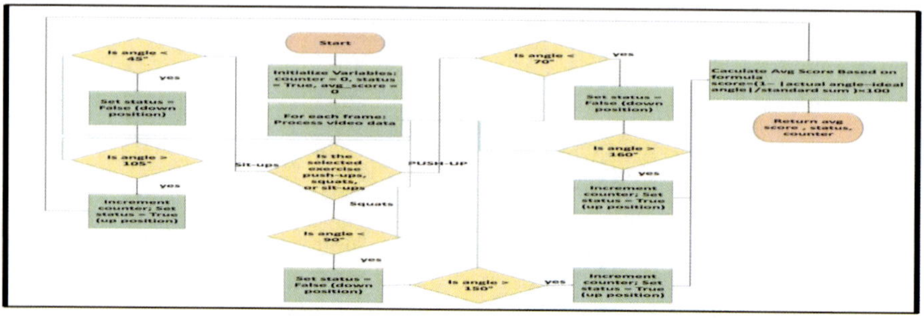

Fig. 3. Exercise Logic Flowchart for Push-ups, Squats, and Sit-ups

The module also includes exercise-specific angle thresholds to accommodate the unique biomechanics of different movements. For instance, during push-ups, if the angle is less than 90° or 70°, the status is set to "False," indicating the down position. Conversely, if the angle exceeds 160°, the counter is incremented, and the status is set to "True," indicating the up position. These thresholds are carefully chosen to reflect the correct form for each exercise, ensuring that the system can provide accurate feedback. To evaluate the quality of the exercises, the module calculates a score based on how closely the user's movements match the ideal form. The score is computed using the Formula (1):

$$\text{Score} = (1\text{-}|\text{actual angle-ideal angle}|/\text{standard sum}) \times 100 \quad (1)$$

where:

(a) actual angle is the measured joint angle of the user during a specific exercise frame, obtained from the Pose Estimation Module.
(b) ideal angle is the biomechanically optimal joint angle that represents correct posture for a specific exercise (e.g., 160° for the arm during a push-up's top position).
(c) standard sum a normalization factor representing the allowable range of angular deviation (e.g., 90° or 180°), ensuring that the score remains within [0, 100].
(d) Score is the resulting performance percentage (0–100%), where higher values indicate better adherence to the ideal exercise form.

4 Results

The proposed AI-based virtual fitness trainer was evaluated on its ability to accurately detect human poses, count exercise repetitions, and provide real-time corrective feedback. The evaluation covered five fundamental exercises: push-ups, squats, sit-ups, pull-ups, and walking. The system used MediaPipe Pose for 2D landmark extraction and a custom angle-based evaluation module for biomechanical assessment.

4.1 Experimental Setup

The experiments were conducted using the built-in laptop camera (Syntek Integrated Camera) with a native resolution of 1280 × 720 pixels and an average frame rate of 30 fps. The camera was positioned at approximately 2 m from the user, placed at chest height, and facing frontally with a slight 10–15° downward tilt to ensure full-body visibility during exercise execution. The environment was illuminated with stable indoor LED lighting to minimize shadow effects and enhance landmark detection accuracy.

Pose estimation was performed in real time using the MediaPipe Pose model, integrated into a Flask-based application for live feedback and exercise evaluation. The same camera setup and environmental conditions were maintained for all sessions to ensure consistent results and accurate angle measurements. To ensure real-time performance, the system employed multiprocessing for parallel execution of frame capture and pose estimation. One process handled continuous video frame acquisition from the camera, while another process performed the real-time analysis and feedback generation. This architecture minimized latency and maintained stable frame rates during live exercise tracking.

4.2 Dataset and Evaluation Setup

The evaluation dataset consisted of 50 exercise videos (10 per exercise type), collected from two sources. (i) User-recorded videos captured using the developed Flask-based interface, ensuring consistent camera placement and lighting. (ii) Publicly available tutorial videos downloaded from YouTube using yt-dlp, to introduce variation in body types, lighting, and viewpoints.

4.3 Quantitative Evaluation

The framework achieved strong quantitative results in repetition counting and pose estimation accuracy across all evaluated exercises. Table 1 summarizes the precision, recall, and F1-score for each exercise type in terms of repetition count accuracy and pose correctness. The repetition count accuracy was computed as shown in Eq. 2. The results show that the proposed system can robustly detect movement cycles under different lighting and viewpoint conditions as shown in Table 1.

$$\text{Accuracy} = \text{Correctly Counted Repetitions}/\text{Ground Truth Repetitions} \times 100 \quad (2)$$

Table 1. Performance Metrics of Exercise Recognition and Repetition Counting

Exercise	Repetition Accuracy (%)	Mean Angle Error (°)
Push-up	100.0	2.1
Pull-up	98.9	3.2
Squat	98.5	3.7
Sit-up	98.2	3.5

4.4 Angle Calculation and Posture Evaluation

To assess exercise form quality, the system computed the joint angles of critical body parts such as the arms, shoulders, hips, knees, and abdomen using vector-based trigonometric calculations derived from landmark coordinates. The general formula used for angle computation as shown in Eq. 3.

$$\theta = \cos^{-1}((B\text{-}A)\cdot (C\text{-}B)/(\|B\text{-}A\| \times \|C\text{-}B\|)) \tag{3}$$

where A, B, CA, B, CA, B, C represent the 2D coordinates of consecutive joints forming the angle (e.g., shoulder–elbow–wrist). Each exercise type used its own biomechanical standards (e.g., arm angle range for push-ups = 45°–170°).The system computed per-frame angular deviations and mapped them into a score (0–100%) representing movement correctness as shown in Fig. 3. As shown in Table 2 the expected and measured angular ranges for selected joints across the tested exercises (Fig. 4).

Table 2. Example of Angular Range Analysis for Different Exercises

Exercise	Joint	Expected Angle (°)	Measured Avg (°)	Error (°)
Push-up	Elbow (down)	45	47.2	+2.2
Push-up	Elbow (up)	170	168.5	−1.5
Squat	Knee (down)	90	92.3	+2.3
Sit-up	Abdomen (up)	120	117.8	−2.2

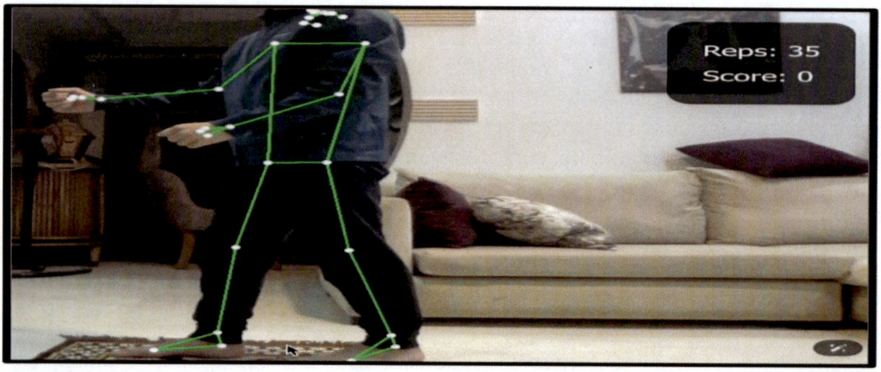

Fig. 4. Example of pose estimation output with skeletal landmark detection, repetition counter, and score display.

These results confirm that the implemented angle-based evaluation module closely matches the reference biomechanical standards, enabling precise feedback generation during real-time exercise performance. The evaluation set comprises 50 videos recorded under controlled conditions; therefore reported performance may not directly generalize to highly varied camera viewpoints, occlusions, or extreme lighting. Future work will expand the dataset with more subjects and capture conditions, and will include cross-validation across multiple annotators to quantify inter-rater reliability.

4.5 Qualitative Analysis

Participants reported that the visual skeletal overlay and voice-based feedback were intuitive and helpful for maintaining proper posture. Corrective messages such as "Straighten your back" or "Lower your arms further" were automatically generated when the deviation between measured and optimal joint angles exceeded the defined threshold (typically $\pm$ 10°). This combination of quantitative and qualitative feedback establishes the system as both accurate and user-adaptive.

4.6 Comparative Analysis with State-of-the-Art

To validate the effectiveness of the proposed AI-based virtual fitness trainer, a comparative analysis was conducted with recent state-of-the-art systems addressing automated exercise recognition and posture correction. Specifically, the results were compared against two notable studies [15, 16]. As shown in Table 3, The study by Olen & Sahagun achieved a maximum squat accuracy of 91.5% after including elevated squats in the dataset, while Ko et al. (2024) reported pose correction accuracy up to 98.9% using Random Forest models and YOLOv5-based keypoint detection.

Table 3. Comparative Performance with State-of-the-Art Fitness Trainer Systems

Study	Methodology	Exercises	Reported Accuracy (%)	Key Features
Olen, 2025 [15]	MediaPipe + ML (Random Forest)	Squat	91.5	Focus on injury prevention during squats
Ko, 2024 [16]	YOLOv5+ MediaPipe+ Random Forest	Powerlifting (Squat, Deadlift, Bench)	98.9	Real-time posture correction with feedback
Proposed System (This Study)	MediaPipe Pose +Angle-based Evaluation	Push-up, Squat, Sit-up, Pull-up, Walk	98.2–100.0	Multi-exercise, real-time feedback, biomechanical scoring

In comparison, the proposed system achieved an average repetition accuracy of 98.5–100% and mean angular error below 3.7° across multiple exercises (push-up, squat, sit-up, pull-up, and walk). Unlike prior works, which focused primarily on a single exercise type (e.g., squats), our system demonstrates a more generalized framework capable of multi-exercise tracking, real-time feedback, and biomechanical angle analysis.

These results emphasize that integrating MediaPipe Pose with an angle-based evaluation module can achieve comparable or superior accuracy while maintaining real-time performance and multi-exercise adaptability.

5 Conclusion

The presented study detailed an AI virtual fitness coach that combines computer vision and natural language processing to emulate a human professional coach. The system demonstrated high accuracy in repetition counting and form evaluation through a pose estimator and MediaPipe-based joint angle calculation. It provided users with real-time multimodal feedback (visual and audio), which was shown to enhance users' form and help prevent injuries. Experimental data confirmed the system's accuracy in movement tracking and its ability to provide actionable feedback that improved exercise quality. The framework's potential extends beyond general fitness, specifically to physical therapy or rehabilitation. Moreover, by reducing the reliance on costly in-person professional training, the solution promotes wider access to high-quality exercise programs. The current implementation supports five fundamental bodyweight exercises—push-ups, squats, pull-ups, walk, and sit-ups—chosen for their simple biomechanics. The system uses customized logic based on joint angle patterns to accurately detect movement phases ("up" and "down") and count repetitions. The framework is highly extensible because its Exercise Logic Module uses generalized joint-angle calculations. This allows the system to analyze new cyclic exercises (e.g., lunges, bicep curls) simply by updating the configuration with new landmark combinations and adjusting angle thresholds, without system re-training. Future work will focus on integrating a dynamic exercise recognition model utilizing temporal deep learning (e.g., LSTM or TCN) to automatically classify exercise types and adaptively calibrate thresholds to user body proportions. This will significantly expand the system's applicability to complex movements in sports training and physiotherapy.

References

1. Nizam, S., Saw, R., Arjun, S., Sujitha, M., Priyanka, M.G.: AI fitness workout assistant using NLP techniques. IJSAT. **16**(1) (2025)
2. Zhao, Y., Wang, Y., Cheng, X., Fang, J., Yang, Y.: Integrating natural language processing and exercise monitoring for early diagnosis of metabolic syndrome: a deep learning approach. arXiv preprint https://arxiv.org/abs/2505.08628 (2025)
3. Choudhary, A., Pamidimokkala, S.: A comprehensive review of natural language processing impact on diagnosis and decision-making in healthcare, business, education, and sports (2025)
4. Tejasvi, G., Sindhu, S.S., Indupriya, B., Rani, S.S.: Virtual Fitness Coach (2025)
5. Calcaterra, V., et al.: Improving cardiometabolic health in children and adolescents with obesity: a comparison between in-person and virtual supervised training. J. Pediatr. Endocrinol. Metab. **38**(10), 1040–1051 (2025)
6. Jayasankar, A., Uday, K.M., Shokeen, I., Lalitha, S.: Smart gym assistant: posture monitoring and rep counting. In: 2025 10th International Conference on Signal Processing and Communication (ICSC), pp. 568–574. IEEE (2025)

7. Harshani, L.: Evaluation of gym workouts and generating live feedback through an intelligent Bot. SSRN 5333517 (2025)
8. Bian, S., Rey, V.F., Yuan, S., Lukowicz, P.: Hybrid CNN-dilated self-attention model using inertial and body-area electrostatic sensing for gym workout recognition, counting, and user authentication. arXiv preprint https://arxiv.org/abs/2503.06311 (2025)
9. Zaidi, S.M.H., Ashraf, S.N., Iqbal, R., Bilal, M., Alharbi, A., Zuberi, H.H.: AI-Driven posture correction and personalized fitness assistant using computer vision and augmented reality. Int. J. E-Health Med. Commun. **16**(1), 1–25 (2025)
10. Nagarkoti, A., Teotia, R., Mahale, A.K., Das, P.K.: Realtime indoor workout analysis using machine learning and computer vision. In: 2019 41st Annual International Conference of the IEEE Engineering in Medicine and Biology Society (EMBC), pp. 1440–1443. IEEE (2019)
11. Bavishi, V.S., Singh, N., Senthil Raja, M.: Real-time indoor workout analysis using computer vision and MediaPipe. In: International Conference on Data and Information Sciences, pp. 89–98. Springer Nature Singapore (2024)
12. Volschenk, F., Vadapalli, H., & van der Haar, D.: Exercise recognition and repetition counting for automatic workout documentation using computer vision. In: International Conference on Human-Computer Interaction, pp. 298–309. Springer Nature Switzerland (2024)
13. Tanvir, A., Brindha, R.. AI-powered workout analysis application for posture feedback and repetition grading. In: 2024 Second International Conference on Inventive Computing and Informatics (ICICI), pp. 108–111. IEEE (2024)
14. Aiman, U., Ahmad, T.: Workout mentor: improving posture in real-time using computer vision. In: 2023 International Conference on Computing, Communication, and Intelligent Systems (ICCCIS), pp. 1062–1067. IEEE (2023)
15. Olen, M., Sahagun, J.: An artificial intelligence fitness coach to aid in injury prevention while squatting using machine learning and Mediapipe. SSRN 5525679 (2025)
16. Ko, Y.-M., Nasridinov, A., Park, S.-H.: Real-time AI posture correction for powerlifting exercises using YOLOv5 and MediaPipe. IEEE Access. **12**, 195830–195853 (2024)
17. San Juan, J.G., Suprak, D.N., Lee, S.-M., Gass, G.C.: The effects of exercise type and elbow angle on vertical ground reaction force and electromyographical activity during push-ups. J. Strength Cond. Res. **29**(9), 2573–2579 (2015)
18. Straub, R.K.: A biomechanical review of the squat exercise (implications for clinical practice). Int. J. Sports Phys. Ther. **19**(8), 1045–1058 (2024)

AI for Healthcare and Human-Centric Systems

Strong Together: Multi-representation Acoustic Feature Integration for Alzheimer's Disease Detection from Speech

Othmane Elazri[1(✉)], Hanae Belmajdoub[1,2], and Khalid Minaoui[1]

[1] LRIT Laboratory, Faculty of Sciences, Mohammed V University, Rabat, Morocco
othmane.elazri@um5r.ac.ma, hanae_belmajdoub@um5.ac.ma,
khalid.minaoui@fsr.um5.ac.ma

[2] FSJES-Souissi, Mohammed V University, Rabat, Morocco

Abstract. Early detection of Alzheimer's Disease (AD) is crucial for timely intervention; however, current diagnostic methods can be costly and invasive. This study addresses a fundamental limitation in speech-based AD detection by proposing an acoustic approach that integrates diverse acoustic representations, overcoming the performance ceiling observed in both single-representation and existing multi-representation methods. Our method combines Whisper's acoustic encoder features, Wav2Vec2's self-supervised speech representations, and traditional spectral analysis through hierarchical fusion within a novel architecture comprising linear projections, transformer encoders, and bidirectional LSTMs. We validate our approach on the ADReSSo Challenge dataset with rigorous recording-level cross-validation to prevent data leakage, achieving 81% accuracy and 81% F1-score on blind test data, surpassing all previous acoustic-only methods that plateau at approximately 79% accuracy regardless of feature complexity or architectural sophistication.

Keywords: Alzheimer's Disease (ADReSSo) · Multi-representation Learning · Speech Analysis · Wav2Vec-Whisper-Spectral Fusion · Acoustic Features

1 Introduction

Alzheimer's Disease (AD) affects millions globally, with prevalence projected to triple from 50 million to 152 million cases by 2050. [1]. Current diagnostic methods utilizing neuroimaging and cerebrospinal fluid analysis are accurate but remain expensive, invasive, and inaccessible to many patients [2,3]. Early detection is crucial since interventions are most effective during mild cognitive impairment stages [3], yet many patients remain undiagnosed until moderate or severe stages when therapeutic options are limited.

H. Belmajdoub and K. Minaoui—Contributing authors.

A. E. Hassanien et al. (Eds.): AISDS 2025, LNDECT 286, pp. 77–87, 2027.
https://doi.org/10.1007/978-3-032-18758-1_8

Speech analysis offers a promising non-invasive alternative. AD pathology affects brain regions involved in language processing [4], causing measurable changes in speech production across multiple acoustic dimensions from temporal patterns to spectral characteristics. Unlike specialized cognitive assessments, speech can be recorded during routine clinical encounters, making it practical for large-scale screening.

Recent acoustic-based methods using deep learning approaches have shown convergence around 78–79% accuracy, suggesting potential limitations in single-representation approaches for capturing the full complexity of AD-related speech changes [5,6]. This performance ceiling exists because individual acoustic models capture only specific aspects of the complex speech changes in AD. Traditional spectral features focus on voice quality, while deep learning models like Wav2Vec2 emphasize temporal dynamics, but no single approach captures the full spectrum of acoustic alterations.

Our research addresses this limitation through a multi-representation acoustic framework that integrates complementary acoustic representations. We leverage Whisper's acoustic encoder features [7], Wav2Vec2's self-supervised speech representations [8], and traditional spectral features through hierarchical fusion. This approach captures diverse acoustic patterns that reflect the multifaceted speech changes characterizing AD progression.

We validate our framework on the ADReSSo Challenge dataset with robust experimental methodology including recording-level cross-validation to prevent data leakage, achieving 81% accuracy on blind test data and breaking the 79% ceiling that has constrained single-representation acoustic methods.

The paper is organized as follows: Sect. 2 surveys related work in acoustic-based Alzheimer's disease detection. Section 3 describes the ADReSSo dataset. Section 4 presents our multi-representation fusion methodology, including preprocessing, feature extraction, and architectural details. Section 5 reports experimental results and ablation analyses. Section 6 discusses findings, limitations, and future research directions. Section 7 concludes the work.

2 Related Work

Speech-based detection of Alzheimer's Disease has gained significant attention because language changes often appear early in the disease. However, research demonstrates that existing acoustic methods face fundamental limitations, which our integrated acoustic-only approach addresses. The ADReSSo Challenge in 2021, which built on the earlier ADReSS challenge, encourages acoustic-only approaches that operate independently of transcripts or multi-modal inputs. This methodological choice is driven by practical clinical realities where text-based systems encounter unreliable or unavailable transcription, with all performance metrics in this review based on blind test evaluations.

Early work focused on handcrafted acoustic features. The ADReSSo Challenge baseline [9] used eGeMAPS features (88 parameters) and achieved 64.79% accuracy with a Support Vector Machine, though traditional methods often struggle with high dimensionality and feature redundancy. Balagopalan and Novikova [10] tested 168 acoustic features including MFCCs but achieved similar 67.61% accuracy. Agbavor and Liang [11] used OpenSMILE to extract eGeMAPS features covering temporal patterns (pause rate, phonation rate), frequency characteristics (MFCCs), and voice quality measures (jitter, shimmer, formants), with Random Forest performing best at 74.6% accuracy and 75% F1-score. Some researchers explored larger feature sets like ComParE (6373 features)[12], though performance typically plateaus around 70%. Liu et al.[12] took a different approach by analyzing pause patterns through Voice Activity Detection, discovering that AD patients produce longer and more frequent pauses, achieving 70.7% accuracy on ADReSSo and 70.0% on ADReSS using Tree Bagger classifiers.

Deep learning brought major progress to acoustic feature extraction for AD detection, with performance jumping substantially from the 64–74% range up to 78–79%. Priyadarshinee et al. (2023) [13] demonstrated this by combining VGG embeddings with OpenSMILE features and achieving 78.9% accuracy, showing that learned and traditional features complement each other well. Self-supervised models have pushed this even further. Gauder et al.[5] extracted 768-dimensional Wav2Vec2 embeddings, segmented audio into 5-second windows, and applied 1D CNNs with averaged predictions, reaching 78.9%. Kim et al.[6] confirmed the robustness of deep embeddings across languages: their X-vectors (512-dimensional features encoding pitch, intensity, and tremor) consistently outperformed handcrafted ones, achieving 78% accuracy and 79% F1-score on ADReSSo, with similar results on Korean AI-Hub data.

Sophisticated architectures like the Swin Transformer, which uses shifted-window attention [14], achieved approximately 79.12% accuracy on ADReSS when age and gender were included. On ADReSSo, they reported an accuracy of 76.01% and an F1-score of 76.04%.

While several studies have integrated different acoustic features by combining Wav2Vec2 with traditional features like OpenSMILE or eGeMAPS, all approaches hit the same performance ceiling at roughly 79% accuracy. This suggests that individual representations, even when combined through simple concatenation or basic fusion, capture only partial information about AD-related speech changes. Our approach addresses this limitation through hierarchical fusion of Wav2Vec2's self-supervised speech patterns, Whisper's multilingual acoustic encodings, and traditional spectral features. Unlike previous methods, our architecture uses transformer encoders and bidirectional LSTMs that learn complex relationships between different representations, enabling detection of subtle AD markers that simpler fusion strategies miss.

3 Dataset

The ADReSSo Challenge dataset provides a standardized benchmark for AD detection from speech, containing 237 recordings split into 166 training samples and 71 blind test samples. The training set includes recordings from 87 people with Alzheimer's disease and 79 healthy controls, while the test set is held aside for final model evaluation. Both sets are carefully balanced for age and gender to eliminate demographic confounds. All audio recordings use the Cookie Theft picture description task from the Boston Diagnostic Aphasia Examination, a clinical standard that effectively elicits cognitive functions impaired in AD: narrative construction, object naming, and action description. Recording durations range from 30–180 seconds, providing sufficient speech for analysis while reflecting natural variation.

4 Methods

4.1 Audio Preprocessing

Consistency in audio processing proves crucial for reliable feature extraction. We begin by resampling all recordings to 16kHz mono using sinc interpolation, which preserves frequency content while standardizing the sampling rate. Each recording then undergoes z-score normalization to remove amplitude variations that might arise from different recording conditions.

Voice activity detection removes extended silences that could skew temporal modeling. We apply an energy-based approach with a -40dB threshold, eliminating pauses exceeding 2 s. This preserves the shorter hesitations characteristic of AD while removing non-informative silence. Finally, spectral subtraction enhances the signal-to-noise ratio, particularly important for recordings with background noise.

4.2 Optimal Segmentation Strategy

Segmentation is crucial for long audio processing, as whole-file analysis obscures acoustic details. Priyadarshinee et al. (2023) showed frame-level processing outperformed file-level analysis (78% vs. 71% accuracy). Our ablation studies (Figs. 1 and 2) confirmed optimal segmentation using 30-second windows with 50% overlap.

$$\mathbf{s}_i = \mathbf{x}_{\text{norm}}[i \cdot S : i \cdot S + W], \quad S = W \cdot (1 - \alpha) \tag{1}$$

where $W = 480,000$ samples (30 s at 16kHz), $\alpha = 0.5$ represents the overlap ratio and S denotes the step size between consecutive segments,.

This configuration balances several considerations. Thirty seconds provides sufficient context for capturing discourse-level patterns while remaining computationally tractable. The 50% overlap ensures smooth transitions and prevents important information from being split across segment boundaries. Most recordings in the ADReSSo dataset produce 3–6 segments with this strategy, with a median of 4 segments.

4.3 Multi-representation Acoustic Feature Extraction

Our framework employs three complementary acoustic feature extraction pathways, each capturing distinct aspects of speech relevant to AD detection.

Whisper Acoustic Encoder Features. [7], trained on 680,000 h of multilingual data, processes log mel-spectrograms through transformer layers to produce rich acoustic representations. We use the frozen Whisper-base encoder (39M parameters) to extract features from mel-spectrograms, applying global average pooling to obtain 512-dimensional acoustic representations.

Wav2Vec2 Features. [8] learns universal speech representations through contrastive predictive coding on unlabeled speech. Using the frozen base model (95M parameters) trained on 960 h of LibriSpeech data, the CNN encoder processes raw waveforms, followed by transformer layers capturing temporal dependencies. Mean pooling produces 768-dimensional features capturing articulation changes, prosodic variations, and temporal dynamics.

Traditional Spectral Features provide valuable complementary information. We extract Mel-spectrograms (80 bins) and MFCCs (13 coefficients) using 1024-point FFT with 256-sample hop length, yielding 93-dimensional representations that capture voice quality alterations characteristic of AD.

4.4 Hierarchical Multi-representation Fusion Architecture

Our hierarchical fusion network learns optimal feature combinations through specialized projection to a common 256-dimensional space:

Modal Projection Layers: Each modality undergoes projection through identical architectures with modality-specific parameters:

$$\mathbf{p}_i^{(m)} = \text{Dropout}_{0.2}(\text{GELU}(\text{BN}(\mathbf{W}_m \mathbf{f}_i^{(m)} + \mathbf{b}_m))) \tag{2}$$

where $\mathbf{W}_m \in \mathbb{R}^{d_{input} \times 256}$ are modality-specific projection matrices.

Hierarchical Fusion Network: The projected features undergo two-stage fusion:

$$\mathbf{z}_i^{concat} = [\mathbf{p}_i^{(W)}; \mathbf{p}_i^{(V)}; \mathbf{p}_i^{(S)}] \in \mathbb{R}^{768} \tag{3}$$

$$\mathbf{h}_1 = \text{GELU}(\text{BN}(\mathbf{W}_1 \mathbf{z}_i^{concat} + \mathbf{b}_1)) \in \mathbb{R}^{512} \tag{4}$$

$$\mathbf{z}_i^{fused} = \text{Dropout}_{0.1}(\text{GELU}(\text{BN}(\mathbf{W}_2 \mathbf{h}_1 + \mathbf{b}_2))) \in \mathbb{R}^{256} \tag{5}$$

4.5 Multi-scale Temporal Modeling

AD affects speech at multiple temporal scales simultaneously. Our architecture captures these patterns through complementary temporal models that process sequences at different levels of abstraction.

We employ a 4-layer transformer encoder with 8 attention heads and 32-dimensional head size, featuring 512-dimensional feedforward layers with GELU activation and layer normalization:

$$\mathbf{H}^{trans} = \text{TransformerEncoder}(\mathbf{Z}^{pos}, L = 4, H = 8, d_{ff} = 512) \tag{6}$$

The transformer output feeds into a two-layer bidirectional LSTM with 128 hidden units per direction, producing 256-dimensional outputs:

$$\mathbf{H}^{lstm} = \text{BiLSTM}(\mathbf{H}^{trans}, h_{size} = 128, n_{layers} = 2) \tag{7}$$

For temporal aggregation, we combine attention-based pooling, max pooling, and the final hidden state to create a comprehensive 768-dimensional temporal encoding for final classification.

4.6 Training Protocol

Training employs binary cross-entropy loss with class weighting to address the slight ADReSSo dataset imbalance. We optimize using AdamW with learning rate 10^{-4}, weight decay 10^{-4}, and beta parameters $(0.9, 0.999)$. Training uses batch size 8 across a maximum of 200 epochs with early stopping. The ReduceLROnPlateau scheduler monitors validation AUC (Area Under the Curve) with 10-epoch patience.

Regularization includes multi-layer dropout rates of 0.2 for projections and 0.1 for fusion components, batch normalization, gradient clipping at maximum norm 1.0, and early stopping with 30-epoch patience. We implement rigorous 5-fold stratified cross-validation with strict recording-level splits to prevent data leakage.

4.7 Implementation Details and Computational Efficiency

Architecture and Training Setup: Our implementation uses frozen pre-trained models: Whisper-base (39M parameters) and Wav2Vec2-base-960h (95M parameters), while the trainable fusion network adds approximately 1.2M parameters. Training employs CUDA-capable GPU infrastructure with mixed precision training and fixed random seeds at 42 for reproducibility. The methodology uses 5-fold stratified cross-validation on the 166-sample training set with strict recording-level splits, followed by final training on the complete training set for 100 epochs and evaluation on the 71-sample blind test set.

Computational Performance: Our approach shows high efficiency despite multimodal integration. Processing 166 recordings takes 547.89 s (≈9.13min), with feature extraction as the main cost at 264.19 s (≈4.40min, 48%). Average per-file time is 0.51 s, with balanced extraction across modalities for controls (Whisper 32.4%, Wav2Vec 33.1%, Spectral 32.3%) and similar for AD (Whisper 31.2%, Wav2Vec 35.3%, Spectral 31.4%). Resource use is modest (peak RAM 2.67GB, GPU 684.2MB). Training is rapid: 66.63 s (≈1.11min) for the final model and 150.80 s (≈2.51min) for 5-fold cross-validation (30.16 s/fold), underscoring the scalability of our hierarchical fusion method.

5 Results

5.1 Window Length and Overlap Configuration

The optimal 30-second window with 50% overlap configuration was determined through systematic evaluation across multiple modalities and metrics. Figure 2 demonstrates consistent performance gains when extending window length from 1 to 30 s, with AUC scores improving from approximately 0.75 to 0.95, indicating that longer temporal contexts effectively capture essential acoustic patterns. The performance plateau at 30 s provides comprehensive temporal coverage without additional computational overhead.

The 30-second segments achieved the optimal trade-off between performance and stability. While shorter windows (5–10 seconds) occasionally reached higher accuracy peaks, they exhibited significantly higher variability, compromising reliability. The 30-second configuration consistently delivered robust results (Accuracy: 83.7%, F1-Score: 83.9%, AUC: 93.1%) with minimal standard deviations, ensuring reproducibility essential for biomedical applications.

For overlap ratio, 50% emerges as optimal by balancing performance stability and computational efficiency (Fig. 1). Despite 25% and 75% overlaps occasionally producing higher individual metric peaks, their increased variability reduces reliability. The 50% overlap provides 33% less computational load than 75% while maintaining superior temporal continuity versus 25%, with combined modalities achieving AUC scores above 0.9. This configuration ensures consistent high predictive power while maintaining computational efficiency critical for practical deployment.

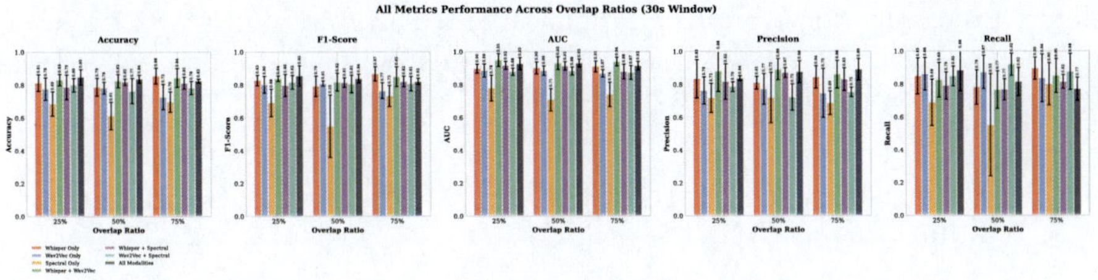

Fig. 1. Performance metrics across overlap ratios with 30 s window.

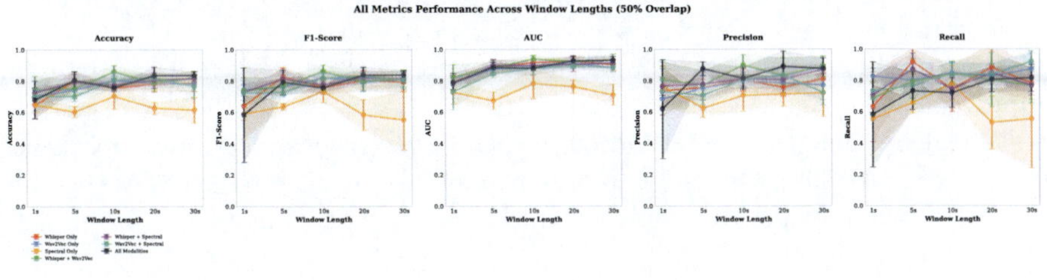

Fig. 2. Performance metrics across window lengths with 50% overlap.

5.2 Comprehensive Ablation Study and Performance Analysis

We systematically evaluated each component through comprehensive ablation study using recording-level 5-fold cross-validation (Table 1).

Table 1. Comprehensive Ablation Study Results: multi-representation AD Detection Performance across Different Feature Combinations (mean ± std across 5 folds)

Configuration	Accuracy	AUC	Precision	Recall	F1-Score
Individual Modalities					
Whisper Only	78.9±5.4	90.2±3.7	81.1±3.9	78.2±10.5	79.3±6.2
Wav2Vec Only	78.3±3.6	88.6±2.9	77.3±9.4	87.4±10.1	80.9±1.6
Spectral Only	61.4±8.3	70.9±6.8	72.1±15.3	55.2±31.2	54.9±19.1
Dual Modalities					
Whisper + Wav2Vec	82.5±4.0	93.1±3.9	**89.2±6.4**	76.7±11.4	81.7±5.4
Whisper + Spectral	81.9±1.9	91.4±2.6	87.2±3.0	76.9±6.4	81.5±2.9
Wav2Vec + Spectral	76.5±7.6	88.3±2.5	72.4±8.0	**92.1±6.7**	80.6±5.3
Complete Integration					
All Modalities	**83.7±2.4**	**93.1±2.4**	87.7±6.7	81.6±8.6	**83.9±2.7**

Whisper leads individual modalities at 90.2% AUC, while feature combinations optimize specific metrics: Whisper+Wav2Vec achieves peak precision (89.2%), and Wav2Vec+Spectral features maximize recall (92.1%). Full integration delivers superior overall performance with 93.1±2.4% AUC, 83.7% accuracy, and 83.9±2.7% F1-score, demonstrating exceptional stability and consistent detection reliability across diverse clinical conditions.

5.3 Comparison with State-of-the-Art Methods and Blind Test Validation

We evaluated our model on the independent ADReSSo blind test set containing 71 samples and compared our results with existing state-of-the-art methods. The blind test accuracy of 81% closely matches our cross-validation results, demonstrating excellent generalization without overfitting (Table 2).

Our method achieves 81% accuracy, representing the first meaningful improvement beyond the long-standing 79% barrier. Early methods using traditional features achieved 64–71% accuracy, while recent deep learning approaches plateaued at 78–79%. Our multi-representation approach demonstrates that the limitation was not due to insufficient model complexity, but rather incomplete acoustic representation.

Table 2. Acoustic-based Alzheimer's Disease Detection Results on ADReSSo Blind Test Set

Reference	Year	Accuracy	Precision	Recall	F1-Score
Luz et al. (baseline) [9]	2021	64.79	78.8	78.8	78.87
Balagopalan & Novikova [10]	2021	67.61	63.64	80.00	70.89
Perez-Toro et al. [15]	2021	67.61	-	75	67.0
Gauder et al. [5]	2021	78.87	-	-	-
Agbavor & Liang [11]	2022	74.6	77.1	73.00	75.00
Liu et al. [12]	2023	70.7	-	-	-
Priyadarshinee et al. [13]	2023	78.87	77.78	80.0	78.84
Pan et al. [14]	2024	76.01	76.52	76.11	76.04
Kim et al. [6]	2024	78.0	82.0	76.0	79.0
Our Method	**2025**	**81.00**	**84.00**	**78.00**	**81.00**

6 Discussion

Breaking the Acoustic Performance Limitation and Clinical Impact: Our results demonstrate that the 79% accuracy threshold stems from the inability of single-representation approaches to capture the full complexity of AD-related acoustic changes. The advancement to 81% accuracy with 81% F1-score represents meaningful clinical progress, with balanced recall (78%) ensuring fewer missed AD cases while high precision (84%) minimizes false diagnoses.

Limitations: Several limitations remain: multiple model computations increase computational cost, cross-validation variability suggests instability, and validation is limited to English speakers on ADReSSo dataset requiring broader demographic evaluation. Ethically, clinical deployment must position this as a screening tool complementing diagnostic protocols, with established patient privacy, informed consent procedures, and equitable access safeguards.

Methodological Rigor and Future Directions: Our implementation addresses critical methodological concerns through strict recording-level cross-validation and complete architectural specifications ensuring reproducibility.

Future research should investigate additional acoustic representations and fusion strategies, expand validation to larger diverse datasets with multiple languages, and conduct longitudinal studies tracking disease progression. Clinical validation through prospective studies in real healthcare settings represents the most critical next step.

7 Conclusion

This work demonstrates that the longstanding 79% accuracy threshold in acoustic-based AD detection reflects fundamental limitations of single-representation approaches. Our multi-representation acoustic framework advances beyond this threshold by synergistically combining Whisper's acoustic features, Wav2Vec2's representations, and spectral features through hierarchical fusion, achieving 81% accuracy and 81% F1-score with rigorous validation methodology.

The clinical impact warrants further investigation: improving detection accuracy enables earlier intervention when treatments remain most effective. By proving that intelligent integration of diverse acoustic representations overcomes single-representation limitations, we establish a new paradigm for speech-based AD detection. Our complete implementation details ensure reproducibility, facilitating further research advancement in this critical domain.

References

1. World Health Organization: Dementia: Number of People Affected to Triple in Next 30 Years. WHO Health Topics, Geneva (2017). https://www.who.int/news/item/07-12-2017-dementia-number-of-people-affected-to-triple-in-next-30-years
2. McKhann, G.M., Knopman, D.S., et al.: The diagnosis of dementia due to alzheimer's disease: recommendations from the national institute on aging-alzheimer's association workgroups on diagnostic guidelines for alzheimer's disease. Alzheimer's Dementia **7**(3), 263–269 (2011)
3. Jack, C.R., Jr., et al.: Nia-aa research framework: toward a biological definition of alzheimer's disease. Alzheimer's Dementia **14**(4), 535–562 (2018)
4. Fraser, K.C., Meltzer, J.A., et al.: Linguistic features identify alzheimer's disease in narrative speech. J. Alzheimers Dis. **49**(2), 407–422 (2016)
5. Gauder, L., Pepino, L.: Alzheimer disease recognition using speech-based embeddings from pre-trained models. In: Proc. Interspeech, pp. 3795–3799 (2021)
6. Kim, T.M., et al.: Comparison of ai with and without hand-crafted features to classify alzheimer's disease in different languages. Comput. Biol. Med. **180**, 108950 (2024)
7. Radford, A., Kim, J.W., et al.: Robust speech recognition via large-scale weak supervision. Preprint at (2022). https://arxiv.org/abs/2212.04356
8. Baevski, A., Zhou, Y.: wav2vec 2.0: a framework for self-supervised learning of speech representations. In: Advances in Neural Information Processing Systems, vol. 33, pp. 12449–12460 (2020)
9. Luz, S., Haider, F.: Detecting cognitive decline using speech only: the adresso challenge. In: Proc. Interspeech 2021, pp. 3780–3784., Brno, Czech Republic (2021)
10. Balagopalan, A., Novikova, J.: Comparing pre-trained and feature-based models for prediction of alzheimer's disease based on speech. Front. Aging Neurosci. **13**, 635045 (2021)
11. Agbavor, F., Liang, H.: Predicting dementia from spontaneous speech using large language models. PLOS Digital Health **1**(12), 0000168 (2022)
12. Liu, J., et al.: Efficient pause extraction and encode strategy for alzheimer's disease detection using only acoustic features from spontaneous speech. Brain Sci. **13**(3), 477 (2023)
13. Priyadarshinee, P.: Alzheimer's dementia speech (audio vs. text): Multi-modal machine learning at high vs. low resolution. Appl. Sci. **13**(7), 4244 (2023)

14. Pan, Y.: Swin-bert: a feature fusion system designed for speech-based alzheimer's dementia detection. In: Proceedings of the 6th ACM International Conference on Multimedia in Asia Workshops (2024)
15. Pérez-Toro, P.A.: Influence of the interviewer on the automatic assessment of alzheimer's disease in the context of the adresso challenge. In: Proceedings of Interspeech 2021, pp. 3785–3789 (2021)

Artificial Intelligence and Inclusive Recruitment: Reducing Bias and Promoting Female Leadership in Morocco and Sub-saharan Africa

Imane Abdelfadel(✉)

International University of Casablanca, Morocco Research Laboratory in Organizational Management, Law and Diversity (MODD) Doctoral Researcher in Human Resources, Casablanca, Morocco
abdelfadel.imane@gmail.com

Abstract. This study adopts a comparative qualitative approach conducted in Morocco and three Sub-Saharan African countries (South Africa, Rwanda, and Kenya) to examine the role of Artificial Intelligence (AI) in recruitment within the banking and financial sector. Based on 23 semi-structured interviews carried out between October 2024 and August 2025 with HR managers, recruiters, and data scientists from ten financial institutions—five in Morocco and five in Sub-Saharan Africa—the findings reveal that AI can mitigate certain biases but also reproduces inequalities linked to data quality and institutional contexts. Its effectiveness depends on integration into strong gender parity policies and inclusive organizational reforms.

Keywords: artificial intelligence · inclusive recruitment · gender bias · female leadership · organizational inclusion

1 Introduction

The growing use of Artificial Intelligence (AI) in recruitment is often presented as a step toward greater objectivity; however, it remains deeply dependent on data and organizational rules shaped by pre-existing inequalities, which may reconfigure—rather than eliminate—biases [1]. In the banking and financial sector, this tension directly affects women's access to decision-making and leadership positions.

This study adopts a comparative qualitative approach conducted in Morocco and three Sub-Saharan African countries (South Africa, Rwanda, and Kenya) to explore AI's inclusive potential across different institutional environments. It draws on 23 semi-structured interviews with HR managers, recruiters, and data scientists from ten financial institutions (five in Morocco and five in Sub-Saharan Africa).

Two hypotheses guide this research: (H1) AI's effectiveness *would* depend on the quality of data and organizational practices; (H2) its impact *would* vary according to sociocultural and regulatory contexts.

A. E. Hassanien et al. (Eds.): AISDS 2025, LNDECT 286, pp. 88–96, 2027.
https://doi.org/10.1007/978-3-032-18758-1_9

We argue that AI should not be seen as an autonomous solution; rather, its inclusive potential depends on its alignment with ambitious gender-equality policies and long-term organizational transformations.

2 Theoretical Framework and Literature Review

2.1 Inclusive Recruitment and Gender Bias

Inclusive recruitment goes beyond regulatory compliance to promote genuine equality of access to professional opportunities. Yet, gender bias continues to influence selection decisions. Bertrand and Mullainathan [1] demonstrated that identical résumés received fewer callbacks when associated with minority names, confirming the persistence of implicit judgments [2].

The introduction of artificial intelligence (AI) into human resource practices has been presented as a driver of objectivity, but research shows that it can also reproduce the very forms of discrimination embedded in training data. Noble [3] highlights how information systems perpetuate stereotypical representations, while Benjamin [4] warns against algorithmic invisibility that reinforces pre-existing inequalities. AI, therefore, cannot be regarded as a neutral tool; its impact depends on the quality of data and the organizational governance that regulates its use.

2.2 Female Leadership: Structural Barriers and Regional Disparities

Female leadership refers to women's ability to access and exercise strategic decision-making roles, yet it remains constrained by structural barriers such as the glass ceiling, occupational segregation, and gender stereotypes [5, 6]. In Africa, women hold about 21% of board seats but only 9% of CEO positions, with significant disparities: 32% in South Africa, 26% in Rwanda, and 24% in Kenya [7].

These gaps stem less from skill shortages than from organizational and institutional dynamics that perpetuate inequality and hinder women's advancement to strategic leadership roles [7, 9]. In Morocco, female representation remains low, with only 12.8% of companies led by women, mostly in the service sector, according to the Haut-Commissariat au Plan [10].

2.3 AI and Recruitment: Opportunities, Risks, and African Contexts

Artificial intelligence (AI) is transforming recruitment practices through automation and predictive analytics. Its effectiveness depends on ethical and inclusive design to ensure fairness and transparency in decision-making [11, 12]. When developed within such a framework, AI can expand the talent pool and reduce human bias [13]. However, several studies indicate that algorithms may reproduce existing inequalities if not supported by responsible governance and diverse design teams [14, 15]. In Africa, nearly a quarter of AI systems still exhibit gender- or ethnicity-related bias [16].

Contextual disparities remain: Morocco is progressively aligning its institutional framework with international diversity standards, while Rwanda, South Africa, and Kenya display contrasting dynamics—strong political parity but limited inclusion in the private sector [7].

2.4 Literature Synthesis and Conceptual Framework

Recent studies highlight the ambivalence of artificial intelligence (AI): while designed to enhance the objectivity of recruitment, its effectiveness depends on data quality, algorithmic governance, and the institutional environments in which it operates [1, 3, 4, 15]. At the same time, research on female leadership shows that gender inequalities in decision-making roles stem from enduring structural and cultural barriers within organizations [5, 6].

AI should therefore be viewed as a conditional lever for inclusion, whose impact relies on its alignment with gender-equality policies and inclusive managerial practices. These findings justify a comparative approach between Morocco and Sub-Saharan Africa and support an integrative conceptual framework combining algorithmic justice, gendered organization theory, and the institutional approach.

2.4.1 Algorithmic Justice: Towards Measurable Fairness

Recent work in computer science and the social sciences stresses that AI is not inherently neutral, but can be governed through algorithmic fairness metrics [17]. Hardt, Price, and Srebro (2016) introduced the Equal Opportunity principle, which assesses whether candidates with comparable qualifications face the same selection chances across groups [18]. In recruitment, this implies that an algorithm must guarantee candidates with equal skills the same chances of selection, regardless of gender. This approach makes the fairness of automated recruitment systems auditable and measurable, but requires strong organisational commitment to continuous monitoring mechanisms. The legitimacy of algorithmic systems therefore rests on transparency, traceability of decisions, and regular audits.

2.4.2 Organisations and Gender: Mechanisms of Inequality Reproduction

Acker's theory of gendered organisations [8] provides a critical perspective on why inequalities persist, even when inclusive policies are formally adopted. It demonstrates that criteria such as merit, time availability, and career progression are historically constructed on a male model and thus tend to exclude women or limit their advancement [8]. This perspective is particularly relevant for Morocco and several Sub-Saharan countries, where female representation in corporate governance remains modest [9, 10]. Even when AI tools are deployed, they are trained on data reflecting these organisational realities, which leads to an implicit reproduction of bias. AI must therefore be understood in conjunction with deeper transformations in management practices and internal norms.

2.4.3 The Institutional Approach: The Weight of National Contexts

In Morocco, despite the adoption of Law No. 19.20 (2019) progressively introducing gender quotas in the boards of joint-stock companies, including financial institutions, its concrete impact remains limited: only 12.8% of companies are actually led by women [10]. This measure reflects institutional willingness to move towards parity, but its effects remain confined to formal governance bodies, without deep transformation of executive positions.

By comparison, South Africa relies on more binding legislation, notably the Broad-Based Black Economic Empowerment Act and the Companies Act, which require listed companies to publish detailed reports on the diversity of their governing bodies. This obligation has contributed to higher female representation (32% of board seats) [9]. Yet, even here, increased female presence on boards does not automatically translate into access to CEO or general management roles, underscoring the persistence of a glass ceiling.

This comparison demonstrates that while legal tools are essential, they remain insufficient unless supported by organisational and cultural change, so that AI in recruitment can truly foster gender equality.

3 Methodology

3.1 Sample Selection

Ten financial institutions operating within the banking and financial sector were selected to ensure sectoral homogeneity and minimise structural bias. The sample includes five organisations based in Morocco and five in Sub-Saharan Africa (South Africa, Rwanda, Kenya). Selection criteria included the declared use of digital or algorithmic tools in recruitment, as well as the existence of practices related to diversity and inclusion [19, 20].

The study involved 23 participants: HR managers (n = 10), recruiters (n = 7), and data scientists (n = 6). Thematic coding was conducted in two stages—a descriptive phase followed by an analytical one—o identify relationships between algorithmic bias, organisational norms, and institutional frameworks. The coding process followed a three-level hierarchical structure (main themes, subthemes, and analytical dimensions) to ensure conceptual depth and cross-case coherence.

3.2 Data Collection Methods

Twenty-three semi-structured interviews were conducted between October 2024 and August 2025 with HR managers, recruiters, and data scientists involved in the deployment or design of AI-based recruitment systems. Interviews lasted approximately 60 min and were designed to combine organisational and technical perspectives to enhance the reliability of the collected data [19].

3.3 Data Collection and Analysis Instruments

The interview guide was structured around three core themes: (i) the use of AI in recruitment, (ii) the presence of gender bias and inclusive practices, and (iii) the articulation between technological tools and diversity policies. All interviews were fully transcribed and analysed following the principles of qualitative inquiry outlined by Lincoln and Guba [21], and thematic analysis as described by Braun and Clarke [22]. Iterative coding was used to refine and structure the main themes and subthemes in alignment with the research hypotheses.

3.4 Validation and Reliability of Results

Methodological robustness was ensured through:

(1) triangulation across participant roles and national contexts, as recommended by Denzin [23] and Silverman [24];

(2) independent double-coding with reconciliation of discrepancies;

(3) targeted member-checking with participating organisations to confirm the factual accuracy of interpretations [25].

4 Results

4.1 Perceived Reduction of Bias through AI

Participants perceive AI as a tool that enhances recruitment objectivity. HR managers in Morocco, Rwanda, and South Africa noted that standardized criteria help reduce subjective judgments: "The algorithm focuses on skills, not gender" (HR, Morocco). Others emphasized the expansion of candidate pools: "AI attracts profiles we would not have considered before" (HR, Kenya). These insights partially support Hypothesis H1, suggesting that AI may help mitigate bias at the initial screening stage.

4.2 Persistent Biases and Identified Limitations

Despite these positive perceptions, many participants pointed to the persistence of biases. The quality of training data strongly conditions the outcomes of automated systems.

As one recruiter in South Africa explained: *"When the company's history favours male profiles, AI simply reproduces that trend, even if it is presented as neutral."*

A Moroccan HR manager confirmed: *"We noticed that female applications are often under-represented in the automatically generated shortlists. This comes from past data, not just the tool itself."*

Others emphasised the impact of proxy variables: *"Criteria such as mobility or time availability indirectly disadvantage women, even if gender is not explicitly included."* (HR, Rwanda)

Another participant pointed out the repetitiveness of results: *"Without recalibration, the tool keeps putting forward the same types of candidates."* (Data scientist, Kenya)

A Moroccan recruiter concluded: *"If we don't check the system regularly, AI simply reproduces our old recruitment patterns."*

These findings reinforce the idea that AI is not a universal solution and may reproduce discrimination if not accompanied by corrective mechanisms.

4.3 Perceptions of Female Leadership in the Digital Era

The study also highlights a tension between expectations associated with AI and the reality of organisational practices.

A Kenyan data scientist observed: *"AI can identify female talent that might have been overlooked, but if the organisation is not ready to promote them, nothing changes."*

A South African executive added: *"We now see more women in the final shortlists thanks to AI, but their access to decision-making positions still depends on managerial culture."*

Other participants stressed the importance of complementary support: *"Without mentors or sponsors, women selected by AI still struggle to progress."* (Executive, Morocco)

An HR manager in Rwanda echoed this concern: *"Leadership criteria are still defined according to a male model, even if AI expands the candidate base."*

A South African career committee member summed it up: *"AI opens the door, but without parity objectives, promotions do not follow."*

These insights illustrate the conditional role of AI, which cannot generate lasting inclusive effects without broader organisational transformation.

4.4 Morocco/Sub-saharan Africa Comparison

The comparison between Morocco and Sub-Saharan Africa reveals significant divergences.

A Rwandan recruiter testified: *"The state imposes parity rules, but private companies do not always follow them, and AI reflects these contradictions."*

A Moroccan executive noted: *"Here, AI is still seen as a trend. As long as organisational culture does not evolve, its impact will remain limited."*

In South Africa, the legal framework is particularly influential: *"Mandatory diversity reporting pushes companies to audit their AI tools as well."* (HR, South Africa)

In Kenya, perceptions remain more cautious: *"We are still in a pilot phase: progress is visible, but budgets and data are still lacking."* (HR, Kenya)

Finally, in Morocco, the scarcity of reliable data was highlighted: *"Without robust databases, AI cannot genuinely reduce biases."* (HR manager, Morocco)

These differences confirm that the effectiveness of AI as a lever for inclusion depends largely on institutional and socio-cultural environments, thus supporting hypothesis H2.

4.5 Summary Table of Results

The following table presents the dominant themes, illustrative quotations, and corresponding analytical implications (Table 1).

These results confirm that AI functions as a conditional lever for inclusion, whose effectiveness depends on data quality, organizational culture, and the institutional environment in which it operates [11, 17, 26].

Table 1. Synthesis of empirical results: themes, quotations, and analytical implications

Core Themes	Main Findings	Illustrative Quotations	Analytical Implications
1. AI and Perceived Objectivity	Automation is perceived as a way to limit human subjectivity and strengthen recruitment neutrality.	"The algorithm does not consider gender; it focuses on skills." (HR Manager, Morocco)	AI acts as an initial standardization tool, promoting fairer preselection processes [1, 5].
2. Persistent Biases and Data Dependence	Algorithms reproduce existing imbalances found in historical datasets.	"If the past favors male profiles, AI tends to replicate that pattern." (Recruiter, South Africa)	Without regular auditing, AI systems risk amplifying inequalities rather than correcting them [15, 16].
3. Female Leadership and Organizational Barriers	AI can help identify female talent, but access to decision-making roles remains constrained by managerial culture.	"AI opens the door, but culture still decides who walks in." (Senior Executive, South Africa)	The impact of AI depends on transforming underlying organizational norms [7, 8].
4. Influence of the Institutional Framework	The effects of AI vary according to the maturity of regulations and national equality cultures.	"AI reflects what each country allows or demands." (Compliance Officer, Rwanda)	Inclusive systems are more effective within structured institutional environments [12, 26].

5 Discussion of Results

5.1 AI and the Question of Neutrality

The findings of this research confirm that artificial intelligence cannot be regarded as a neutral tool. While some participants consider that "AI eliminates human bias by standardising selection" (HR interview, Morocco), the empirical data reveal that these systems reproduce asymmetries embedded in historical data. This observation echoes the critiques of Noble [5] and Benjamin [6], who emphasise the risk of algorithmic invisibility. In the studied context, women perceive AI as a double-edged instrument: "It feels as if it evaluates us more objectively, but in reality, it reproduces existing patterns" (Executive interview, Kenya).

5.2 Structural Barriers and Female Leadership

Beyond technology, the discussion highlights the persistence of organisational and socio-cultural barriers. The qualitative results show that gender bias manifests upstream of AI use, for instance in the pre-selection criteria defined by companies. As one participant noted: "Even if AI sorts the CVs, it is still the same male profiles that are valued" (Recruitment manager interview, South Africa). These findings confirm Acker's theory of gendered organisations [8], which demonstrates that performance and merit criteria are historically constructed on male models, making them difficult to neutralise through technology alone.

5.3 Theoretical and Practical Implications

Artificial intelligence does not eliminate bias; it reshapes it through technical, organizational, and institutional dynamics [27]. Integrating algorithmic justice, gendered organizational structures, and institutional frameworks offers a comprehensive lens for understanding inclusive female leadership in the digital era [8, 15, 18, 26].

6 Conclusion and Implications

This study underscores the dual nature of AI: it can mitigate certain biases while reproducing inequalities when implemented without adequate safeguards [15, 16, 26]. Its effectiveness depends on ethical governance, continuous auditing, and coherent gender equality policies [12, 26]. AI should therefore be regarded not as an autonomous solution but as a conditional lever for inclusion.

6.1 Limitations and Future Research

This study presents three key limitations. First, the sample includes only ten organisations within the banking and financial sector, which restricts the generalisability of the findings to other industries with different AI maturity levels. Second, while the qualitative approach provides rich contextual insights, it does not allow for statistical inference. Third, the comparative scope remains limited to four African countries, preventing broader institutional generalisation.

Future research should adopt mixed or quantitative methods to measure the prevalence of these mechanisms on a wider scale, extend the analysis to additional sectors and countries, and use longitudinal or experimental designs to evaluate how algorithmic fairness evolves and how corrective interventions may reduce gender bias in practice.

References

1. Raghavan, M., Barocas, S., Kleinberg, J., Levy, K.: Mitigating bias in algorithmic hiring: evaluating claims and practices. In: Proceedings of the 2020 Conference on Fairness, Accountability, and Transparency (FAT'20), pp. 469–481. ACM (2020)
2. World Bank: Profiting from Parity: Unlocking the Potential of Women's Businesses in Africa. World Bank, Washington, DC (2019)
3. Bertrand, M., Mullainathan, S.: Are Emily and Greg more employable than Lakisha and Jamal? Am. Econ. Rev. **94**(4), 991–1013 (2004)

4. Greenwald, A.G., Banaji, M.R.: Implicit social cognition: attitudes, self-esteem, and stereotypes. Psychol. Rev. **102**(1), 4–27 (1995)
5. Noble, S.U.: Algorithms of Oppression: How Search Engines Reinforce Racism. NYU Press (2018)
6. Benjamin, R.: Race After Technology: Abolitionist Tools for the New Jim Code. Polity Press (2019)
7. Eagly, A.H., Carli, L.L.: The female leadership advantage: an evaluation of the evidence. Leadersh. Q. **14**(6), 807–834 (2003)
8. Acker, J.: Hierarchies, jobs, bodies: a theory of gendered organizations. Gend. Soc. **4**(2), 139–158 (1990)
9. UN SSE and IFC: Gender equality in corporate leadership – Africa. IFC Report (2023)
10. Haut-Commissariat au Plan: Rapport national sur les entreprises dirigées par des femmes au Maroc. HCP, Rabat (2023)
11. IBM Institute for Business Value: Women in leadership in the age of AI. IBM Global Report (2024)
12. UN Women: Artificial intelligence and gender equality. United Nations, New York (2024)
13. UN Women: Facts and figures: women's leadership and political participation. UN Women (2025)
14. Financial Times: Why AI is the key to unlocking leadership opportunities for women. FT Special Report (2024)
15. Raji, I. D., Buolamwini, J.: Actionable Auditing: Investigating the Impact of Publicly Naming Biased Performance Results of Commercial AI Products. In: AAAI/ACM Conference on AI, Ethics, and Society, pp. 429–435 (2019)
16. O'Neil, C.: Weapons of Math Destruction. Crown (2016)
17. World Bank: AI, inclusion and gender in Africa. Technical Paper (2025)
18. Hardt, M., Price, E., Srebro, N.: Equality of opportunity in supervised learning. In: Advances in Neural Information Processing Systems, vol. 29, pp. 3315–3323 (2016)
19. Miles, M.B., Huberman, A.M., Saldaña, J.: Qualitative Data Analysis: A Methods Sourcebook, 3rd edn, Sage (2014)
20. Yin, R.K.: Case Study Research and Applications: Design and Methods, 6th edn, Sage (2018)
21. Lincoln, Y.S., Guba, E.G.: Naturalistic Inquiry. Sage (1985)
22. Braun, V., Clarke, V.: Using thematic analysis in psychology. Qual. Res. Psychol. **3**(2), 77–101 (2006)
23. Denzin, N.K.: The Research Act: A Theoretical Introduction to Sociological Methods. AldineTransaction (2012)
24. Silverman, D.: Interpreting Qualitative Data, 6th edn, Sage (2020)
25. Birt, L., Scott, S., Cavers, D., Campbell, C., Walter, F.: Member checking: a tool to enhance trustworthiness or merely a nod to validation? Qual. Health Res. **26**(13), 1802–1811 (2016)
26. OECD: Gender Equality in Africa: Progress and Challenges. OECD Publishing (2022)
27. Royaume du Maroc: Loi n° 19-20 modifiant et complétant la loi n° 17-95 relative aux sociétés anonymes. Bulletin Officiel, n° 6807 (2019)

Smart Dairy: Agentic AI-based System for Everyday Dairy Operation

Rajesh Neupane[1](✉), Toya Acharya[2], Bhuwan Shrestha[1], Sushil Paudyal[1], Annamalai Annamalai[2], and Mohamed Chouikha[2]

[1] Department of Animal Science, Texas A and M University, College Station, Texas, USA
rajeshneupane7@gmail.com

[2] Department of Electrical and Computer Engineering, Prairie View A and M University, Prairie View, Texas, USA

Abstract. SmartDairy is an open-source, LLM-powered (Large Language Model) chatbot designed specifically for dairy producers, integrating Retrieval-Augmented Generation (RAG) and agentic Artificial Intelligence (AI) workflows to provide accurate, domain-specific recommendations. The system allows users to upload authentic sources such as PDFs, Word documents, and manuals, which are processed through text chunking and embedding models to build an embedded dataset stored in a vector database. In addition to user-provided materials, the system incorporates a curated collection of highly cited research articles, ensuring responses are grounded in reliable and scientific knowledge. A web tool is embedded within the platform to capture recent developments, while a shared working directory enables farm-specific queries such as estimating milk production per cow or identifying sick animals based on available data. Through an intuitive web interface, producers interact with the chatbot, which delivers actionable, context-aware insights to enhance productivity, optimize herd management, and support data-driven decision-making

Keywords: Artificial Intelligence (AI) · Agentic AI · Chatbot · Dairy · LLM

1 Introduction

1.1 Agentic Artificial Intelligence

Artificial Intelligence (AI) has been evolving since its inception, reshaping industries and economies across the globe. Recent advancements in computing infrastructure, particularly innovations in parallel computing chips, have accelerated AI development and made it an integral part of daily life. The integration of AI across diverse sectors has demonstrated significant improvements in efficiency, speed, and automation, driving its rapid adoption as the "new electricity" powering modern society [1, 2]. However, being trained in the historically published articles, internet scrapping, and inherently predicting in autoregressive manner, these artificial intelligence models or large language models struggle to act to real time data and make real time decisions [3]. This traditional model

A. E. Hassanien et al. (Eds.): AISDS 2025, LNDECT 286, pp. 97–106, 2027.
https://doi.org/10.1007/978-3-032-18758-1_10

often lacks autonomy to dynamically adapt to complex situations, and this shortcoming has led to new type of workflow in artificial intelligence called agentic AI. These systems are capable of independently making decisions, interacting with the complex environment and working without any human interaction or intervention. Thes systems incorporate the fundamentals of human type learning such as web search, self-learning, multi agent collaboration and finally making data informed decisions.

1.2 Basic Dairy Operations

Dairy farming is inherently a multifaceted operation involving multiple stakeholders, including producers, nutritionists, milking system operators, and dairy processing facilities. Modern dairy management relies heavily on advanced technologies and record-keeping systems. For instance, specialized herd management software is widely used to track animals from their entry into the herd, beginning with early ear tagging using Radio Frequency Identifier (RFID) numbers. These systems record detailed information on health status, disease incidence, milk yield, and reproductive parameters [4].

Milking technologies also play a critical role. Robotic and conventional machine milking systems not only measure milk yield at each session but also capture key milk quality indicators such as fat, protein, and somatic cell counts [5]. In addition, automated activity monitoring systems, including accelerometers, pedometers, and rumen boluses are increasingly used to track feed and water intake, detect behavioral changes, and assess the physiological status of animals, often providing early warnings for potential health interventions [6].

Beyond individual animal monitoring, emerging technologies are being developed to optimize resource use at the system level. For example, pasture growth monitoring tools can recommend optimal grazing times to maximize land-use efficiency [7]. While dairy farming may appear simple from a consumer perspective, the integration of diverse technologies has transformed it into a highly complex, data-driven operation.

1.3 Problem Statement

Despite the rapid adoption of technology in dairy farming, a major challenge lies in the lack of compatibility and seamless integration across different software platforms and monitoring systems. For example, herd management tools are often not fully interoperable with activity monitoring systems, leading to fragmented datasets and difficulties in consolidating animal health and performance information. As a result, producers are frequently forced to switch between multiple platforms simply to identify sick or at-risk animals, creating inefficiencies and increasing the risk of missed interventions. The problem is further compounded by the continuous introduction of new technologies such as automated lameness detection cameras that generate additional streams of valuable data but often lack integration with existing infrastructure. This siloed approach to data management hinders the full potential of precision dairy technologies, ultimately limiting their effectiveness in supporting timely and informed decision-making on farms.

1.4 Objectives and Uniqueness of Study

This study aims to propose novel approaches for addressing interoperability challenges in precision dairy farming by exploring the application of advanced AI techniques such as Agentic AI. Specifically, the study seeks to investigate AI-driven data harmonization methods for integrating heterogeneous data streams from herd management software, activity monitoring systems, and emerging technologies such as automated lameness detection cameras; to examine how RAG based tool, web search tool and automated local data analytics help producers. By emphasizing the scalability and adaptability of these approaches, the study underscores their potential to transform currently isolated technologies into a cohesive, interoperable ecosystem for sustainable dairy production.

2 Methodology

2.1 Foundation Model Description

Several open-source language models have become central to the development of agentic AI systems, offering a balance between scalability, reasoning ability, and deployment flexibility. DBRX (132B total, 36B active parameters) provides strong multi-step reasoning and structured tool use within a 32 k token context window, making it well suited for agentic ai applications. Meta's Llama family (7B, 13B, and 70B parameters) supports up to 32 k tokens and is widely adopted due to its versatility, strong fine-tuning ecosystem, and suitability for both local assistants and multi-agent frameworks. Falcon, Mistral, Mixtral, and many more open-source models are available for agentic AI workflow.

2.2 RAG Based Agentic AI

Retrieval-Augmented Generation (RAG) has become a foundational paradigm for enabling agentic AI, as it combines the generative capacity of large language models with dynamic access to external knowledge sources. In this framework, an agent retrieves relevant information from structured or unstructured databases, knowledge graphs, or document collections, and integrates it into the reasoning process of the model to generate contextually accurate and up-to-date responses [8]. Unlike static models that rely solely on pre-trained parameters, RAG-based systems can ground decisions in domain-specific data, reduce hallucinations, and improve transparency by linking outputs to verifiable sources [9]. When extended to agentic settings, RAG enables autonomous agents not only to answer queries but also to plan multi-step actions, execute tool calls, and iteratively refine their strategies using external memory. This makes RAG particularly valuable in high-stakes applications such as healthcare, finance, and scientific discovery, where accuracy, explainability, and adaptability are critical.

For our study we used Llama-7B model which offers several practical advantages over larger language models that make it particularly well suited for agentic AI applications. First, its relatively small parameter size allows it to run locally on modest hardware, requiring only ~7 GB of GPU memory, which significantly lowers the computational and infrastructure barriers to deployment. Second, this compact footprint enables fast response times, a critical factor for real-time interaction in agentic systems.

Finally, despite its smaller scale, Llama-7B retains strong instruction-following and reasoning capabilities, making it a cost-effective and efficient alternative to larger models for scenarios where speed, resource efficiency, and on-device operation are prioritized (Fig. 1)

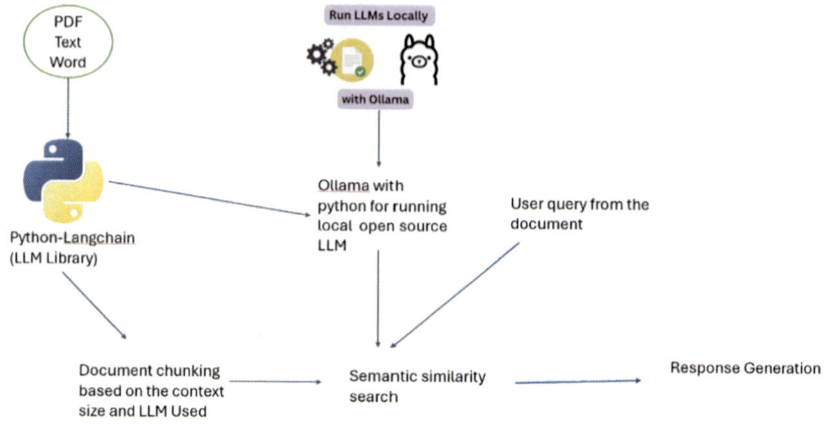

Fig. 1. RAG process applied in this study

Leveraging a modular system architecture, Smart Dairy platform was designed to process heterogeneous data sources, including PDF documents, spreadsheets, and unstructured textual reports. Document parsing was performed using PyPDF for PDFs, while other file types were standardized into a textual representation prior to downstream processing. To preserve semantic integrity during text segmentation, we implemented a recursive text chunking algorithm with a sliding window approach, incorporating a 512-token overlap between adjacent chunks to mitigate the risk of context fragmentation. Each chunk was embedded using the nomic-text-embed-v1.5 model, generating dense vector representations optimized for semantic retrieval tasks. These embeddings were indexed in ChromaDB to support low-latency vector search, while FAISS (Facebook AI Similarity Search) was integrated for GPU-accelerated approximate nearest neighbor search, thereby enhancing scalability and retrieval efficiency under large document corpora. The inference pipeline employed the llama 7b language model, deployed through a FastAPI backend to manage retrieval-augmented generation (RAG) workflows. On an NVIDIA GeForce RTX 4060 (8GB) GPU, for end-user interaction, a React-based frontend was developed to enable multimodal engagement, including natural language querying, document uploads, and structured user feedback. To support longitudinal system evaluation, PostgreSQL was used to log chat sessions with timestamped metadata, enabling detailed analysis of query patterns, user interactions, and model performance over time.

To provide most accurate responses to users top 5 highly cited review articles from respective dairy domain, Breeding, Nutrition, Management, Reproduction and Precision dairy were selected and stored in vector database, and users were asked if they would like to add any extra resources such as pdf manual from suppliers (equipment), protocols from veterinarians, local and state level polices which were relevant to their questions.

2.3 Web Search Based Agentic AI

The proposed system implements a multi-layered RAG-based AI agent designed to intelligently answer user queries by integrating retrieval, validation, and web search. When a user submits a question, the RAG agent first attempts to formulate an answer using its internal knowledge base, which may include previously stored documents, and embeddings. Once an initial response is generated, a secondary evaluation agent assesses the quality and completeness of the answer. If the evaluation agent determines that the response is insufficient, the system automatically triggers a web-search-capable agent, which retrieves relevant information from external sources, such as DuckDuckGo. The newly retrieved content is then stored in a local database for future queries and combined with the initial context to generate a refined answer as demonstrated in Fig. 2. This multi-agent architecture ensures that the system produces accurate, up-to-date, and contextually relevant responses, while optimizing efficiency by avoiding unnecessary web searches when the internal knowledge base suffices. The workflow allows for scalable, autonomous reasoning, combining RAG, quality verification, and agentic web search in a seamless pipeline.

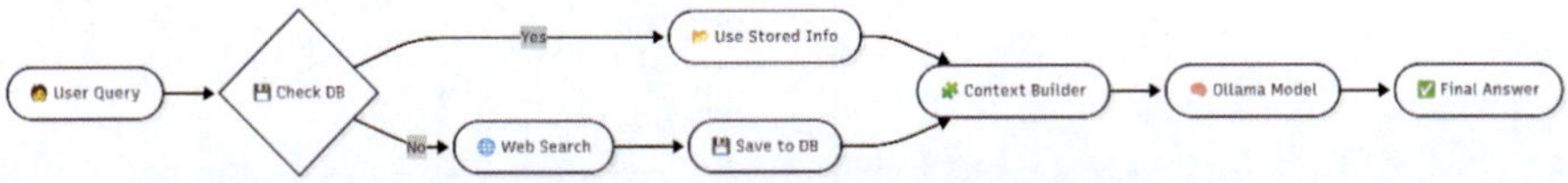

Fig. 2 Web search-based agent

2.4 Local csv/excel File Evaluation Agentic AI

The main use of this system relies on its ability to read specific csv/excel files that have been saved to common directory, where herd management software, precision sensor software would routinely write their data into csv files, when user queries about some information, agentic AI would look into the shared directory, load the file based on key words, generate the python script based on the user query, run the python script on the loaded data, and give output as demonstrated in Fig. 3.

This solves the two main problems, compatibility and integration issues between the software and risk of data leakage when used API based LLM, since data is stored locally in a system, and producer is the owner.

2.5 Information Protection

The agentic AI system prioritizes dairy producers data privacy through a dedicated local-first deployment model, ensuring data sovereignty. The core RAG knowledge is derived only from public, peer-reviewed articles, with a web search tool used only for

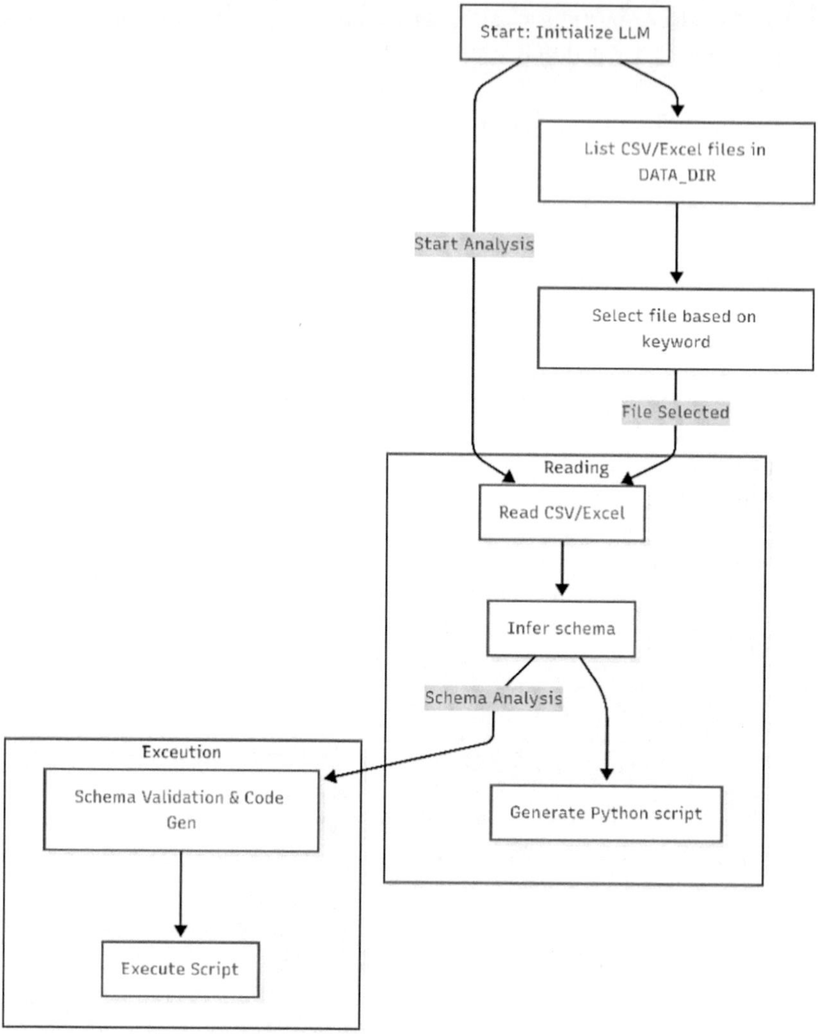

Fig. 3. Local csv/excel reading agentic workflow

general knowledge gaps. Crucially, private, proprietary farm records (e.g., cow milking data from local CSV/Excel files) are accessed and processed entirely on the farmer's hardware. This strict on-premises design means no farm-specific data is ever transmitted to the cloud or external servers, thereby eliminating risks associated with data leakage, unauthorized access, and third-party data violation issues

3 Experiments

We developed a smart dairy chat interface to help dairy producers to make informed decisions Fig. 4. As a preliminary test, we asked some questions related to dairy farming, as illustrated in Fig. 5, as it represents real world question producers could ask for their everyday dairy operation. To answer this question our system approximately took around 15 ~ 20 s. However, the response time depends on the type of model uses, hardware capacity, & the vector database used.

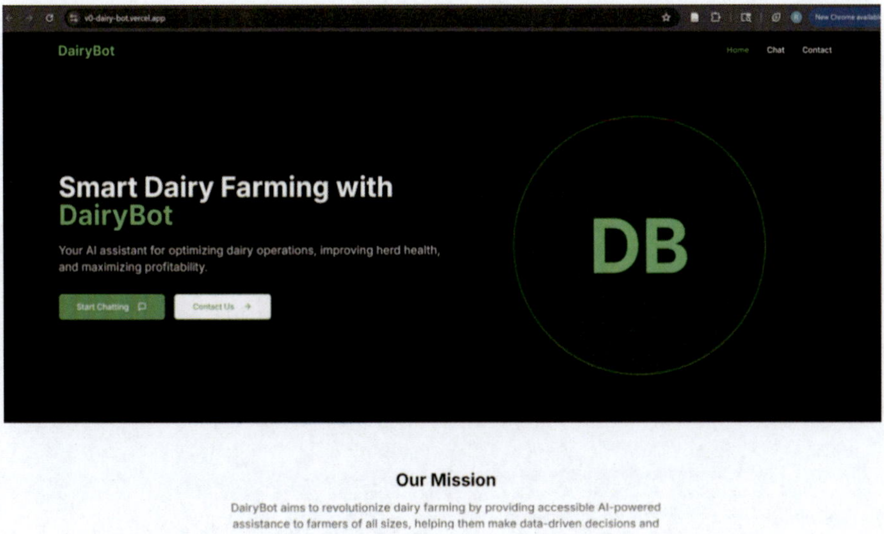

Fig. 4. Smart dairy bot chat interface, designed to run locally.

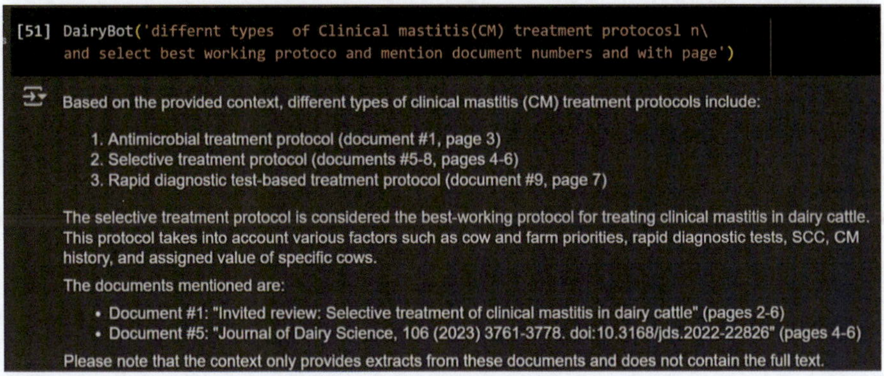

Fig. 5 Sample Response generated by the system when asked a domain specific question.

When we asked about the most popular dairy technology, it searched for its Chroma dB database for the matching keyword, if keyword overlap is less than the user defined threshold, it triggered for the web search tool, and searched for top 5 websites, and generated response based on the content of website (Fig. 6).

```
question = "which is the most popular dairy technolgy used in everyday farming,?"
answer = agent_answer(question)
print("\n🤖 Final Answer:\n", answer)
✓ 4.4s                                                                 Python

Keyword overlap: 1
Doing web search...

🤖 Final Answer:
Based on the latest technological trends in dairy farming, I can conclude that one of the most popular dairy technologies used in everyday farming is Autom

This technology has become increasingly common on modern dairy farms due to its ability to optimize milk production while also promoting animal welfare and
```

Fig. 6 Web search-based tool implementation.

When a real-time dairy management query was submitted, the system accessed synthetic farm data stored in a shared directory as a CSV file. This dataset contained approximately 5,000 records across 20 unique cow IDs, with columns including cow ID, milk yield, fat percentage, and protein percentage. Upon receiving a request regarding the average production of each cow, the system's CSV/Excel evaluation module was automatically triggered. It identified and loaded the relevant file, executed the query, and returned the calculated averages for each individual cow. This example highlights the system's ability to autonomously integrate data retrieval, processing, and analysis to provide actionable, farm-specific insights (Fig. 7).

```
question = "What is the average milk production for each cow?"
result = agent_answer(question)
print("\n🤖 Final Answer:\n", result)
[95]  ✓ 0.9s

cow_id
0      9.767582
1      9.445007
2      9.735094
3      9.645850
4      9.981165
5     10.176560
6      9.708028
7     10.026985
8     10.038964
9      9.703181
10     9.759811
11     9.358437
12     9.966342
13     9.764467
14     [illegible]
15     9.967757
16    10.170426
17    10.014831
18    10.298662
19     9.102033
Name: milk_yield, dtype: float64
```

Fig. 7. Automated local csv file query-based answers, demonstrating a real-world question by producers

This agentic AI system's core novelty lies in its real-world application for everyday dairy farming decisions while strictly maintaining data sovereignty for the farmer. By utilizing a local-first RAG framework, the system translates complex operational data, such as milk yield, breeding cycles, and treatment costs stored in local files, into actionable advice via a natural language interface. This enables a farmer to simply ask, "Which cows should I breed next week?" or "List potential culls due to low productivity?" and receive a data-driven recommendation, effectively making cutting-edge AI analytics immediately accessible and impactful for critical tasks like breeding, culling, and productivity tracking. This integration of an autonomous, agentic workflow with on-premises data ensures the system is not just an advisory tool, but a private, operational decision-support assistant.

However, as a primary conceptual study, the research faces an inherent limitation in the current lack of robust, standardized performance metrics or comparative accuracy findings. The system's actual response quality and speed are currently dependent on variable factors, including the farm's specific hardware configuration (CPU/GPU) and the inherent variability in the farmer's local dataset quality and structure. To overcome this, the future involves running a pilot study where the entire system will be packaged into a Dockerized container. This containerization ensures the software environment is reproducible across all participating farms, standardizing the foundation for testing and finally allowing for the collection of validated, generalizable metrics on accuracy and performance, thereby moving the research from a conceptual demonstration to an empirically supported, deployable solution.

4 Conclusions

Thus, we have demonstrated that agentic AI workflow is a novel innovation for a everyday dairy operations. Future developments should focus on building more robust agentic AI workflows. For example, incorporating automated task orchestration would allow the system to chain multiple steps such as retrieving herd health records, cross-validating them with veterinary guidelines, and generating a ranked list of interventions, without requiring manual prompts. Similarly, integrating multimodal inputs (e.g., images of cows for lameness detection, body condition scoring) could enable richer, context-aware recommendations. Strengthening the agentic framework in these ways would allow SmartDairy to evolve from a decision-support tool into a proactive management assistant capable of addressing complex, dynamic challenges in modern dairy farming.

Acknowledgements. Authors would like to acknowledge the support from Department of Animal Science Texas A&M University, and SECURE center at Prairie View A & M University.

References

1. Siemon, D., Strohmann, T., Michalke, S.: Creative potential through artificial intelligence: recommendations for improving corporate and entrepreneurial innovation activities. In: Association for Information Systems (2022)

2. Cheng, L., Yu, T.: A new generation of AI: a review and perspective on machine learning technologies applied to smart energy and electric power systems. Int. J. Energy Res. **43**(6), 1928–1973 (2019)
3. Reznikov, R., Turlakova, S.: Data Science methods and models in modern economy (2024)
4. Wallace, R.: Computer-assisted herd mastitis monitoring (2003)
5. Vik, J., Stræte, E.P., Hansen, B.G., Nærland, T.: The political robot—the structural consequences of automated milking systems (AMS) in Norway. NJAS-Wagening. J. Life Sci. **90**, 100305 (2019)
6. Tullo, E., Fontana, I., Gottardo, D., Sloth, K.H., Guarino, M.: Validation of a commercial system for the continuous and automated monitoring of dairy cow activity. J. Dairy Sci. **99**(9), 7489–7494 (2016)
7. Fox, S.A.: Linear programming applied to a forage based cattle and sheep production system on a demonstration farm (1983)
8. Jeong, C.: A study on the implementation method of an agent-based advanced rag system using graph. arXiv preprint https://arxiv.org/abs/2407.19994 (2024)
9. Gao, Y., et al.: Retrieval-augmented generation for large language models: a survey. arXiv preprint https://arxiv.org/abs/2312.10997 (2023)

The Product Manager's Compass: Navigating Responsible Generative AI in Consumer Health and Telecom Services

Supraja Chinthala(✉)

Cumming, GA, United States
Supraja.chinthala@ieee.org

Abstract. Deployment. of generative AI on customer-facing products of the pharma and telecom industry is a leap forward in the. user experience. The blog is a try to solve the paralyzing task. of the product manager to implement such integration. The blog revolves around the ethics and security ratio of creating such kinds of applications such as med chatbots and smart voice assistants, wherein the consumer's trust should be at the top. Top priority is the reduction of hallucination risk, compliance with regulations, and an assurance guarantee on the life cycle of a product. A simulated test suite has been run on a 'Consumer AI Interaction (CAII-2025)' of artificially simulated 100,000 user interactions. Python libraries such as scikit-learn for tracking performance metrics and matplotlib/seaborn for plotting have been utilized in data work within an emulated testbed of their own design. Impacts are significant and exhibit extremely high negative correlation between user trust and AI hallucination ratings, and extremely high positive impact of transparent regulation-compliant attributes on adoption. Studies reveal a 'Responsible AI Product Lifecycle Framework' as product manager's guidebook to bake safety, transparency, and accountability into product DNA. It claims that where profitable goods are viable, responsible development is not a constraint but an obligation and a duty for sensitive cases.

Keywords: Responsible AI · Generative AI · Trust · Pharma · Telecom · Digital Products

1 Introduction

Emerging generative AI models are a paradigm shift in digital product development, concludes [3] research. In consumer-facing sectors like telecom and pharma, this technology opens up unprecedented options for personalization of experiences, ease of communication, and round-the-clock customer service, corroborated by research work undertaken by [11]. AI chatbots in the pharma sector can provide real-time information about medicine and medical-related issues, revolutionizing patient treatment, a boost to the arm of [4]. IVAs are used in telephony to enable network services, fault detection, and customer

S. Chinthala—Independent Researcher

A. E. Hassanien et al. (Eds.): AISDS 2025, LNDECT 286, pp. 107–118, 2027.
https://doi.org/10.1007/978-3-032-18758-1_11

account management in a very user-friendly program, a program that was spearheaded by [6]. Power, on the other hand, has an exponentially higher cost to bear, sought by study research conducted by [9]. The operation of the very essence of generative AI itself—fresh, unforeseen content generation—is replete with dubious ethics and security concerns to which product managers need to be attentive with extreme vigilance, a concern highlighted by [10]. The point where ethics meet innovation is the product manager since it is him who crosses the middle ground upon which ethical considerations must be dealt with in an attempt to create AI products, according to arguments brought by research conducted by [1]. He is not just product production; it is being a hero of faith by consumers, an experience that comes with [5]. When a user comes to a telemedicine chatbot or telecom IVA with a grievance, he is trusting the system with his safety, reliability, and accuracy, and that is something which remains true as per [7]. AEmployment of the same type of trust as disinformation (hallucinations), intrusion into data privacy, or unbalanced sources is disastrous in terms varying from physical damage in medicine to full economic or personal destruction in the telecommunication sector, on an impression brought to spotlight by [12]. A response to four pillars of significance as much as the product management is concerned is as follows. It first restates trust as the currency of success as mandated in research by [8]. Without trust from the users, however well-quality AI program, it will never get use and sustainability, something testified by [2]. Second, it restates the danger of hallucination, where the AI provides rational but false or deceptive information, as cited in research used by [1]. It is not merely a technical shortcoming in the product manager's arsenal but also a product shortcoming inherent in the product that calls for drastic mitigation strategies, e.g., retrieval-augmented generation (RAG) and rigorous human-in-the-loop validation, an approach borrowed from [9]. Third, unification of regulations is also talked about in the paper. Applying such future models like Health Insurance Portability and Accountability Act (HIPAA) to health care or data protection legislation like GDPR to the telecommunication industry is ancillary to law but a reasonable product mandate towards data rule, model training, and user interface development, as one research study by [10] has accounted for. Finally, end-to-end system for ethical product life cycle management is what the following research advises. From birth to deployment, tracking, and retirement, moral concerns must be established at every step since one can make money out of already done work on [11]. I.e., asking hard questions at every step: Are we handling our data in an unbiased manner when training? Have we been properly protected from misuse? How are we providing users with data transparency and control? How do we keep lines of accountability open when the AI gets something wrong? As a product manager, in defining these challenges, this paper aims to establish an operating agenda for the build of generative AI products that are not only useful and smart but also equitable, secure, and worthy of customer trust, as explained in conclusions by [12]. They are elaborated in detail below through literature review, optimal practice with simulated data set, and careful results analysis with conclusion drawn for AI product leaders.

2 Review of Literature

The debate on AI had witnessed a paradigm shift in the last decade from technology promise towards increasing realization of its social as well as ethical implications, with proof being presented by [3]. Early studies of AI product management tended to focus less on what machine learning was being used for in terms of process automation, recommendations, and predictions, for instance, by [11]. The most important key performance indicators of interest would typically be predominantly efficiency, accuracy, and user experience that were measured through measures such as click-through rates and conversion funnels, a model by [12]. The "black box" quality of the AI was relished but in mindset as a performance compromise, a mindset rationalized by [2]. The business needs would be met by the product manager as technology needs for data science teams with little regard for models' own explainability or fairness, as argued in [5]. As that many AIs are currently put in consumer products, a new field of study in the psychology of trust and human-AI collaboration, i.e., [6] was born.

Researchers began to investigate whether such measures like explainability, predictability, and perceived benevolence actually produced a user to trust an AI system, i.e., employed by [7]. The research observed that trust is not a virtue but an evolving notion that keeps getting constructed as it keeps getting repeated and repeated and repeated with experience-bound knowledge, as identified in [9]. This follows the observation that human beings would judge failure of AI systems less stringently than failure by human beings, particularly for high-risk cases of use, as indicated by experiments by [4]. This new design of product from being to fear to simply try to maximize to produce precise outputs to designing systems that fit well with uncertainty with good boundary reporting and recourse terms in the event of failure, a challenge first addressed by [8]. Generative AI as a new player is with a still greater order of complexity that is challenged to duplicate the paradigms of the time, as set by studies of research of [10].

Generative models create new content, predictive models predict or classify from new created content, and this is accompanied by the issue of "hallucinations," an issue addressed by [1]. Text therefore arrives with the new algorithmic flaw, and its even more sinister back-end that the output produced is believable, eloquent, and realistic written but completely fabricated, a scenario already addressed by research utilized by [3]. It destroys the user trust foundation itself, as suggested by [7]. And so the work has shifted to building mitigation methods such as retrieval-augmented generation (RAG), where it trains its AI response against a corpus of verified facts, and strong content safety guardrails such as tested by [2]. But responsible AI has been a deeper and longer-running space with more fairness, accountability, and transparency, something which has been the aspiration of work by [11].

Algorithmic bias problem, or the presence of unwarranted conclusions due to biased training data, has been the general setting, subsequent to [5]. Additionally, product managers have an obligation to track the provenance and composition of datasets and utilize fairness toolkits to have faith that demographic prejudice in models can be identified, according to [9]. Accountability has further been challenged as to whom one blames when autonomous agent has caused damage, which has been challenged in the literature by [6]. This has necessitated AI control structures and provision to make sure that there is a "human-in-the-loop" in consequential decision-making in practice, as argued by [8].

Regulatory weather also forms one of the cool trends with regulations like GDPR and upcoming ones on AI requiring product teams to get data protection and privacy into a product's DNA right from the start, now also referred to as "privacy by design" through [10] research. To product managers in general, in total, this literature is referring to a broader trend: not producing what may be produced, but producing what should be done in a responsible manner and with purpose, something with which [12] agrees and with which sentiments employed by [4].

3 Methodology

Quantitative method simulation has been used in this study to measure the impact of consumption of generative AI by consumers on customers' telecommunication services and health information. This study has been constructed through simulating an estimate of a product manager's preference and control in the case of an influence in a product life cycle decision. In applying the simulation of creating a simulated data set, 'Consumer AI Interaction (CAII-2025)' is used as the original data set. It simulates 100,000 customers into two basic simulated applications: 'HealthBot', the pill robot to respond to the questions of the patients about the medicine, and 'ConnectAI', the telecommunication smart voice assistant to recognize network failure. A user group, intent, and question difficulty level schema-trained transformer language model was utilized in the generation process in an effort to create a realistic scenario set. The initial solution involved developing and progressively augmenting four consecutive generations of the respective central AI model of the respective application using a coordinated test environment developed with Python. Version 1.0 was a basic generative model with comparatively lesser emphasis on the safety aspect. Version 2.0 included algorithmic debiasing methods with specific emphasis on training data re-weighting for fairness across simulated user sub-populations. Version 3.0 also included a robust Human-in-the-Loop (HITL) test plan with a holdout set of high-stakes queries for human-simulated judgment and response-augmented generation (RAG) architecture for response grounding on a stable knowledge base. Version 4.0 also included extra transparency features such as AI response confidence scores and system limitation disclaimers on self-labeling systems. interest metrics were monitored at all levels: model correctness, F1 measure, latency, and very likely more importantly, hallucination rate, as model response to an at-random subset of correct responses to 10,000 indiscriminate test queries. To determine further specifically the influence of the users, the second test was also carried out to provide user sentiment scores (1–100), adoption (%), and an algorithm-derived cumulative 'User Trust Score' based on task success rate, sentiment score, and escalation to human representative. We used pandas and scikit-learn libraries for calculations and plots were created using matplotlib and seaborn libraries to show model improvement, measures, and user trust results graphically.

4 Data Description

Standard Consumer AI Interaction (CAII-2025)' data were synthetically generated for purposes of research here to present a sound and ethical foundation for analysis without requiring real-user data collection. The data set generates 100,000 artificially simulated user interactions and divides them in half into 50,000 each of the pharma chatbot scenario and 50,000 each of the telecom intelligent voice assistant scenario. The interaction record contains an artificially simulated user query, user demographic fields (skill level, age group), query difficulty level, AI response, best ground-truth response, and other secondary user-dependent values like task done status and value of sentiment. Data set has been obtained from an internal simulation environment built in a large language model of anonymized, aggregate human behavior patterns through publicly released industry case studies of IVA and chatbot implementations.

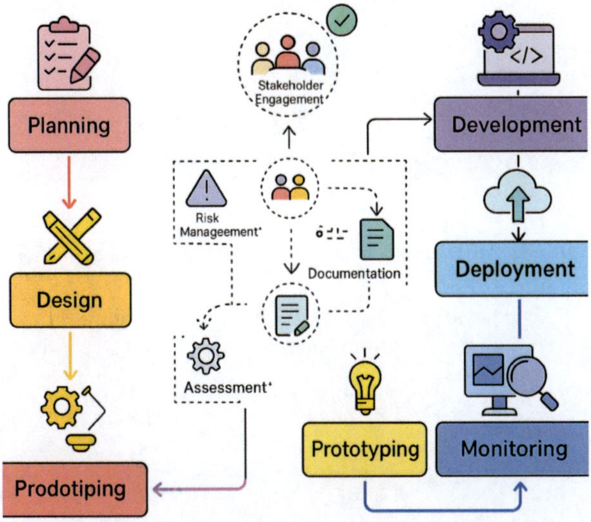

Fig. 1. Responsible AI Product Lifecycle Framework

Figure 1 depicts the Responsible AI Product Lifecycle Framework with an infinitely rigorous process towards AI system design, development, and deployment openly responsibly. It begins in Planning, where intent and ethical boundaries are defined, followed by Design, where system design and equity policy are defined. Next Prototyping, where function testing is feasible at a basic level, before additional development to Development, where models are defined and ready for production. Tested, the system moves on to Deployment, where the AI solution is being manufactured. Rolled out, Continuous Monitoring enables ongoing monitoring of performance, bias, and adherence to regulation compliance requirements. Risk Management, Assessment, and Documentation enabler building blocks it stages to try to make it transparent, minimize vices' scope, and maximize auditability. Stakeholder Engagement is that glue which binds users, regulators, and developers of the AI life cycle together to get technology working as per the requirement of society. Briefly, the frameworkspecifies a bottom-up design framework for building ethical, sustainable deployable, and responsible AI systems.

5 Results

Results from the simulation test make the case compelling and coherent for infusing responsible practice in AI into consumer-facing generative AI product development. Experimentation, following four rounds of iterative development on the 'HealthBot' (Pharma) and 'ConnectAI' (Telecom) models, showed good technical advancement and user-controllable limits as progressively stronger safety and ethical guardrails were made available. Baseline baseline model (Version 1.0) results captured inherent risk of uncontrolled generative AI. Hallucination was 12.4% for HealthBot and 9.8% for ConnectAI. Categorical cross-entropy loss function is given as:

$$L(\theta) = -\frac{1}{N}\sum_{i=1}^{N}\sum_{j=1}^{C} y_{ij}\log\left(\hat{y}_{ij}(\theta)\right) \tag{1}$$

(1)

Table 1. Model Performance Metrics Across Product Versions

Metric	Version 1.0 (Baseline)	Version 2.0 (Bias Mitigation)	Version 3.0 (HITL Validation)	Version 4.0 (Enhanced Guardrails)
Accuracy (%)	78.5	81.2	95.8	98.2
F1 Score	0.76	0.80	0.94	0.97
Hallucination Rate (%)	12.4	11.8	1.5	0.9
Latency (ms)	850	920	1250	1100
User Satisfaction (%)	48	55	86	91

Table 1 illustrates the development of the most significant performance indicators of the HealthBot application throughout four consecutive development cycles, as a means of demonstrating the effect of integrating responsible AI features. Columns are successive releases of products, and rows are some key performance measures. Numbers itself quite clearly indicates the way in improvement both in model quality as well as mood of the users. F1 Score and Accuracy, standard performance measures applied on the models, continue to rise from baseline model's 78.5% and 0.76 to record high of 98.2% and 0.97 for the current release with the additional guardrails. Most striking of all, maybe, is that the most important of all the measures in question here, the Hallucination Rate, has an incredible drop, from 12.4% to as low as less than 1%. This suggests the quality of performance with which HITL and RAG systems work in V3.0. Ironically enough, Latency does actually increase when one includes these other additional more sophisticated validation stages (peaks at V3.0), an ad-hoc product performance v. security trade-off which a product manager would rightly need to bear. But column 4, User

Satisfaction, reins in strategy in constraint of this trade-off since satisfaction increases from 48% to 91%, which attests that users mostly concern themselves with correctness and reliability, and not with incremental responsiveness enhancement, particularly in a health setting. Bayesian inference for updating trust can be expressed as:

(2)

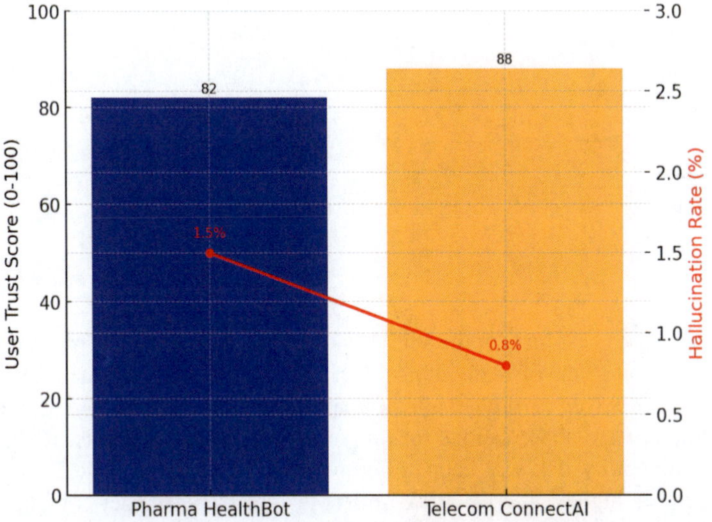

Fig. 2. AI hallucination rate and consumer confidence in pharma and telecom

In Fig. 2 use Cases Overview comparison view is represented in the form of the bar line graph, reflecting the strong relationship between customer sentiment and AI model performance in telecom (Telecom) and pharmaceutical (Pharma) use cases. Colour bars and left Y-axis are tracking 'User Trust Score' on 0 to 100 scale by metric of task completion as far as achievement, user satisfaction, and trust towards AI. Blue bar of Pharma HealthBot at User Trust Score of 82 and orange bar of Telecom ConnectAI at 88 in second position behind them. Both are using the optimised V4.0 versions of the study. The secondary vertical axis is properly and the line plot superimposed show the 'Hallucination Rate (%), or the percent by which factually wrong or fictional information had been produced by the AI. Blue color marking of the line is the fact that the Pharma application had a consistent hallucination rate of 1.5%, and orange marking for Telecom application was 0.8%. The storyline illustrates the ideal inverse correlation pictorially: as the user trust increases, the AI hallucination decreases. This infographic is a grim reminder from the viewpoint of both the research scientist and the product manager that even on those areas where low error rates are achieved on high-stakes application areas like healthcare, a single error can be a net user trust loss. Pharma bot's lesser confidence with relatively lower hallucination rate implies that the cost of one error in a health-care setting is far greater and virtual absolute precision is all the better. F1 score with precision and recall components will be

(3)

Table 2. Comparative analysis of ethical risks and mitigation strategies

Ethical Risk Category	Risk Probability (%)	Risk Impact (1–5)	Mitigation Effectiveness (%)	Residual Risk (%)
Data Privacy Breach	15	4	95	0.75
Algorithmic Bias	25	3	90	2.50
Misinformation (Hallucination)	30	5	97	0.90
Lack of Transparency	40	2	85	6.00
Accountability Gap	20	4	80	4.00

The Table 2 is an estimate of the V4.0 HealthBot application risk to the V4.0 HealthBot application in terms of likely ethical harm and control effectiveness to prevent it. Briefings are employed as proactive risk management business by product managers. Briefing columns are abbreviations for five broad types of ethical risk. Risk Probability (%) is probability risk will happen before control can be exercised, and 'Lack of Transparency' is highest at 40%. Risk Impact is 1–5, worst impact is highest impact; 'Misinformation (Hallucination)' would need to have highest impact rating since highest risk of immediate bodily damage is in the health care setting. Column 'Mitigation Effectiveness (%)' calculates duty AI action success—i.e., encryption as mitigation to make information secure, RAG to avoid hallucination, and bias audits to make it unbiased—to lower initial risk. For instance, misinformation prevention measures are 97% effective. Finally, the Residual Risk (%) column calculates the risk exposure after using the formula: (Risk Probability * (1 - Mitigation Effectiveness)). The final figure is the one product leadership will be interested in since there can never be a hundred percent safe control system. There are strict controls but always a very rare supply risk in each category that is sensitive and small but has to be removed from always, user feedback paths, and incident response plan well defined. Residual risk quantification model is:

(4)

Figure 3 represents a four-panel chart of projected quarterly user adoption of Pharma HealthBot and Telecom ConnectAI solutions, the latter with and without some regulatory alignment and transparency features. Four-quarter (Q1-Q4) on the x-axis and the 'User Adoption Rate (%))' on the y-axis. Four lines: a blue solid line for Pharma with features, a blue dashed line for Pharma without features, an orange solid line for Telecom with features, and an orange dashed line for Telecom without features. The graph indicates quite well the use of apps with moral AI attributes—i.e., easy data privacy conditions, easy consequences, and AI limit reminder—are being used more and more. They both put their thicker lines far above thinner ones. Pharma app and together have experienced 45% uptake through to date latest Q4 compared to 30% without app. And a rich-featured Telecom app scored 52% against its less feature-honesty twin at 40%. Excellent product

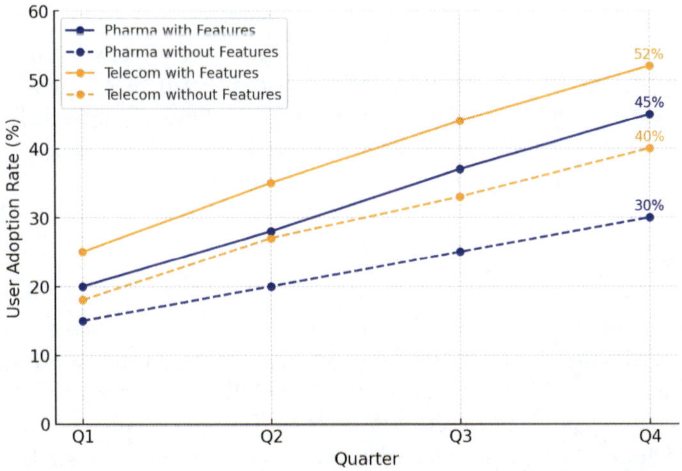

Fig. 3. Four quarters user adoption regulatory alignment feature impact subsequent

lesson: moral transparency and regulation as an issue of law isn't merely important, it's an incredibly powerful trust catalyst for markets and consumers both. End-users will re-use and utilize AI capabilities more comfortably if they are guaranteed with certainty that they are safe, secure, and that their data rights are being respected. Shannon entropy for model output uncertainty will be:

(5)

They were a result of User Trust Scores of 45 and 52 out of 100 and very high human agent escalation rates (35% for HealthBot). Later releases were considerably better. Bias counteracting features introduced in Version 2.0 were becoming progressively more effective at improving simulated populations but with little impact on total trust. The largest leap was Version 3.0 which added Human-in-the-Loop (HITL) validation and retrieval-augmented generation (RAG) to the platform. Hallucination scores decreased to 1.5% for HealthBot and 0.8% for ConnectAI. The all-time lowest percentage of disinformation was the largest improvement to elevate User Trust Scores to 82 for HealthBot and 88 for ConnectAI. The transparency and regulation compliance feature was also confirmed by the study. Figure 3 shows the adoption of both apps was much higher when both apps also possessed alignment feature like responsible AI values (i.e., transparent data usage policy, accuracy disclaimer).

The large gap between the 'with alignment features' cluster and 'without' cluster at four-quarter simulated time was due to a maximum of 15 points in HealthBot adoption, and 12 points in ConnectAI adoption. This means that customers make a moral decision to adopt the cleaner and safer services. Table 1 illustrates world view of model performance trend. Both, in all the remaining tests, F1 Score as well as Accuracy both were increasing with reducing hallucination rate and latency simultaneously. Health Bot accuracy was improved from 78.5% for V1.0 to 98.2% for V4.0, which exhibits the synergistic capability of moral AI action. Comparative estimation of risks is presented in Table 2, highlighting probability and magnitude of various ethical risks. It illustrates the nefarious power of disinformation within the pharma sector (5/5 Impact Score) when compared to the telecom sector (3/5). Above everything, the results are that despite extremely effective counter-measurement (more than 90% effectivity on average), no plan could eradicate risk completely and there exists a ‘Residual Risk’ that will need to be monitored and tackled by product managers. Overall, findings are formal quantitative proof that such a product policy to trust, openness, and safety results in more performing products and more patient consumers.

6 Discussion

The research findings forecast a true picture of the problem and demand arising for product managers with the advent of generative AI. The conclusions firmly affirm the thesis that by design responsible is not just a matter of CSR but a general success driver of consumer-facing AI products in sensitive markets. The below ‘Responsible AI Product Lifecycle Framework’ (Fig. 1) is not an ideal exercise but an action guide, which was directly applied to remedy issues faced within a simulation. The initial ‘Ideation & Ethical Framing’ stage, for instance, forces a product team to cast day one the physical cost of misinformation in a pharma context (Table 2‘s metric) and outline the whole development route. Most difficult to see in the mind’s eye is high negative correlation between user trust and hallucination rate in Fig. 2. It is most interruptible by a product manager’s metric. Experiment led us to discover that with 1.5% hallucination, HealthBot’s 82 trust score was lower than telecom bot’s 88 (0.8% hallucination rate). It indicates how much hallucination “costs” differently by domains.

A medical error would be far more devastating as a blunder than a debugging network error. To that extent, pharma, a hazardous products business unit with high-risk character, product managers can never select industry standards of accuracy measures; they will have to select a virtual zero hallucination rate. Those are at true cost, as is observable from a higher latency in Table 1 with HITL validation. It is the product manager who will be forced to explain this compromise, and explain that safety expenditure is one way of maintaining at arm’s length loss of confidence on the part of the masses one single failure to do so would entail. Outcomes, in Fig. 3, as opposed to non-concern on the part of consumers regarding where a product sits on the ethical scale are. Epidemic growth curves in open-app adoption are yet more people-driven digital citizenship on a global scale. They are not consumers of technology but in fact determining what is respectful and safe for their information themselves.

That is a solid business justification for product managers to be spending money on moral AI. They are “value drivers,” not “cost centers,” and can be leveraged to create long-term competitive tailwinds. Whether to reveal data use policies explicitly, whether to reveal output confidence scores on the answer, or whether to reveal shallow escalation paths to human representatives is an engineering trade that establishes long-term trust with the customer. Third, iterative optimization such as Table 1 and risk analysis such as Table 2 illustrate the dynamic AI product management process. An AI product is never complete. That very first V1.0 baseline model, while genius, barely was stable enough to ship to cut free on the wild. Every subsequent subsequent iteration de-risking the product but still in development adding to its value proposition by introducing some thoughtfully added responsible AI interventions. The idea in Table 2‘s ‘Residual Risk’ comes particularly into use. It encourages product managers to abandon the fantasy of a completely safe system and work towards a state of vigilant mindfulness instead. That means instituting robust post-launch monitoring controls, open feedback loops to fix AI errors, and retraining or even retirement access if a feature is found to be more wicked than useful. Last but not least, the topic of this essay is the new product manager paradigm: delivering features and running giants of socio-technical systems in which user trust is the payment currency.

7 Conclusion and Future Work

The leadership of the product manager is firmly at the front in this essay due to the ultimate duty of moral AI generative production and shipping in consumer telecommunication and telemedicine products. The 'HealthBot' and 'ConnectAI' program simulation testing of this research gives us qualitative evidence that safety and ethical practice should be the minimum of product design. The strongest assumption is that users' trust is not a soft measure but a hard requirement and that trust is inversely linear with respect to AI hallucination rate. Even reductions in such hallucinations, through processive methods like Human-in-the-Loop verification and retrieval-augmented generation, optimally expected high user satisfaction and trust ratings. The study also credits liable AI as an actual value force to commerce. More usage of product releases that have initiated compliant regulation functionality in the open will see customers go out of their way to buy and pay for products online which they will consider safe and secure. That is not saying that any such observation that ethical aspects would serve to stifle innovation but are an additional boost to design green, friendly products. The variation between version-to-version in measures of performance that are fluctuating also means that the ethical development of AI is not something done once but something repeatedly accomplished and reiterated over and over again in an iterative cycle. Product managers will be able to make level-set trade-offs, e.g., accepting a little extra latency for the deep cut in malicious disinformation. The current risk estimate is that even with optimal-of-the-art mitigation, there are risks at play that require an infinite modesty and watchfulness of product management culture. Simply put, good and well-intentioned products for generative AI product managers are not on the algorithmic track but on that of relentless endeavor with openness, responsibility, and an infinite one on user well-being. While this study is fundamentally exploratory, there are a few probable avenues of future research that are worth noting. Longitudinal study, if feasible on actual user data, would be the Holy Grail. User trust and adoption rate over time can inform us regarding the users' perception of AI worthiness over time after usage and when and whether ever the trust is regained (permanently) after AI to induce failure. Second, follow-up studies must be highly sensitive to socio-demographic and cultural specificity in trusting AI. A best-case population model of consumers was assumed for analytical purposes in this study as a working hypothesis, but depending on AI, particularly medicine, would be hugely age-sensitive, digitally literate level-sensitive, culture-sensitive, and technology familiarity-sensitive. Variations have to be accounted for while developing inclusive and unbiased AI products. Another study is developing consumer-choicereadable, standardized "AI nutrition labels." Activity can involve designing and prototyping various forms of such labels to set best practice in communicating model strengths and limitations, data, and bias to the non-technical consumer. Finally, the tech R&D agenda for the detection of real-time hallucinations is in development. One such potential area of employment would be to create light-weight models or AI-born systems that can predict probable hallucinations as robust warnings early on before response generation back to the user, from active prevention model to post-hoc audit model (i.e., HITL). That would be a giant leap towards designing safer generative AI systems by design.

References

1. Ahmad Husairi, M.A., Rossi, P.: Delegation of purchasing tasks to AI: the role of perceived choice and decision autonomy. Decis. Support. Syst. **179**, 114166 (2024)
2. Aldausari, N., Sowmya, A., Marcus, N., Mohammadi, G.: Video generative adversarial networks: a review. ACM Comput. Surv. **55**(2), 1–25 (2023)
3. André, Q., et al.: Consumer choice and autonomy in the age of artificial intelligence and big data. Cust. Needs Solut. **5**(1–2), 28–37 (2018)
4. Ashfaq, M., Yun, J., Yu, S., Loureiro, S.M.C.: I, chatbot: Modeling the determinants of users' satisfaction and continuance intention of AI-powered service agents. Telematics Inform. **54**, 101473 (2020)
5. Castelo, N., Bos, M.W., Lehmann, D.R.: Task-dependent algorithm aversion. J. Mark. Res. **56**(5), 809–825 (2019)
6. Choi, J.K., Ji, Y.G.: Investigating the importance of trust on adopting an autonomous vehicle. Int. J. Hum. Comput. Interact. **31**(10), 692–702 (2015)
7. Fan, Y., Liu, X.: Exploring the role of AI algorithmic agents: the impact of algorithmic decision autonomy on consumer purchase decisions. Front. Psychol. **13**, 1009173 (2022)
8. Gaggioli, A., Riva, G., Peters, D., Calvo, R.A.: Positive technology, computing, and design: shaping a future in which technology promotes psychological Well-being. In: Emotions and Affect in Human Factors and Human–Computer Interaction, pp. 477–502. Academic (2017)
9. Schneider, S., Nebel, S., Beege, M., Rey, G.D.: The autonomy-enhancing effects of choice on cognitive load, motivation and learning with digital media. Learn. Instr. **58**, 161–172 (2018)
10. Walczak, K., Cellary, W.: Challenges for higher education in the era of widespread access to generative AI. Econ. Bus. Rev. **9**, 71–100 (2023)
11. Davenport, T., Guha, A., Grewal, D., Bressgott, T.: How artificial intelligence will change the future of marketing. J. Acad. Mark. Sci. **48**, 24–42 (2020)
12. Abd Wahab, N., Hassan, L.F.A., Shahid, S.A.M., Maon, S.N.: The relationship between marketing mix and customer loyalty in hijab industry: the mediating effect of customer satisfaction. Procedia Econ. Finance. **37**, 366–371 (2016)

Artificial Intelligence in Obstetric Outcome Prediction: A Comprehensive Review

Ezaccani Salma[1](✉), Bousselham Abdelmajid[2], Barakat Oussama[3], Mottet Nicolas[4], Chollet Justine[4], and Bouattane Omar[1]

[1] EEIS Laboratory, ENSET Mohammedia, Hassan II University of Casablanca, Mohammedia, Morocco
salma.ezaccani-etu@etu.univh2c.ma

[2] 2IACS Laboratory, ENSET Mohammedia, Hassan II University of Casablanca, Mohammedia, Morocco

[3] SINERGIES Laboratory, University of Bourgogne-Franche-Comté, Besançon, France

[4] Mother-Child Department, CHU Besançon, UR SINERGIES, University of Marie and Louis Pasteur, Besançon, France

Abstract. Artificial Intelligence (AI) is increasingly transforming healthcare, including maternal and fetal medicine. This systematic review examines recent applications of Machine Learning (ML) and Deep Learning (DL) in predicting critical obstetric outcomes. The reviewed studies address tasks such as delivery mode prediction, fetal weight estimation, preterm birth risk assessment, preeclampsia and fetal distress detection, and labor onset forecasting. Relevant publications from 2021 to 2025 were retrieved from IEEE Xplore, BMC, Scientific Reports, and MedRxiv. Out of 110 studies screened, 20 met the inclusion criteria. Each was analyzed based on AI model type, dataset characteristics, input features, prediction targets, and performance metrics. The findings highlight strong predictive potential but also methodological limitations, including data imbalance, small sample sizes, and limited external validation. Future research should emphasize larger and more diverse datasets, explainable AI approaches, and ethical integration into clinical practice to ensure safe and equitable maternal care.

Keywords: Artificial intelligence · Machine learning · Deep learning · Maternal healthcare · Predictive modeling

1 Introduction

The rapid evolution of technology has transformed our way of life and shows promising results, particularly in healthcare where AI; including Machine Learning, Deep Learning, and Internet of Things (IoT) has a significant positive impact by facilitating personalized decision-making, improving patient care, reducing medical errors, and optimizing diagnostic time and costs. (Figure 1), illustrates the various roles of AI in healthcare.

In recent years, several ML and DL-based methods have been proposed to predict various diseases; diabetes, COVID-19, different types of cancer, and adapt treatments

A. E. Hassanien et al. (Eds.): AISDS 2025, LNDECT 286, pp. 119–130, 2027.
https://doi.org/10.1007/978-3-032-18758-1_12

for more precise medicine. Shatnawi et al. [1] used ConvNeXt and CNN for lung cancer classification, achieving 100% accuracy. Chikkala et al. [2] applied a Bidirectional Recurrent Neural Network (BRNN) for histopathological breast cancer image classification. Fakieh and Saleem [3] proposed a machine learning-based framework using KNN and SVM for COVID-19 prediction. Dihmani et al. [4] developed a CAD system based on thermal imaging and Explainable AI, achieving 98.27% accuracy for breast cancer detection. In another contribution, Bousselham et al. [5] enhanced ischemic stroke MRI segmentation using thermal data and a U-Net model trained on over 3,000 images.

This review focuses on AI applications for predicting critical childbirth parameters including delivery mode, preterm birth, fetal weight, and labor timing - crucial public health issues requiring simultaneous protection of maternal and fetal wellbeing. However, challenges persist: healthcare access inequalities, limitations in monitoring tools, and difficulties in anticipating complications (hemorrhages, infections, hypertension, prematurity, etc.). Through enhanced monitoring and early warning systems, AI could improve anomaly detection and tailor management for each case, thereby reducing risks and optimizing maternal and neonatal outcomes.

The article is structured as follows: Sect. 2: Study Methodology" Presents the research methodology. Section 3: "Overview of Selected Articles" Provides a thematic analysis of included studies. Section 4: "Comparative Analysis" Summarizes key findings. Section 5: "Discussion" Interprets results and implications. Section 6: "Conclusion" Synthesizes findings and future directions.

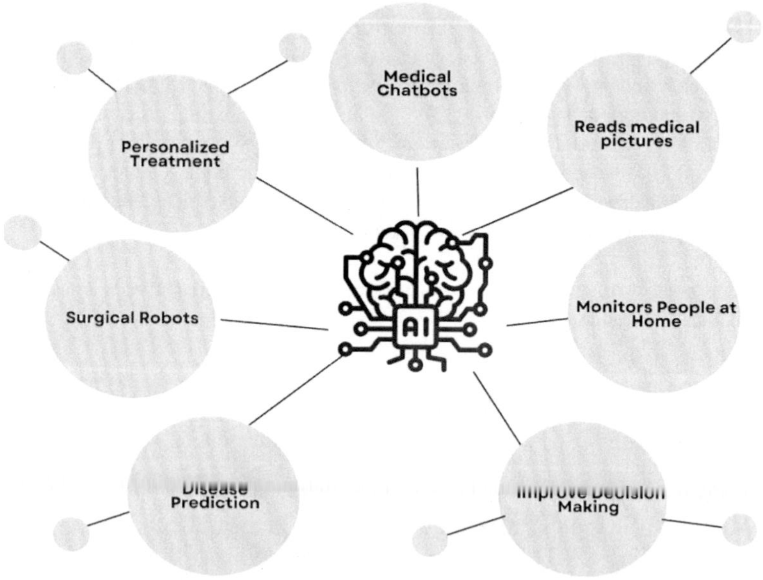

Fig. 1. Roles of AI in Healthcare. Original.

2 Study Methodology

2.1 Search Strategy

This review examined studies published between 2021 and 2025, obtained from various sources such as IEEE Xplore, BMC, Web of Science, Scientific Reports, and MedRxiv. The dataset included different publication types, including review papers, journal articles, and doctoral theses. Our comprehensive search strategy combined artificial intelligence—related terms ("machine learning," "deep learning," "neural networks," "AI") with obstetric prediction parameters ("delivery mode prediction," "fetal weight estimation," "preterm birth risk," "labor onset prediction," "Adverse Pregnancy Outcomes Prediction," and "maternal health risk").

2.2 Inclusion and Exclusion

To ensure the relevance and quality of the studies analyzed, strict criteria were applied. **Inclusion criteria** comprised: (a) studies using AI techniques; (b) recent publications (2021–2025); (c) English-language articles; (d) papers with clear methodology and quantifiable results. **Exclusion criteria** eliminated: (a) older studies (pre-2021); (b) non-AI approaches; (c) non-obstetric outcomes; (d) non-English publications; (e) abstract-only records without full-text articles. The summary of the search and selection of final articles are illustrated in Fig. 2.

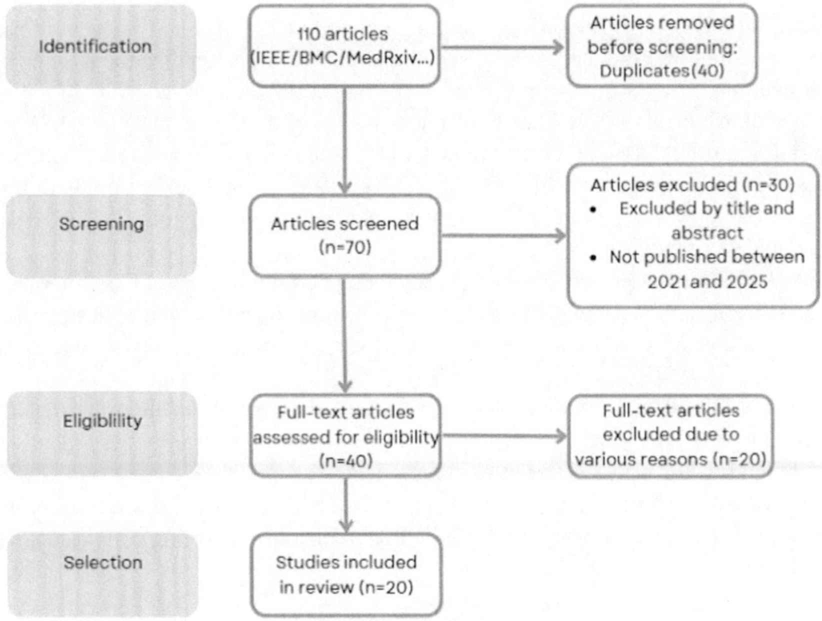

Fig. 2. Flowchart illustrating the study selection process. Original.

2.3 Data Extraction and Analysis

For each study included in the review, we systematically extracted three categories of information: (1) general characteristics (authors, publication year, country, and study design), (2) technical specifications of AI models (algorithms used, e.g., Convolutional Neural Network (CNN), Random Forest; input data types; and validation methods), and (3) reported performance metrics (accuracy, Area Under the Curve (AUC), and sensitivity). These elements were compared to evaluate the effectiveness of different approaches for each obstetric prediction task, with particular attention to data quality and diversity, clinical applicability, and methodological strengths and limitations. The results were synthesized in comparative tables highlighting performance patterns across studies, as well as the advantages and limitations of each method.

3 Overview of Selected Articles

In this section, we provide a comprehensive analysis of selected studies examining artificial intelligence applications in obstetric care. The review is organized by clinical prediction tasks, highlighting key methodologies and findings from each work.

3.1 Delivery Mode Prediction

Studies [6–10] investigate AI models for predicting cesarean versus vaginal delivery:

Nguyen [6] Introduces a biomechanical framework using Generative Adversarial Networks (GANs), Long Short-Term Memory networks (LSTMs), and physics-informed neural networks to simulate delivery scenarios, enabling personalized and safer birth route planning.

Chowdhury et al. [7] validated their Random Forest (RF) model on 1,659 Electrohysterogram (EHG) recordings using an 80/20 holdout split and cross-validation, addressing class imbalance with SMOTE. The model achieved perfect prediction performance (AUC = 1.00) for normal-weight women, demonstrating robust results despite the initial data imbalance.

Devika et al. [8] reported a 99% accuracy using an Support Vector Classifier (SVC) model with data splits of 70/30 and 80/20. However, the unspecified sample size and the absence of external validation limit the generalizability of these otherwise exceptional results.

Islam et al. [9] applies multiple machine learning models to thousands of birth records; the Stacking Classifier (F1-score 97.9%) proves most effective, identifying critical features like prior cesareans and complications.

Kuanar et al. [10] implements a Deep Neural Network that combines various biometric indicators to predict delivery mode with 99% precision, supporting clinical decisions through integrated analysis.

3.2 Fetal Weight Estimation

Studies [11–15] explore AI approaches for fetal/birth weight prediction:

Wang et al. [11] developed a deep neural network to predict fetal weight using selected maternal and neonatal features. The model showed better accuracy than the traditional Hadlock method, with a lower RMSE. Its simplicity and strong performance make it practical for clinical use.

Jeyalakshmi and Bhavani [12] introduced an enhanced recurrent neural network optimized with bio-inspired techniques. It achieved a high prediction accuracy and could detect growth abnormalities early. The model outperformed standard methods by a wide margin.

Płotka et al. [13] presented BabyNet, combining 3D ResNet and Transformer modules to analyze ultrasound (US) videos. It delivered performance close to expert sonographers and automated plane detection. This approach helps streamline and improve fetal weight estimation.

Płotka et al. [14] proposed a hybrid CNN-Transformer model to directly estimate fetal weight from ultrasound videos. It surpassed expert accuracy and eliminated the need for manual measurements. The model was highly consistent and reliable.

Yao et al. [15] introduced FetalAI, an RNN-based tool for monitoring fetal weight over time. It integrates various clinical features to assess risks like macrosomia or LGA. The tool enables continuous and proactive pregnancy management.

3.3 Preterm Birth Prediction

Studies [16–18] explore how machine learning (ML) can improve risk assessment:

Kloska et al. [16] an optimized Support Vector Machine (SVM) model achieved 84% F1-score, identifying CRP, hematocrit, platelet levels, and clinical factors like parity and education level as significant risk markers, though the small sample size limits generalizability.

Zhang et al. [17] AdaBoost emerged as the most effective method, with an AUC of 0.91, highlighting placenta praevia and fetal distress as critical risk factors for preterm birth.

Belaghi [18] logistic regression achieved 80.5% AUC with 97% negative predictive value, outperforming neural networks (AUC 68.8%) and demonstrating the continued relevance of simpler, interpretable models in clinical practice.

Table 1. Summary of Comparative Analysis.

Scope	Reference	Algorithms/Methods
Delivery Mode Prediction	[6]	GAN – LSTM – PINN
	[7]	RF
	[8]	SVC – CNN – KNN – NB
	[9]	SC – RF – KNN – DT – SVM
	[10]	DNN – ANN – KNN – RF ...

(*continued*)

Table 1. (*continued*)

Scope	Reference	Algorithms/Methods
Fetal Weight Estimation	[11]	DNN
	[12]	AURNN
	[13]	BabyNet
	[14]	CNN + Transformers
	[15]	RNN
Preterm Birth Prediction	[16]	SVM
	[17]	AdaBoost
	[18]	LR
Labor Onset Prediction	[19]	AE-LSTM
Adverse Pregnancy Outcomes Prediction	[20]	NB
	[21]	1D-CNN-LSTM, 2D-CNN
Embryo Selection for IVF	[22]	ResNet-CNN
Maternal Health Risk Prediction	[23]	MLP
	[24]	Stacking Ensemble (stratified sampling), CatBoost, XGBoost, RF, DT
	[25]	Random Forest

3.4 Labor Onset Prediction

Basavaraj et al. [19] developed an AE-LSTM (Autoencoder Long Short-Term Memory) model using Oura Ring data (n = 91) to predict labor onset. It achieved 79% accuracy for predicting spontaneous delivery 7–10 days ahead ($R^2 = 0.93$), with lower performance for induced labor ($R^2 = 0.68$). Larger clinical validations are needed for implementation.

3.5 Adverse Pregnancy Outcomes Prediction

This section highlights AI tools for early detection of pre- eclampsia and fetal distress.

Munyao et al. [20] a wearable system using biomedical sensors and Naïve Bayes classification was developed for real-time pre-eclampsia risk detection, enabling remote monitoring.

Asfaw et al. [21] proposed an innovative deep learning model, but its sensitivity was limited (20%), mainly due to severe class imbalance (0.9% severe cases versus 99.1% normal cases) and the heterogeneity of historical data collected over two decades, which may have affected annotation consistency.

3.6 Embryo Selection for IVF

AI is revolutionizing embryo selection in In Vitro Fertilization (IVF) by offering more accurate assessments, as shown by Huang et al. [22], whose ResNet-CNN model trained on time-lapse images achieved an AUC of 0.968.

3.7 Maternal Health Risk Prediction

Studies [23–25] explore how machine learning (ML) can improve risk assessment using the same clinical dataset of 1,014 pregnant women from Bangladesh:

Pi et al. [23] evaluated six ML algorithms, with the multilayer perceptron (MLP) achieving the highest overall accuracy of 82%, and 91% accuracy for high-risk cases.

Malde et al. [24] developed a stacking ensemble model combined with stratified sampling, outperforming models like CatBoost and XGBoost, reaching 87.2% accuracy.

Tzimourta et al. [25] compared six classifiers, with Random Forest performing best at 88.03% accuracy, but facing challenges in classifying the mid-risk category due to class imbalance. Table 1 provides a summary of the comparative analysis.

4 Comparative Analysis

This section synthesizes and compares AI applications across key obstetric prediction domains, focusing on datasets used, methodological choices, and clinical relevance.

4.1 Delivery Mode Prediction

Several studies explored AI models to distinguish between vaginal and cesarean delivery. The approach in [6] combined GANs, LSTMs, and physics-informed networks to simulate delivery scenarios, enhancing prediction but with limited clinical practicality due to system complexity. Researchers in [7] applied a Random Forest on EHG signals and maternal BMI, showing perfect performance for normal-weight women (AUC 1.00) but lower for obese women (AUC 0.86), reflecting dependence on signal quality. According to [8] used SVM on Electronic Health Record (EHR) with 99% accuracy, outperforming CNNs but lacking model transparency. Study [9] applied a stacking ensemble on 6,000+ records, achieving 97.9% F1-score and identifying 32 key features, though interpretability remains a concern. In [10] used a DNN with clinical inputs from 101 women (99% accuracy), but its small dataset limits generalization.

4.2 Fetal Weight Estimation

AI models also improved fetal weight prediction. Study [11] used a DNN with 10 features on 34,824 records, achieving RMSE 189.64 g, outperforming the Hadlock formula. Researchers in [12] used an augmented RNN trained on ultrasound and medical history (accuracy 95.4%), enabling early growth issue detection. As discussed in [13] introduced BabyNet (3D ResNet + Transformer) with 7.5% error from ultrasound (US) videos, near expert level. In [14] used a CNN-Transformer hybrid achieving 2.59% error—better than sonographers—without traditional biometric input. Study [15] proposed FetalAI, an RNN tool integrating ultrasound and maternal data for dynamic risk monitoring, especially for macrosomia.

4.3 Preterm Birth Prediction

In [16], an optimized SVM model (F1-score 84%) identified biomarkers like CRP and hematocrit in a small high-risk cohort. According to [17] used AdaBoost (AUC 0.91) on 5,411 records, highlighting placenta praevia and fetal distress, though excluding comorbidities may bias results. Study [18] showed logistic regression (AUC 80.5%, NPV 97%) outperformed neural networks in a cohort of 145,846 women, underlining the utility of simpler, interpretable models.

Table 2. Characteristics and Results of Reviewed Studies.

Reference	Study Objectives	Data Type	Main Output
[6]	Delivery Mode	CT images, simulation data, time-series signals, 3D models	IoU = 93.68%–99.63% for skeleton segmentation; RMSE = 0.103–0.845 mm for tissue prediction; real-time PINN model with RMSE < 0.01 m;
[7]	Delivery Mode	EHG signals	AUC = 1.00 (normal weight), AUC = 0.86 (obese);
[8]	Delivery Mode	EHR data	Accuracy = 99%
[9]	Delivery Mode	Birth records	F1-score = 97.9%
[10]	Delivery Mode	Biometrics	Pr = 99%
[11]	Fetal Weight	EHR	RMSE = 189.64 g
[12]	Fetal Weight	Temporal clinical	accuracy = 95.4%
[13]	Fetal Weight	US videos	Error = 7.5%
[14]	Fetal Weight	US videos	MAPE = 2.59%
[15]	Fetal Weight	Longitudinal US	MAE = 292.5 g; sens: 96.77% (macrosomie), 75% (LGA)
[16]	Preterm Birth	Biomarkers	F1 = 84%
[17]	Preterm Birth	EHR	AUC = 0.91
[18]	Preterm Birth	Birth registry	AUC = 80.5%
[19]	Labor Onset	Wearable time-series data	accuracy = 79%
[20]	Pre-eclampsia	IoT sensors	rap = 80%,pre = 77%
[21]	Fetal Distress	CTG traces	Sens = 20%; spec = 95%
[22]	IVF Selection	Time-lapse images	AUC = 0.968
[23]	Maternal Health Risk	clinical data	accuracy = 82%
[24]	Maternal Health Risk	clinical data	accuracy = 87.2%
[25]	Maternal Health Risk	clinical data	accuracy = 88.03%

4.4 Labor Onset Prediction

Researchers in [19] used an AE-LSTM model with wearable sensor data to predict labor up to 10 days in advance ($R^2 = 0.93$), opening new possibilities for remote monitoring, though large-scale validation is still needed.

4.5 Adverse Pregnancy Outcomes

In [20] proposed a wearable IoT-based pre-eclampsia detection system, where Naïve Bayes outperformed other models for analyzing vital signs in real time, though clinical validation remains pending. Study [21] analyzed 51,449 Cardiotocography (CTG) traces with a hybrid 1D-CNN-LSTM + 2D-CNN model, achieving 20% sensitivity at 95% specificity (PAUC = 0.20) despite severe class imbalance.

4.6 Embryo Selection for IVF

Study [22] applied ResNet-CNN to time-lapse embryo imaging, achieving AUC 0.968 and surpassing subjective visual assessments by embryologists, supporting better embryo selection in IVF.

4.7 Maternal Health Risk Prediction

Study [23] applied an MLP achieving 91% accuracy for high-risk pregnancies, benefiting from GPU-accelerated processing. Study [24] developed a stacking ensemble model with stratified sampling (accuracy 87.2%), outperforming CatBoost and XGBoost on imbalanced data. Study [25] used Random Forest, reaching 88.03% accuracy, with challenges in the mid-risk class due to class imbalance.

Table 2 summarizes key models, data types, and performance across all thematic areas for comparative reference.

5 Discussion

5.1 Implications

The integration of artificial intelligence in obstetrics marks a significant step toward predictive and personalized maternal-fetal care. The reviewed models demonstrate promising performance in anticipating key outcomes such as preterm birth, fetal weight, labor onset, and complications like pre-eclampsia or fetal distress. Particularly, deep learning, ensemble methods, and multimodal signal processing improve clinicians' ability to tailor monitoring and interventions. Furthermore, wearable and connected technologies are reshaping traditional prenatal follow-up by enabling continuous, remote monitoring of maternal health.

5.2 Challenges and Limitations

Despite these advances, several limitations remain. Many studies rely on retrospective, single-center datasets, which restrict generalizability. Some technically advanced models still fall short in terms of clinical readiness, due to insufficient sensitivity or specificity, as seen in fetal distress detection.

A critical limitation is the narrow scope of input data. Nearly all studies focus exclusively on maternal features, neglecting paternal data. Information such as the father's medical history, genetic background, or age could influence outcomes like birth weight, delivery mode, or inherited conditions. The exclusion of paternal factors represents a methodological gap that may limit the comprehensiveness of current predictive models.

In addition, the complexity and lack of interpretability in some AI systems—particularly deep learning—pose barriers to clinical integration. In medical practice, explainability is key to building trust and ensuring informed decision-making.

5.3 Future Directions

To enhance clinical impact, several directions should be explored. First, training and validating models on multicenter, demographically diverse datasets is essential. Incorporating longitudinal and multimodal data—such as biomarkers, genetic profiles, environmental, and behavioral factors—can further improve predictive accuracy. Importantly, future research should aim to integrate paternal data. A holistic view that includes both maternal and paternal contributions would better reflect the multifactorial nature of pregnancy outcomes. This could improve predictions related to fetal development, congenital risks, and obstetric complications. Finally, the development of explainable AI will be crucial. Models that offer transparent reasoning will facilitate adoption by healthcare providers, improve clinical trust, and support responsible integration into care protocols.

5.4 Ethics and Fairness Considerations

The use of AI in obstetrics raises ethical and fairness concerns. Ensuring patient privacy and data confidentiality is essential, especially for sensitive maternal and fetal information. Many studies used anonymized datasets, but few reported ethical approval or consent procedures. Bias may arise from limited or unrepresentative datasets, affecting performance across demographic groups. AI should serve as a decision-support tool, with clinicians retaining final responsibility, to ensure safe, equitable, and accountable integration into clinical practice.

6 Conclusion

Artificial intelligence is transforming obstetrics by improving prediction of pregnancy outcomes and supporting clinical decisions. While machine learning and deep learning models show high performance in predicting fetal weight, preterm birth, and labor onset, challenges remain, including limited external validation, poor interpretability, and exclusion of paternal data. Incorporating both parental data, multimodal inputs, and transparent, clinically validated models is crucial for accurate, fair, and ethical AI in maternal and neonatal care.

Acknowledgment. This work was carried out with support co-funded by the CNRST and the MTNRA as part of the "PhD-ASsociate Scholarship – PASS" program for doctoral teaching assistants in the field of digital technology, in line with the 2027-2030 National Digital Morocco Strategy.

Data Availability Statement. The data analyzed in this study come from publicly available sources (IEEE Xplore, BMC, Scientific Reports, MedRxiv) and confidential hospital datasets.

Conflict of Interest. The authors declare that they have no conflict of interest.

References

1. Shatnawi, M.Q., Abuein, Q., Al-Quraan, R.: Deep learning-based approach to diagnose lung cancer using CT-scan images. Intell.-Based Med. **11**, 100188 (2025). https://doi.org/10.1016/j.ibmed.2024.100188
2. Chikkala, R.B. et al.: Enhancing breast cancer diagnosis with bidirectional recurrent neural networks: a novel approach for histopathological image multi-classification. IEEE Access. **13**, 41682–41707 (2025). https://doi.org/10.1109/ACCESS.2025.3542989
3. Fakieh, B., Saleem, F.: COVID-19 from symptoms to prediction: a statistical and machine learning approach. Comput. Biol. Med. **182**, 109211 (2024). https://doi.org/10.1016/j.compbiomed.2024.109211
4. Dihmani, H., Bousselham, A., Bouattane, O.: A new computer-aided diagnosis system for breast cancer detection from thermograms using metaheuristic algorithms and explainable AI. Algorithms. **17**, 462 (2024). https://doi.org/10.3390/a17100462
5. Bousselham, A., Bouattane, O., Youssfi, M., Raihani, A.: Towards an accurate MRI acute ischemic stroke lesion segmentation based on bioheat equation and u-net model. Int. J. Biomed. Imaging. **2022**, 5529726 (2022). https://doi.org/10.1155/2022/5529726
6. Nguyen, D.: Advanced modeling of the childbirth system using deep learning methods: from fetal skeleton segmentation to real-time soft tissue deformation. Doctoral Dissertation, Centrale Lille Institute (2024). https://tel.archives-ouvertes.fr/tel-04831150
7. Chowdhury, R.H., Hossain, Q.D., Ahmad, M.: Predicting the birth delivery mode: uncovering the link between EHG topological features and BMI. IEEE Access. **13**, 6855–6868 (2025). https://doi.org/10.1109/ACCESS.2024.3525358
8. Devika, P., Bhuvaneswari, S., Rithika, R., Swarna, S., Vijitha, N.: Prediction of modes of child birth using SVC. In: Sharmila, V., Kannan, R., Kannadhasan, R., Sivakumar, P. (eds.) Challenges in Information, Communication and Computing Technology, pp. 726–731. CRC Press, London (2024). https://doi.org/10.1201/9781003559085-125
9. Islam, M.N., Mahmud, T., Khan, N.I., Mustafina, S.N., Islam, A.K.M.N.: Exploring machine learning algorithms to find the best features for predicting modes of childbirth. IEEE Access. **9**, 1680–1692 (2021). https://doi.org/10.1109/ACCESS.2020.3045469
10. Kuanar, A., Akbar, A., Sujata, P., Kar, D.: Deep neural network (DNN) modelling for prediction of the mode of delivery. Eur. J. Obstet. Gynecol. Reprod. Biol. **297**, 241–248 (2024). https://doi.org/10.1016/J.EJOGRB.2024.04.012
11. Wang, Y. et al.: Fetal weight estimation based on deep neural network: a retrospective observational study. BMC Pregnancy Childbirth. **23**, 560 (2023). https://doi.org/10.1186/s12884-023-05819-8

12. Jeyalakshmi, C., Bhavani, G.: Prediction of clinical risk factors in pregnancy using optimized neural network scheme. Placenta. **163**, 33–42 (2025). https://doi.org/10.1016/J.PLACENTA.2025.03.003
13. Płotka, S. et al.: BabyNet: residual transformer module for birth weight prediction on fetal ultrasound video. In: Wang, L., Dou, Q., Fletcher, P.T., Speidel, S., Li, S. (eds.) Medical Image Computing and Computer Assisted Intervention – MICCAI 2022, pp. 350–359. Springer Nature Switzerland, Cham (2022)
14. Płotka, S.S. et al.: Deep learning for estimation of fetal weight throughout the pregnancy from fetal abdominal ultrasound. Am. J. Obstet. Gynecol. MFM. **5**, 101182 (2023). https://doi.org/10.1016/j.ajogmf.2023.101182
15. Yao, Y., Liu, R., Zhang, B.: FetalAI: a deep learning web-based application for predicting birthweight from prenatal ultrasound measurements. Inform. Med. Unlocked. **49**, 101555 (2024). https://doi.org/10.1016/J.IMU.2024.101555
16. Kloska, A., Harmoza, A., Kloska, S.M., Marciniak, T., Sadowska-Krawczenko, I.: Predicting preterm birth using machine learning methods. Sci. Rep. **15**, 5683 (2025). https://doi.org/10.1038/s41598-025-89905-1
17. Zhang, Y. et al.: Establishment of a model for predicting preterm birth based on the machine learning algorithm. BMC Pregnancy Childbirth. **23**, 779 (2023). https://doi.org/10.1186/s12884-023-06058-7
18. Belaghi, R.A.: Prediction of preterm birth in multiparous women using logistic regression and machine learning approaches. Sci. Rep. **14**, 21967 (2024). https://doi.org/10.1038/s41598-024-60097-4
19. Basavaraj, C., Grant, A.D., Aras, S.G., Erickson, E.N.: Deep learning model using continuous skin temperature data predicts labor onset. BMC Pregnancy Childbirth. **24**, 777 (2024). https://doi.org/10.1186/s12884-024-06862-9
20. Munyao, M.M., Maina, E.M., Mambo, S.M., Wanyoro, A.: Real-time pre-eclampsia prediction model based on IoT and machine learning. Discov. Internet Things. **4**, 10 (2024). https://doi.org/10.1007/s43926-024-00063-8
21. Asfaw, D., Jordanov, I., Impey, L., Namburete, A., Lee, R., Georgieva, A.: Multimodal deep learning for predicting adverse birth outcomes based on early labour data. Bioengineering. **10**, 730 (2023). https://doi.org/10.3390/bioengineering10060730
22. Huang, B., Zheng, S., Ma, B., Yang, Y., Zhang, S., Jin, L.: Using deep learning to predict the outcome of live birth from more than 10,000 embryo data. BMC Pregnancy Childbirth. **22**, 36 (2022). https://doi.org/10.1186/s12884-021-04373-5
23. Pi, X., Wang, J., Chu, L., Zhang, G., Zhang, W.: Prediction of high-risk pregnancy based on machine learning algorithms. Sci. Rep. **15**, 15561 (2025). https://doi.org/10.1038/s41598-025-00450-3
24. Malde, A., Prabhu, V.G., Banga, D., Hsieh, M., Renduchintala, C., Pirrallo, R.: A machine learning approach for predicting maternal health risks in lower-middle-income countries using sparse data and vital signs. Future Internet. **17**, 190 (2025). https://doi.org/10.3390/fi17050190
25. Tzimourta, K.D., Tsipouras, M.G., Angelidis, P., Tsalikakis, D.G., Orovou, E.: Maternal health risk detection: advancing midwifery with artificial intelligence. Healthcare. **13**, 833 (2025). https://doi.org/10.3390/healthcare13070833

Integrating Fuzzy Logic with AI in Healthcare: A Comprehensive Review of Applications in Physical and Neurological Health

Abderrazzak Benmoujane[1(✉)], Abdellah Madani[1], and Mohamed Kissi[2]

[1] LAROSERI Laboratory, Chouaib Doukkali University Faculty of Sciences, El Jadida, Morocco
benmoujane.a271@ucd.ac.ma

[2] LIM Laboratory of Intelligence Machine, Hassan II University of Casablanca Faculty of Sciences and Techniques, Mohammedia, Morocco

Abstract. Medical decision-making is often challenged by uncertain information and complex clinical conditions. Artificial intelligence (AI), particularly machine learning (ML) and deep learning (DL), offers advanced capabilities in disease diagnosis and patient monitoring. Fuzzy logic (FL) is an invaluable tool for handling the ambiguity and uncertainty inherent in medical data. This paper provides a comprehensive review of the integration of FL with other AI techniques, emphasizing its role in physical and neurological health. Key applications in the diagnosis of cancer, heart disease, Alzheimer's, and depression are discussed. While some studies report high performance, with accuracy rates up to 99% on specific datasets, these results require critical appraisal regarding their generalizability. By synthesizing recent advancements, this review demonstrates how this integration can improve diagnostic accuracy and patient outcomes. Unlike "black-box" models, FL can offer more interpretable decision-making, which is crucial for building trust with clinicians. Finally, the review outlines key challenges, including the need for rigorous clinical validation and regulatory considerations, and proposes future research directions for developing intelligent, adaptive, and truly explainable healthcare solutions

Keywords: Fuzzy Logic · Artificial Intelligence · Machine Learning · Healthcare · Precision medicine

1 Introduction

The majority of domains in healthcare have explored the usage of artificial intelligence (AI) technologies such as machine learning (ML) and deep learning (DL) to determine their potential benefits [1]. Automated diagnosis and predictive modeling are two well-known research domains in healthcare, both of which

A. E. Hassanien et al. (Eds.): AISDS 2025, LNDECT 286, pp. 131–139, 2027.
https://doi.org/10.1007/978-3-032-18758-1_13

have been fruitful with the use of ML. A number of research works have established the effectiveness of techniques such as decision trees (DT), support vector machines (SVM), artificial neural networks (ANN), and DL in analyzing genetic data [2,3], medical images, and patient reports for the early diagnosis of diseases. Yet, conventional machine learning techniques are plagued by the uncertainties of patient data, inconsistent reports, vagueness in prescriptions, and variations in patients' emotional and physical states at the time of analysis. Combination of fuzzy logic (FL) and machine learning provides a powerful solution to these. ML models are improved pattern recognizers and predictors, whereas fuzzy logic eliminates medicine data uncertainty and simulates human reasoning to allow easier and more comprehensible and flexible decision-making in complex medical cases [3,4]. It allows researchers, doctors, and developers to find new and realistic solutions where knowledge cannot be fully explained in terms of concrete mathematical models.

This article is a comprehensive overview of recent convergence in medicine between fuzzy logic and AI. Section 2 describes our review approach. Section 3 is a brief primer on fuzzy logic basics. Section 4 overviews some of the most promising applications in physical and neurological health. Section 5 provides a critical discussion of the evidence, outlining key challenges and future directions, while Sect. 6 offers a concluding summary of the review's main findings.

2 Methodology

This paper provides comprehensive review of the integration of fuzzy logic (FL) with artificial intelligence (AI) in healthcare. To ensure a structured and thorough analysis, a literature search was performed across prominent academic databases, including PubMed, IEEE Xplore, and Scopus, from 2015 through 2024. The literature investigation was executed by using all of the leading keywords such as "fuzzy logic," "artificial intelligence," "machine learning," "deep learning," "healthcare," "medical diagnosis," "neurological health," and "physical health." The criteria chosen to help identify literature included the following: (1) literature was peer-reviewed from an international research publication or conference publication written in English, (2) research included an understandable application of the combination of fuzzy logic with another form of artificial intelligence (such as machine or deep learning), (3) application was used in the diagnosis or monitoring of a specific physical or neurological health issue. The criteria excluded research literature that was not peer-reviewed or presented an understandable unique viewpoint of current information, literature that did not incorporate fuzzy logic, or research literature related to non-clinical issues of healthcare or healthcare management.

3 Fuzzy Logic

To address uncertainty inherent in human cognition, Lotfi Zadeh designed fuzzy logic in the 1960s. Introduced in 1965 [5], fuzzy logic extends classical binary logic by defining degrees of truth rather than absolute true/false conditions. This capacity to speak of partial truths is particularly relevant in medicine, where clinical data frequently are imprecise, incomplete, or vague. Consequently, it became prominent in the late 1980s [6] as a method for representing the imprecision, and consequently enhancing the accuracy of diagnostic and personalized treatment. This section outlines the fundamental principles of fuzzy logic and its application to computational medicine.

3.1 Fuzzy Set

Fuzzy sets generalize traditional ("crisp") set theory by replacing strict binary membership where a member is either completely in a set (1) or not (0) [7] with a membership degree in the range [0, 1] [8]. In this way, an element can be partially a member of multiple sets simultaneously. For instance, the systolic blood pressure 130 mmHg may be a 0.7 member of the set 'Normal Blood Pressure' and 0.3 member of 'High Blood Pressure'. Such partial membership is useful in representing states of clinical conditions that fall between clear diagnostic cutpoints, such as an illness that is "largely normal but tipping towards high."

3.2 Membership Function

A fuzzy set is defined by its membership function $\mu_A(x)$, which systematically quantifies the degree to which an element x belongs to a set A by assigning it a value in the range $[0, 1]$ [9]. In this range, 0 indicates no membership, 1 indicates full membership, and intermediate values represent partial membership. Figure 1 provides a graphical example for systolic blood pressure. As shown, a reading of 130 mmHg holds partial memberships in two sets: a degree of 0.7 in the 'Normal Blood Pressure' set and 0.3 in the 'High Blood Pressure' set. This ability to model data that falls between crisp diagnostic categories is a core strength of fuzzy logic in medicine.

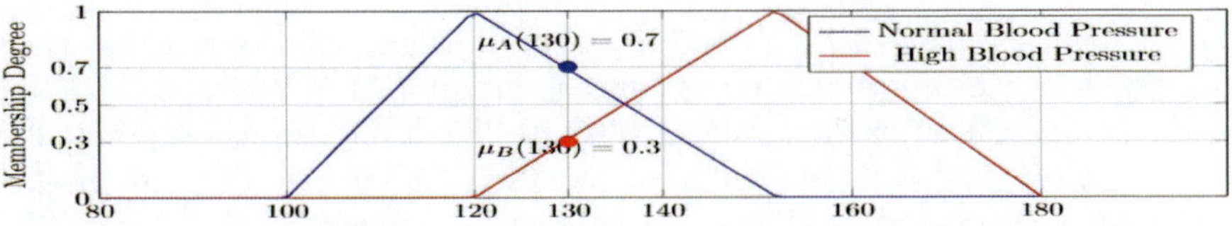

Fig. 1. Membership functions for systolic blood pressure

3.2.1 Linguistic Variables

Linguistic variables are fuzzy logic's core concepts and enable the quantification of qualitative or fuzzy data in terms of words like "low," "normal," or "high" rather than exact numerical values [10]. Membership functions tie fuzzy sets to words, giving an effective description of information when exact values are unavailable or not necessary–e.g., medical screening like blood pressure readings [11].

3.3 Fuzzy Logic System

A Fuzzy Logic model is a rule-based system that uses fuzzy theory to handle vagueness. Figure 2 presents the components of an FL system. The FL system consists of four basic components as follows:

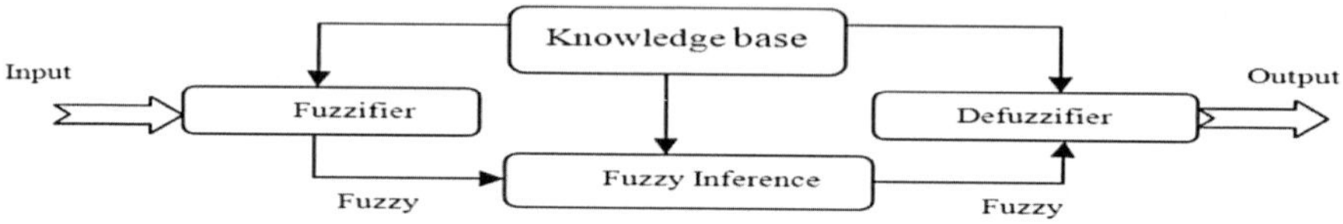

Fig. 2. Fuzzy Logic system components

- **Fuzzifier:** The fuzzifier converts crisp objects into fuzzy sets, assigning membership function grades to linguistic variables in fuzzy sets [12].
- **Inference Engine:** It processes the fuzzified input values based on rules from the knowledge base, acting as the core processing unit of the fuzzy logic system [7].
- **Knowledge Base:** The knowledge base, or rule base, contains "if-then" rules provided by experts that define input-output relationships, guiding decision-making in the system [12].
- **Defuzzifier:** The defuzzifier converts fuzzy outputs into crisp, real-world values, ensuring the output is practical by emphasizing higher membership degrees and minimizing lower ones [11].

4 Applications of AI and Fuzzy Logic in Healthcare

The combination of fuzzy logic and artificial intelligence can solve the "black box" problem, which has become a challenge in using AI in healthcare systems, since it is difficult to explain [13,14]. This combination also has the potential to improve diagnosis and surveillance, since it allows for a strong framework for handling the uncertainties that are present in biomedical information. This review will highlight important applications for neurological and physical conditions.

4.1 AI and Fuzzy Logic in Neurological Health

The human brain remains one of the most complex and least understood systems in biology. Such complexity is heightened by the high prevalence of neurological diseases such as depression, stress, anxiety, and Alzheimer's disease, which have affected millions of individuals worldwide [15]. Alzheimer's specifically is a significant global health issue, according to a 2022 estimate, more than 22% of those 50 years and older fall under the Alzheimer's disease continuum from preclinical to full dementia [16]. Although conventional treatments have used pharmacological and psychotherapeutic interventions, novel technologies like artificial intelligence (AI) and fuzzy logic (FL) are allowing for more targeted and patient-specific diagnostic and therapeutic strategies.

Hybrid AI–FL systems have demonstrated enormous success in the diagnosis of neurodegenerative diseases. Sasikala et al. [17] introduced an MRI-based Computer-Aided Diagnosis (CAD) system that integrates deep learning with fuzzy logic. The proposed model employed the Iteratively Reweighted Adaptive Spatial Fuzzy C-Means (IRW-ASFCM) algorithm to segment brain tissue and a CNN ensemble with 2-Level Selective Wavelet Kernel (2LS-WaveCNN) for feature extraction with 99.54% accuracy. In the same vein, Jayasutha et al. [18] came up with ResNet-augmented 3D–2D Fuzzy Convolutional Neural Network (FCNN) for Alzheimer's diagnosis, using fuzzy logic on 2D layers and initial 3D convolutions to address imaging uncertainty. With preprocessing and dimensionality reduction using PCA, the model attained 99% accuracy, better than traditional approaches in precision as well as resilience.

Chen et al. [19] proposed a comprehensible fuzzy system for diagnosing dementia using dominant sets theory and Adaptive Network-based Fuzzy Inference System (ANFIS). From OASIS imaging data, the system achieved 81.69% accuracy greater than J48, SVM, and Random Forest and at the same time provided transparency through explainable fuzzy rules that enhance clinical explainability.

In dementia, FL–AI integration has been demonstrated to function well with other neurological disorders. Narasimham and Kumar [20] suggested a fuzzy logic-based approach for brain tumor classification from MRI images applying watershed and region-growing segmentation, followed by a Mamdani-type Fuzzy Inference System (FIS). The region-growing algorithm achieved 96.08% accuracy with great precision and recall values.

In psychiatry, Delgado et al. [21] combined fuzzy logic with data mining (decision trees and Apriori algorithm) to derive explainable rules from DASS-21 assessments of 242 patients with 92.98% accuracy in predicting disorder severity. Similarly, Rajawat et al. [22] applied a Fusion Fuzzy Logic (FFL) model with 3D CNNs to detect depression from facial expressions with 94.3% accuracy by managing uncertainty in temporal features.

To handle multimodal data complexity, Han [23] proposed a Heuristic Fuzzy C-Means Clustering Algorithm (HFCA) integrated with CNNs that achieved 96.7% accuracy on a Kaggle stress dataset and demonstrated robustness in the AVEC dataset, indicative of its potential in emotion and stress monitoring.

Further studies identify the versatility of FL in neurological health. Raghavaiah et al. [24] proved how fuzzy deep learning could enhance MRI preprocessing by reducing the uncertainty of data. Likewise, Öztürk et al. [25] used fuzzy logic in linguistic summarization of eye-tracking data in children with autism spectrum disorder (ASD), enabling interpretable analysis of behavioral patterns for clinical insights.

4.2 AI and Fuzzy Logic in Physical Health

The diagnosis of physical diseases a cornerstone in medical decision-making is frequently marred with vagueness and uncertainty in medical information. To deal with such complexity, the union of fuzzy logic (FL) and artificial intelligence (AI) has been widely researched in an enormous array of applications ranging from pandemic monitoring to chronic disease diagnosis and tailored healthcare [26].

Hybrid AI–FL systems have proven to surpass in both diagnostic accuracy and real-time patient observation. For instance, Tayyaba et al. [2] suggested an Internet of Things (IoT)-aided solution that used wearable sensors along with a Rule-Based Fuzzy Logic Classifier (FLCD) to monitor COVID-19 patients. Using real-time physiological signals respiratory rate, heart rate, SpO_2, and body temperature the system precisely classified patients into normal, mild, serious, and critical classes with 95% accuracy. The model handled uncertainty in symptom variability accurately and produced a highly efficient and scalable approach for the management of pandemics.

For chronic disease care and management, Marashi-Hosseini et al. [27] proposed a Clinical Decision Support System (CDSS) using fuzzy logic to help dietitians plan adaptive diets for patients with multiple chronic diseases (MCCs) such as obesity, diabetes, hypertension, and kidney disease. The system, learned from 1,144 Mamdani rules, was 97% concordant with recommendations from expert dietitians and was deserving of trust for personalized nutrition planning. Likewise, Improta et al. [28] proposed an FL-based CDSS for renal function monitoring in kidney transplant patients with proteinuria and glomerular filtration rate (GFR) as major indicators. The system was more than 90% accurate, significantly decreasing drug errors and healthcare expenditures.

In metabolic disease diagnosis, Scrobota et al. [29] proposed an FL-based decision-support system for assessing periodontitis risk in Type 2 Diabetes Mellitus (T2DM) patients. From biomarkers like BMI, glycemia, cholesterol, and triglycerides, the system generated a Periodontitis Cumulative Risk score (PCRisk) with high correlation to the number of periodontal pockets (nrPPs) and demonstrated its utility for early diagnosis and treatment planning specific to the patient.

Aside from individual uses, studies have established the basis for FL's general efficiency in medical health. A meta-review conducted by Özkan Görgülü and Akilli [1] had a total 90% success rate of 71 FL-based medical systems, showing its worth in diagnostics. Likewise, Kaur and Khehra [30] reviewed state-of-the-art fuzzy logic and hybrid FL-based systems for identifying risk of heart disease, where the hybrid systems performed better maximum 99.4% accuracy. Follow-up research has since taken this research to hybrid FL–ML systems for general disease diagnosis, e.g., diabetes diagnosis with greater accuracy [3], and optimization of elementary image processing tasks such as image segmentation and feature extraction using fuzzy approaches [31].

4.3 Comparative Summary of Applications

To give a brief and systematic overview of the primary studies covered here, the following two tables in summary present applications to neurological and physical health, respectively. Table 1 presents solutions to diseases like Alzheimer's and depression, and Table 2 gives an overview of systems designed for illnesses like COVID-19 tracking and heart disease diagnosis. Combined side by side, they enable direct comparison of the methodologies, data sets, and performance measures utilized in each of the two fields and identify parallel trends and domain-related concerns.

Table 1. Summary of AI and Fuzzy Logic Applications in Neurological Health

Reference	Disease/Task	AI + FL Method	Dataset / Size	Key Metrics
Sasikala et al. [17]	Alzheimer's Diagnosis	IRW-ASFCM, 2LS-WaveCNN, Deep Tree Training	MRI Dataset	Acc: 99.54%
Jayasutha et al. [18]	Alzheimer's Diagnosis	ResNet-boosted 3D2D Fuzzy CNN	MRI Images	Accuracy: 99%
Chen et al. [19]	Dementia Diagnosis	Dominant Sets Theory, ANFIS	OASIS Imaging Data	Accuracy: 81.69%
Narasimham & Kumar [20]	Brain Tumor Diagnosis	Mamdani-type FIS, Region-growing Segmentation	Kaggle MRI Dataset	Accuracy: 96.08%
Delgado et al. [21]	Depression Anxiety	FIS, Data mining (Decision Trees, Apriori)	242 patients evaluated with DASS-21 scale	Accuracy: 92.98%
Rajawat et al. [22]	Depression Detection	Fusion Fuzzy Logic (FFL), 3D CNN	Facial Expression Videos	Accuracy: 94.3%
Han, H. [23]	Student Stress Assessment	HFCA, CNN	Kaggle/AVEC.	Accuracy: 96.7%

All studies applied fuzzy logic integrated with AI for improved neurological disorder diagnosis.

Table 2. Summary of AI and Fuzzy Logic Applications in Physical Health

Reference	Disease/Task	AI + FL Method	Dataset / Size	Key Metrics
Tayyaba et al. [2]	COVID-19 Monitoring	IoT, Rule-Based Fuzzy Logic Classifier	Real-time physiological data	Accuracy: 95%
Marashi-Hosseini et al. [27]	Personalized Diet Planning	FL-based CDSS, Mamdani rules	Rule-based (1,144 rules)	Accuracy: 97%
Improta et al. [28]	Renal Function Monitoring	FL-based Clinical Decision Support System (CDSS)	Post-transplant patient data	Precision: +90%
Scrobota et al. [29]	Periodontitis Risk Assessment	FL system with metabolic biomarkers	87 T2DM patients	Significant Correlation

Note: These studies demonstrate the use of hybrid AI-FL systems for diagnosis, risk estimation, and management in physical health.

5 Discussion and Future Directions

While the integration of fuzzy logic and artificial intelligence holds significant promise, careful evaluation uncovers substantial discrepancies between the reported clinical feasibility in research studies and the actual outcomes observed. The superior rates of performance listed in the majority of studies (Tables 1 and 2) have to be taken with care because they typically originate from small, single-center collections and lack independent external verification. This reduces the findings and is a major barrier to demonstrating real-world efficacy. In addition, clinical translation is more than algorithmic precision. The statement that fuzzy logic guarantees explainability can be a self-refuting claim, an interacting system with hundreds of rules can be no more understandable than a "black-box" model. Real value is placed in constructing models with published and clinician-meaningful justification. Follow up effort should therefore focus on moving from simply demonstrating promise to demonstrating clinical usefulness by being thoroughly proven, addressing regulatory compliance and Electronic Health Record (EHR) integration problems, and being interested in developing systems that are actually usable in a clinical context.

6 Conclusion

The combination of artificial intelligence and fuzzy logic has robust potential for managing uncertainties in biomedical information. The applications discussed in this brief review range from applications related to the body and neurological systems, having the potential to enhance surveillance and diagnosis. However, a huge gap exists in the commercial use of test tube accuracy versus the accuracy required for clinical use. The key to the success of the whole field is a massive shift in paradigms from achieving high accuracy on small-sized datasets to experimental validation on a massive scale. To gain the confidence of practicing professionals, it is finally necessary that systems are made hardy enough to pass tests, moral enough to gain acceptance, and interpretable enough to gain trust.

References

1. Gorgulu, O., Akilli, A.: Use of fuzzy logic based decision support systems in medicine. Stud. Ethno Med. **10**, 393–403 (2016)
2. Ilyas, T., Mahmood, D., Ahmed, G., Akhunzada, A.: Symptom analysis using fuzzy logic for detection and monitoring of covid-19 patients. Energies (2021)
3. Arya, L., Lavudiya, N.S., Sateesh, G., Padmanaban, H., Srinivasulu, D.B.V., Rastogi, R.: Fuzzy logic-driven machine learning algorithms for improved early disease diagnosis. Int. J. Adv. Comput. Sci. Appl. (2024)
4. Sahu, B., Sarangi, L., Ghosh, A., Palo, H.K.: Application of Fuzzy Logic to Healthcare Industry (2022)
5. Zadeh, L.A.: Fuzzy sets. Inf. Control **8**, 338–353 (1965)
6. Zadeh, L.A.: Making computers think like people (1984). https://api.semanticscholar.org/CorpusID:59755175
7. Gürsel, G.: Healthcare, uncertainty, and fuzzy logic. Digit. Med. **2**, 101–112 (2016)
8. Kumar, M., Misra, L., Shekhar, G.: A survey in fuzzy logic: an introduction (2015). https://api.semanticscholar.org/CorpusID:32657820
9. Yager, R.R., Zadeh, L.A.: An introduction to fuzzy logic applications in intelligent systems (1992). https://api.semanticscholar.org/CorpusID:110900014
10. Chapter 9 - fuzzy methods. In: Profillidis, V.A., Botzoris, G.N. (eds.) Modeling of Transport Demand. Elsevier (2019)
11. Cirstea, M.N., Dinu, A., Khor, J.G., McCormick, M.: Fuzzy logic fundamentals (2002). https://api.semanticscholar.org/CorpusID:118323159
12. Dweiri, F.T., Kablan, M.M.: Using fuzzy decision making for the evaluation of the project management internal efficiency. Decis. Support Syst. **42**, 712–726 (2006)
13. Liu, C.F., Chen, Z.C., Kuo, S.C., Lin, T.C.: Does AI explainability affect physicians' intention to use AI? Int. J. Med. Inform. **168**, 104884 (2022)

14. Wang, Y., Song, Y., Ma, Z., Han, X.: Multidisciplinary considerations of fairness in medical ai: a scoping review. Int. J. Med. Inform. **178**, 105175 (2023)
15. Vos, G., Trinh, K., Sarnyai, Z., Azghadi, M.R.: Generalizable machine learning for stress monitoring from wearable devices: a systematic literature review. Int. J. Med. Inform. **173**, 105026 (2022)
16. Gustavsson, A., et al.: Global estimates on the number of persons across the Alzheimer's disease continuum. Alzheimer's Dementia **19** (2022). https://doi.org/10.1002/alz.12694
17. Sasikala, T.S.: A cad system design using iteratively reweighted fuzzy c-means and deep tree training for alzheimer's disease diagnosis. Biomed. Sign. Process. Control **88**, 105655 (2024)
18. Jayasutha, D., Divakar, T., Selvam, P.M.S.M., Vigneshkumar, R.A., Kumar, P., S.: Integration of fuzzy logic and deep learning for medical image analysis in neuroimaging. S. East. Eur. J. Public Health (2024)
19. Chen, T., et al.: A dominant set-informed interpretable fuzzy system for automated diagnosis of dementia. Front. Neurosci. **16** (2022)
20. Narasimham, N., KeshavKumar, K.: Fuzzy logic-based system for brain tumour detection and classification. ArXiv:abs/2401.14414 (2024)
21. Delgado, S., Vignola, R.C.B., Sassi, R.J., Belan, P.A., Araújo, S.A.: Symptom mapping and personalized care for depression, anxiety and stress: a data-driven ai approach. Comput. Biol. Med. **182**, 109146 (2024)
22. Rajawat, A.S., Bedi, P., Goyal, S.B., Bhaladhare, P.R., Aggarwal, A., Singhal, R.: Fusion fuzzy logic and deep learning for depression detection using facial expressions. Procedia Comput. Sci. (2023)
23. Han, H.: Fuzzy clustering algorithm for university students' psychological fitness and performance detection. Heliyon **9** (2023)
24. Tanveer, M., et al.: Fuzzy deep learning for the diagnosis of Alzheimer's disease: approaches and challenges. IEEE Trans. Fuzzy Syst. **32**, 5477–5492 (2024)
25. Ozturk, D., et al.: Linguistic summarization of visual attention and developmental functioning of young children with autism spectrum disorder. Health Inf. Sci. Syst. **12** (2024)
26. Ahmadi, H., Gholamzadeh, M., Shahmoradi, L., Nilashi, M., Rashvand, P.: Diseases diagnosis using fuzzy logic methods: a systematic and meta-analysis review. Comput. Methods Programs Biomed. **161**, 145–172 (2018)
27. Marashi-Hosseini, L., Jafarirad, S., Hadianfard, A.M.: A fuzzy based dietary clinical decision support system for patients with multiple chronic conditions (MCCS). Sci. Rep. **13** (2023)
28. Improta, G., Mazzella, V., Vecchione, D., Santini, S., Triassi, M.: Fuzzy logic–based clinical decision support system for the evaluation of renal function in post-transplant patients. J. Eval. Clin. Pract. **26**, 1224–1234 (2019)
29. Ioana, S., et al.: An artificial intelligence-based fuzzy logic system for periodontitis risk assessment in patients with type 2 diabetes mellitus. Bioengineering **12**, 211 (2025). https://doi.org/10.3390/bioengineering12030211
30. Kaur, J., Khehra, B.: Fuzzy logic and hybrid based approaches for the risk of heart disease detection: State-of-the-art review. J. Inst. Eng. (India) Ser. B **103**, 1–17 (2021). https://doi.org/10.1007/s40031-021-00644-z
31. Saxena, P., et al.: Fuzzy-based medical image processing and analysis. Int. J. Intell. Syst. Appl. Eng. (2024)

AI for Optimization and Resource Management

Hybrid Game-Theoretic and Machine Learning Framework for Fair and Efficient Resource Allocation in Cloud Computing

Oumaima Ghandour, Said El Kafhali(✉), and Abdelkrim Haqiq

Faculty of Sciences and Techniques, Computer, Networks, Modeling, and Mobility Laboratory (IR2M), Hassan First University of Settat, B.P. 577, 26000 Settat, Morocco
{o.ghandour,said.elkafhali,abdelkrim.haqiq}@uhp.ac.ma

Abstract. This paper presents a hybrid framework for fast and fair multi-resource allocation in cloud environments by combining the Nash Bargaining Solution (NBS) with Machine Learning (ML). The NBS is used to compute optimal CPU and RAM allocations from real-world Google Cluster traces, serving as training labels for a Random Forest model. The learned model achieves near-optimal performance with a Mean Absolute Error (MAE) = 0.0067 and a coefficient of determination (R^2) = 0.951, but occasionally violates capacity constraints. A lightweight post-processing repair step based on uniform scaling fully restores feasibility without degrading fairness or utilization. This approach significantly reduces computation time compared to solving the NBS repeatedly, enabling real-time deployment in multi-tenant cloud systems.

Keywords: Cloud Computing · Resource Allocation · Nash Bargaining Solution · Machine Learning · Random Forest · Fairness · Multi-Tenant Systems

1 Introduction

Efficient and fair allocation of computing resources is a critical challenge in multi-tenant cloud environments [1], where diverse workloads with varying CPU (Central Processing Unit) and RAM (Random Access Memory) demands compete for limited capacity [2]. Classical optimization approaches, such as the Nash Bargaining Solution (NBS) [3], offer provably fair and efficient allocations but are computationally expensive to solve repeatedly in real time. This limitation becomes significant in large-scale systems with high-frequency allocation decisions [4].

Recent advances in Machine Learning (ML) have shown promise in approximating complex optimization outcomes at a fraction of the computational cost [5]. By learning from optimal allocations produced offline, ML models can deliver near-instant predictions suitable for dynamic cloud environments [6]. However,

A. E. Hassanien et al. (Eds.): AISDS 2025, LNDECT 286, pp. 143–152, 2027.
https://doi.org/10.1007/978-3-032-18758-1_14

these predictions may occasionally violate resource capacity constraints, requiring additional mechanisms to ensure feasibility [7].

In this work, we propose a hybrid framework that combines the fairness guarantees of the NBS with the speed of a Random Forest (RL) model trained on real-world Google Cluster traces. The ML model learns to approximate optimal allocation fractions for CPU and RAM, while a lightweight post-processing repair step ensures strict capacity adherence. Experimental results demonstrate that our method achieves near-optimal fairness and utilization with negligible loss in accuracy after repair, enabling scalable, real-time deployment in multi-tenant cloud platforms.

The main contributions of this paper are:

- A supervised learning approach using RL to approximate Nash Bargaining (NB) allocations in cloud resource allocation scenarios.
- A post-processing repair mechanism that guarantees feasibility without significantly degrading prediction accuracy.
- A comprehensive evaluation on Google Cluster data, showing high fairness, near-optimal utilization, and scalability for real-time applications.

The remainder of this paper is organized as follows. Section 2 reviews relevant literature on game-theoretic and ML approaches for resource allocation in cloud computing. Section 3 presents the reference model based on the NBS and details the proposed methodology. Section 4 reports the experimental results, including predictive accuracy, feature importance, constraint violation analysis, and the impact of post-processing, followed by a detailed discussion of practical implications. Finally, Sect. 5 concludes the paper and outlines potential directions for future research.

2 Related Work

Resource allocation in cloud and edge computing has been extensively studied through game-theoretic and ML approaches. Cooperative Game Theory (CGT) has been widely applied to address fairness and efficiency in resource sharing among multiple service providers. Zafari et al. [8] proposed a CGT-based framework for inter-cloud resource sharing, modeling the problem as an N-player cooperative game with non-transferable utility, and designing an $O(N)$ algorithm that ensures Pareto-optimal allocations from the core. Extending this work to mobile edge environments, Zafari et al. [9] introduced algorithms that guarantee coalition stability and improve application satisfaction.

Another variant [10] focused on multi-objective resource allocation and offered reduced computational complexity compared to Shapley valuebased solutions. Hassan and Alamri [11] adopted the NBS for dynamic virtual machines allocation in multimedia clouds, demonstrating cost reduction and high Quality of Service (QoS) satisfaction. Other notable contributions include Pillai et al. [12], who leveraged coalition formation and the uncertainty principle of game theory to improve utilization, and Du et al. [13], who designed an evolutionary Stackelberg differential game for SDN-based edgecloud systems, optimizing pricing and allocation under dynamic user behavior.

In parallel, predictive modeling has emerged as a powerful tool for resource allocation. Zhang et al. [14] applied supervised learning to auction-based allocation, using linear and logistic regression to predict near-optimal assignments with high social welfare and utilization. Catena et al. [15] proposed distributed LSTM-based prediction in Network Function Virtualization environments, reducing communication overhead while maintaining low prediction error.

While game theory provides fairness guarantees and optimality grounded in cooperative strategies, it often incurs high computational costs in real-time environments. Conversely, ML enables rapid inference but may not inherently enforce fairness or feasibility constraints. Few studies explicitly combine these paradigms. The present work addresses this gap by using NBS-derived allocations as training targets for an RL model, enabling near-instantaneous prediction of fair allocations, followed by a lightweight post-processing step to ensure strict capacity compliance. This hybrid approach bridges the gap between theoretical optimality and practical deployability in large-scale cloud environments.

3 Reference Model and Proposed Methodology

This section describes the theoretical foundation and the step-by-step methodology of our proposed approach. We first present the reference model based on the NBS, which serves as the optimal baseline for fair resource allocation. We then explain how this model is integrated into an ML pipeline for the real-time prediction of allocation fractions.

3.1 The Nash Bargaining Reference Model

We consider a cloud data center environment where multiple independent users (players) compete for two types of limited resources: CPU and RAM. Let CPU_i and RAM_i denote the respective amounts requested by user i. Since physical capacity is limited, the total available CPU and RAM for allocation in a given time slot are restricted to fixed fractions of the total demand:

$$C_{\text{CPU}} = \alpha \cdot \sum_j \text{CPU}_j, \quad C_{\text{RAM}} = \alpha \cdot \sum_j \text{RAM}_j \tag{1}$$

Equation 1 expresses capacity as a ratio α of instantaneous demand to obtain scale-free instances and controllable load regimes. This is equivalent to using fixed physical capacities C_{CPU} and C_{RAM} and writing

$$\alpha = \frac{C_{\text{CPU}}}{\sum_j \text{CPU}_j} = \frac{C_{\text{RAM}}}{\sum_j \text{RAM}_j}.$$

We set $\alpha = 0.6$ to emulate contention (60% offered capacity relative to demand), which forces the bargaining solution to trade off users fairly under scarcity. Using a demand-normalized form makes the optimization invariant to absolute units and dataset-agnostic, while remaining consistent with fixed-capacity systems through the above equivalence.

The allocation problem aims to determine, for each user i, a fraction f_i of their requested demand to allocate to them. The NBS is a cooperative game-theoretic approach that ensures fairness by maximizing the product of all players' utilities, subject to resource constraints. The optimization problem is:

$$\max_{f_i} \sum_i \log(f_i) \tag{2}$$

subject to:

$$\sum_i f_i \cdot \text{CPU}_i \leq C_{\text{CPU}}, \quad \sum_i f_i \cdot \text{RAM}_i \leq C_{\text{RAM}}, \quad 0 < f_i \leq 1 \tag{3}$$

Here, $\log(f_i)$ represents the utility of user i, which increases as their allocated fraction grows, but with diminishing returns. This formulation guarantees:

Fairness: No user can gain more without reducing another's allocation disproportionately.
Pareto optimality: No other allocation can improve one user's share without harming at least one other.

The resulting NBS allocations f_i are treated as the optimal ground truth in our study.

3.2 Dataset Description

To build and validate our predictive framework, we use the Google 2019 Cluster [16] Sample dataset available on Kaggle. This dataset contains real execution traces from a large-scale production data center, including CPU and RAM requests for thousands of tasks.

For consistency, the data is segmented into 5-minute time slots, each representing a snapshot of concurrent workloads and resource demands.

3.3 Training Data Generation

For each time slot, we:

1. Extract the individual CPU and RAM demands for each user.
2. Compute the total CPU and RAM demand in the slot.
3. Solve the NB optimization to obtain the optimal allocation fractions f_i.

This process produces a supervised learning dataset where:

Features: individual and total slot demands.
Target: Optimal allocation fraction f_i from NBS.

3.4 Machine Learning Model

We train a `RandomForestRegressor` using 80% of the dataset and reserve 20% for evaluation. Two main metrics are used:

Mean Absolute Error (MAE): Measures the average prediction error magnitude.
Coefficient of Determination (R^2): Assesses the proportion of variance explained by the model.

Additionally, we perform feature importance analysis to identify which inputs most strongly influence the predicted allocations.

3.5 Post-Processing for Feasibility

Although the RL model achieves high predictive accuracy, its raw predictions may slightly exceed capacity constraints due to regression errors. To address this, we apply a uniform scaling factor γ to all predicted fractions within a slot:

$$\sum_i f_i^{\text{corr}} \cdot \text{CPU}_i \leq C_{\text{CPU}}, \quad \sum_i f_i^{\text{corr}} \cdot \text{RAM}_i \leq C_{\text{RAM}} \tag{4}$$

This repair step ensures:

Feasibility: No resource constraint is violated.
Proportional fairness: Relative allocation ratios between users remain unchanged.

The corrected allocations are then used for performance evaluation in later sections.

4 Results and Discussion

The experimental evaluation assesses the ability of the proposed hybrid MLgame theory framework to predict NB-based resource allocations using real-world cluster data. We examine the predictive accuracy of the RL model, the relative importance of key demand features, and the extent of constraint violations before and after applying a lightweight post-processing repair step. By combining quantitative metrics with visual analyses, the results aim to validate the accuracy and operational feasibility of the approach in realistic multi-tenant cloud environments.

4.1 Predictive Accuracy and Feature Importance

The proposed RL model was trained on 9367 samples (features: individual CPU request, individual RAM request, total CPU demand per slot, total RAM demand per slot) extracted from 493 valid 5-minute slots of the Google Cluster dataset.

The model achieved an MAE of 0.0055 and a coefficient of determination $R^2 = 0.9511$ on the evaluation set, indicating strong predictive performance in approximating the NB optimal allocation fraction (f_{opt}). Over the full dataset comparison of ML predictions against Nash labels, the average MAE was 0.0067 with a median MAE of 0.0013, and the feasibility rate (capacity-respecting allocations) was 61.66% before post-processing.

As illustrated in Fig. 1a, the feature importance analysis revealed that the individual RAM request (`ram_req`) was the most influential predictor, followed by the aggregate CPU demand per slot (`total_cpu_demand_slot`) and the individual CPU request (`cpu_req`). This pattern highlights the dominant role of memory demand in driving the Nash allocation outcomes, suggesting that in realistic multi-tenant environments, RAM pressure is often the critical limiting factor for equitable resource distribution.

The parity plot Fig. 1b shows that predicted allocations align closely with the optimal labels, with most points tightly clustered around the diagonal line. Here, f_{true} denotes the ground-truth allocation fraction obtained from the NBS, and f_{pred} denotes the model-predicted allocation fraction for the same user-slot pair (before applying any feasibility repair, unless stated otherwise). A few minor deviations occur in high-demand situations where aggregated requests are close to the 60% capacity limit.

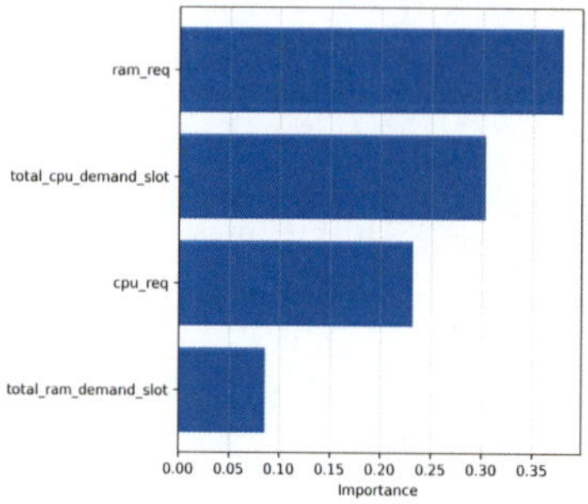

(a) Feature Importance (RL)

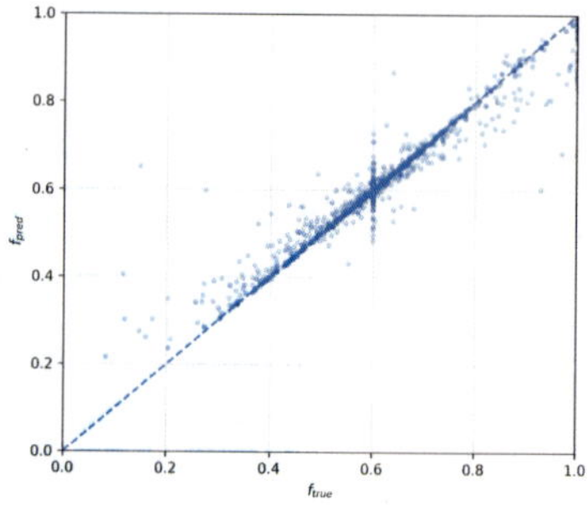

(b) Parity between f_{true} and f_{pred}

Fig. 1. Model Evaluation: Feature Importance of RL and Parity Between Predicted and True Allocations

4.2 Constraint Violations Before Repair

Although the model exhibits excellent predictive accuracy, its outputs are not inherently guaranteed to respect the CPU and RAM capacity constraints. The histograms in Fig. 2a and Fig. 2b illustrate the distribution of CPU and RAM violations in the predicted allocations. A violation occurs when the aggregated predicted allocation exceeds the predefined resource limit (here, 60% of the total capacity) for a given time slot.

The analysis reveals that 38.7% of the slots are infeasible, corresponding to a feasibility rate of only 0.613. Most violations are concentrated very close to the zero boundary, indicating that they stem from slight over-allocations rather than substantial mispredictions. This suggests that the model accurately captures demand patterns but does not explicitly learn the hard constraint boundaries.

From a practical standpoint, this type of infeasibility can be problematic in real-time allocation scenarios, as even marginal overshoots can trigger Service Level Agreement (SLA) breaches or require costly reallocation processes. However, the fact that violations are small and infrequent indicates that a lightweight post-processing or correction mechanism can effectively restore full feasibility without significantly degrading allocation quality.

This observation emphasizes the importance of integrating constraint-awareness into the prediction pipeline, either through post-hoc correction or by embedding resource limits directly into the model's learning process. Doing so would bridge the gap between high prediction accuracy and operational deployability in production-grade cloud environments.

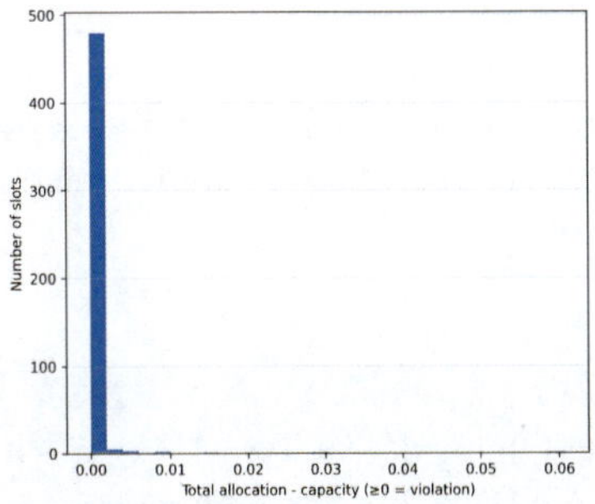

(a) CPU Constraint Violations

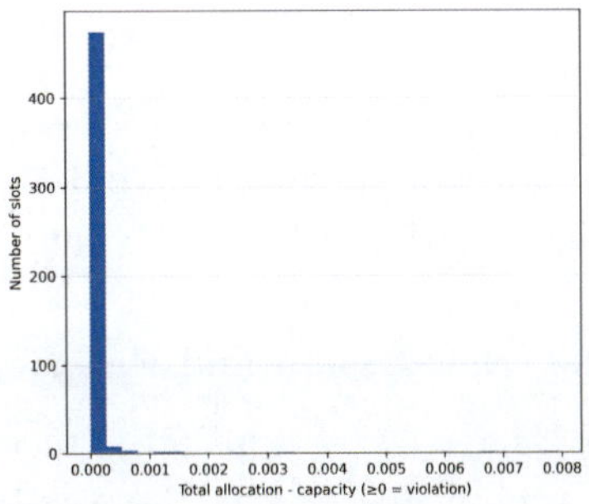

(b) RAM Constraint Violations

Fig. 2. Constraint Violations Before Post-Processing: Distribution of CPU and RAM Allocation Exceedances Across Slots

4.3 Impact of Post-Processing Repair

To ensure that the predicted allocations strictly comply with the CPU and RAM capacity limits, a uniform scaling factor $\gamma \in (0, 1]$ was applied to all predicted fractions within each slot. This post-processing step guarantees feasibility by proportionally reducing allocations whenever the aggregate exceeds the allowed 60% capacity.

The results show that this simple yet effective correction mechanism eliminated all infeasibilities, increasing the feasibility rate from 61.3% to 100% (Table 1). The correction had a negligible impact on predictive accuracy: the MAE increased only slightly, from 0.00665 to 0.00722. Capacity utilization

Table 1. Comparison of key performance metrics before and after applying the repair step.

Metric	True (Nash)	Predicted	After Repair
Feasibility Rate (%)	100.0	61.26	**100.0**
Avg. CPU Utilization	99.83%	99.81%	99.29%
Avg. RAM Utilization	99.35%	99.65%	99.11%
Avg. Jain's Index	0.989	0.991	0.991
Mean Absolute Error	–	0.00665	0.00722

remained extremely high (99.29% for CPU and 99.11% for RAM), indicating that the repair process does not waste significant resources.

Fairness, evaluated using Jain's index, was preserved at a very high level (0.991), identical to the pre-repair value. This means that the proportional scaling maintains the relative distribution among users, ensuring that no player is disproportionately affected by the correction.

From an operational perspective, these findings confirm that a lightweight, uniform-scaling repair step can bridge the gap between high model accuracy and strict constraint adherence without introducing meaningful efficiency losses. This makes the proposed pipeline directly deployable in real-world multi-tenant cloud environments, where hard SLA compliance is critical.

4.4 Discussion and Practical Implications

Experimental results indicate that the proposed hybrid framework correctly learns the distribution patterns derived from the NBS using real cluster traces. The model consistently produces nearly optimal distributions with negligible prediction error, which justifies its ability to capture the NBS model's inherent fairness-efficiency trade-off. The intermittent violation of the constraints observed before post-processing is not surprising, as regression models do not necessarily guarantee feasibility in constrained optimization environments. However, the lightweight repair mechanism described here, expressed as a uniform scaling factor used for all the predicted allocations in a slot, eliminates these violations without sacrificing proportional fairness and utilization efficiency. This ensures that the model predictions are operationally valid at the cost of minimal accuracy loss.

In practice, the approach has great merits in the context of cloud computing environments for real-time resource allocation. By replacing the repeated computation of NBS allocations with one-time training and fast inference after that, the approach radically improves decision latency and computational expense. This reduction in requirements for online processing improves responsiveness and lowers the energy expense of control-plane activity, which can result in significant cost savings on big data centers. Moreover, the capacity to maintain fairness and achieve near-optimal CPU and RAM utilization ensures that QoS goals are achieved without inducing resource underutilization or unfair allocation among tenants.

The ease of the method is the other significant advantage. Although the present work focuses on CPU and memory allocation, the approach is readily extendible to other resources, such as network bandwidth or disk, by choosing an appropriate feature set and constraint. This also makes the approach applicable to multi-objective optimization problem settings where fairness must compete with other factors such as cost-effectiveness or application performance. Moreover, the proposed model can be applied beyond cloud data centers, including edge computing and Internet of Things (IoT) devices, where efficient, fair, and fast heterogeneous resource allocation is required, and high-performance computing (HPC) clusters wherein fairness and efficiency need to be accounted for during scheduling. In each of these domains, combining the inferential flexibility of game-theoretic solutions with the performance of ML inference is an attractive paradigm for fair and scalable resource management.

The experimental evaluation challenges the ability of the proposed hybrid framework of MLgame theory to predict NB-based resource allocations on real-world cluster data. We examine the predictive accuracy of the RL model, the relative importance of leading demand features, and the extent of constraint violations before and after a lightweight post-processing repair step. The outcomes, through a combination of quantitative measurements with qualitative observations, seek to validate the precision and working usability of the methodology in actual multi-tenant cloud environments.

5 Conclusion and Future Work

This research presented a hybrid approach that integrates the fairness guarantees of the NBS with efficient supervised ML inference in cloud computing resource allocation. NBS allocations were used as the training targets, and the model learned well from real cluster traces, the underlying allocation patterns of near-optimal performance and minimal prediction error. The inclusion of a lightweight post-processing correction mechanism ensured strict CPU and RAM capacity constraints without compromising fairness or usage effectiveness. Experiments indicate that the technique offers a practical and computationally affordable alternative to continuously optimizing NBS in real time, enabling rapid decision-making in large-scale and dynamic resource allotment contexts. Beyond cloud data centers, the method can be extended to other computational models, such as edge computing, fog computing, IoT networks, and HPC clusters, and multi-resource and multi-objective issues. Future work will expand the model to cover additional types of resources, dynamic prediction of demands, and adaptive learning algorithms to allow the system to adjust to varying workloads. Moreover, integrating the approach with online monitoring and reinforcement learning could further enhance its adaptability to dynamic environments while guaranteeing fairness, efficiency, and scalability.

References

1. Ghandour, O., El Kafhali, S., Hanini, M.: Computing resources scalability performance analysis in cloud computing data center. J. Grid Comput. **21**(4), 61 (2023)
2. Ghandour, O., El Kafhali, S., Hanini, M.: Adaptive workload management in cloud computing for service level agreements compliance and resource optimization. Comput. Electr. Eng. **120**, 109712 (2024)
3. Binmore, K., Rubinstein, A., Wolinsky, A.: The Nash bargaining solution in economic modelling. Rand J. Econ. 176–188 (1986)
4. Mikram, H., El Kafhali, S., Saadi, Y.: Server consolidation algorithms for cloud computing: taxonomies and systematic analysis of literature. Int. J. Cloud Appl. Comput. (IJCAC) **12**(1), 1–24 (2022)
5. Hafdi, Z.S., El Kafhali, S.: A comparative evaluation of machine learning methods for predicting student outcomes in coding courses. Applied Math **5**(2), 75 (2025)
6. Anbarkhan, S.H.: Optimizing cloud resource allocation with machine learning: strategies for efficient computing. Ingénierie des Systèmes d'Information **30**(1), 1 (2025)
7. Mikram, H., El Kafhali, S.: CHPSO: an efficient algorithm for task scheduling and optimizing resource utilization in the cloud environment. J. Grid Comput. **23**(2), 1–33 (2025)
8. Zafari, F., Leung, K.K., Towsley, D., Basu, P., Swami, A.: A game-theoretic framework for resource sharing in clouds. In: Proceedings of the 2019 12th IFIP Wireless and Mobile Networking Conference (WMNC), pp. 8–15. IEEE (2019)
9. Zafari, F., Leung, K.K., Towsley, D., Basu, P., Swami, A., Li, J.: Let's share: a game-theoretic framework for resource sharing in mobile edge clouds. IEEE Trans. Netw. Serv. Manage. **2**, 2107–2122 (2020)
10. Zafari, F., Li, J., Leung, K.K., Towsley, D., Swami, A.: A game-theoretic approach to multi-objective resource sharing and allocation in mobile edge. In: Proceedings of the 2018 on Technologies for the Wireless Edge Workshop, pp. 9–13 (2018)
11. Hassan, M.M., Alamri, A.: Virtual machine resource allocation for multimedia cloud: a Nash bargaining approach. In: Procedia Computer Science, pp. 571–576 (2014)
12. Pillai, P.S., Rao, S.: Resource allocation in cloud computing using the uncertainty principle of game theory. IEEE Syst. J. **2**, 637–648 (2014)
13. Du, J., Jiang, C., Benslimane, A., Guo, S., Ren, Y.: SDN-based resource allocation in edge and cloud computing systems: an evolutionary Stackelberg differential game approach. IEEE/ACM Trans. Network. **4**, 1613–1628 (2022)
14. Zhang, J., Xie, N., Zhang, X., Yue, K., Li, W.: Machine learning based resource allocation of cloud computing in auction. Comput., Materials Continua no. 1, (2018)
15. Catena, T., Eramo, V., Panella, M., Rosato, A.: Distributed LSTM-based cloud resource allocation in network function virtualization architectures. Comput. Netw. 109111, (2022)
16. Mwiti, D.: Google 2019 Cluster Sample, https://www.kaggle.com/datasets/derrickmwiti/google-2019-cluster-sample. Accessed 16 Aug 2025

Risk-Aware Resource Allocation in Vehicular Fog-Cloud Systems for Emergency Healthcare: An SMDP Approach with Clinical Prioritization

El Mehdi Chtouki[1(✉)], Hibat Eallah Mohtadi[1], Mohamed Hanini[1], and Hicham Ben Alla[2]

[1] Hassan First University of Settat, Faculty of Sciences and Techniques, Computer, Networks, Mobility and Modeling laboratory: IR2M, 26000 Settat, Morocco
{elmehdi.chtouki.doc,h.mohtadi,mohamed.hanini}@uhp.ac.ma

[2] Hassan First University of Settat; Faculty of Sciences and Techniques Settat; LAVETE Laboratory, 26000 Settat, Morocco
h.benalla@uhp.ac.ma

Abstract. Emergency healthcare systems demand rapid and reliable allocation of computational resources, especially in critical cases such as cardiac arrest and trauma. Traditional fog cloud architectures often rely on static allocation strategies that cannot adapt to the fluctuating reliability of mobile nodes or the varying urgency of clinical demands. Consequently, responsiveness and stability may be compromised, with direct implications for patient outcomes. To address this challenge, we propose a dynamic resource allocation framework that integrates vehicular fog computing with cloud infrastructure. The framework is formulated as a semi-Markov decision process (SMDP), providing a rigorous foundation for decision-making under uncertainty. Key contributions include: (i) a real-time reliability scoring mechanism for mobile nodes such as ambulances and drones, (ii) a clinical priorityaware scheduling policy that prioritizes urgent requests, and (iii) an optimization of cost reliability trade-offs through a novel reward function that penalizes unstable nodes. Simulation results show that the proposed SMDP policy significantly improves system performance, achieving higher expected rewards as the arrival rates of service requests increase. This demonstrates the model's ability to efficiently allocate fog resources under varying traffic dynamics.

Keywords: Vehicular fog computing · Emergency healthcaren · SMDP · Resource allocation · Risk-aware scheduling

H. E. Mohtadi, M. Hanini, H. B. Alla—These authors contributed equally to this work.

A. E. Hassanien et al. (Eds.): AISDS 2025, LNDECT 286, pp. 153–163, 2027.
https://doi.org/10.1007/978-3-032-18758-1_15

1 Introduction

The internet of medical things (IoMT) revolutionizes emergency care through real-time patient monitoring using wearable biosensors, connected ambulances, and medical drones that collect critical physiological data requiring immediate processing. While cloud computing provides substantial processing capabilities, its inherent latency, bandwidth constraints, and operational costs present significant challenges for time-sensitive medical applications where response time directly impacts patient survival outcomes.

The emergence of fog computing has brought computational resources closer to data sources [1]. In healthcare, fog-based systems show potential for real-time analytical processes including remote patient surveillance. However, current implementations depend on stationary fog nodes, failing to account for the mobility and unpredictability of emergency medical units. These mobile platforms exhibit variable connectivity, face energy limitations, and are susceptible to environmental factors that impact their reliability [2]. Contemporary resource allocation strategies lack mechanisms for evaluating these dynamic risk factors while maintaining cost-effectiveness.

This research presents a risk-sensitive resource allocation framework for hybrid fog-cloud emergency medical systems. Our methodology introduces key innovations including a sophisticated reliability assessment model that continuously evaluates mobile fog nodes (ambulances and drones) forming a dynamic mobile local cloud infrastructure. This model incorporates real-time parameters such as velocity, battery status, and environmental conditions to quantify failure probabilities.

The technical foundation utilizes a semi-Markov decision process (SMDP) formulation that models medical service requests, dynamic availability of mobile fog resources, and temporal requirements of emergency response. Our reward mechanism integrates three dimensions: (i) medical priority differentiation, (ii) dynamic risk assessment penalties, and (iii) cost considerations for cloud utilization. We employ an advanced continuous-time value iteration algorithm to derive optimal policies capable of adapting to system perturbations.

The remainder of this paper is organized as follows. Section 2 reviews related work in fog computing and resource allocation for healthcare systems. Section 3 describes our system framework, including the reliability model and SMDP formulation. Section 4 presents our numerical results and analysis. Finally, Sect. 5 concludes the paper and discusses future research directions.

2 Related Work

The integration of fog computing and the internet of medical things (IoMT) has garnered significant attention for enhancing emergency care. Foundational work by [1] established fog computing as a solution to reduce latency by bringing computation closer to data sources. Subsequent studies have explored its application in healthcare: [3] demonstrated effective patient monitoring using static fog architectures. More recently, [4] proposed a machine learning-based data segmentation approach using random forest technique for fog-cloud healthcare systems, while [5] introduced a deep reinforcement learning approach to jointly optimize offloading decisions and resource allocation in mobile edge computing. However, these approaches assume predictable network conditions and focus on stationary scenarios, overlooking the dynamic challenges posed by mobile units like ambulances and medical drones.

Recent work has begun addressing mobile fog environments. The authors in [6] developed a task offloading model incorporating basic mobility and energy metrics for vehicular fleets. While this approach demonstrates innovative use of mobility patterns, it neglects critical factors for emergency healthcare. Similarly, other contemporary studies in mobile fog computing, including [7], overlook essential environmental factors such as weather impacts on reliability. Collectively, these existing solutions lack comprehensive real-time risk assessment and clinical prioritization mechanisms necessary for emergency medical scenarios.

Our work advances the state of the art by addressing these limitations through three key contributions: (i) a dynamic reliability model integrating real-time speed, battery, and weather conditions; (ii) a clinical prioritization mechanism distinguishing urgent and normal requests; and (iii) a reward function balancing risk and cost while discouraging both unstable node usage and excessive cloud offloading.

3 System Framework

3.1 System Description Model

The proposed model depicts a vehicular mobile cloud (VMC) architecture for emergency healthcare, where smart ambulances and medical drones act as mobile fog nodes within a hybrid fog-cloud computing ecosystem. These vehicles are equipped with onboard computational resources, called Resource Units (RUs), enabling real-time processing of medical tasks such as patient monitoring, injury assessment, and emergency triage. Vehicles are dynamically classified into safe and risky nodes based on a real-time reliability score that incorporates their speed, battery level, and surrounding weather conditions. Safe vehicles offer stable performance, while risky ones subject to potential failures are penalized in the decision-making process. Upon the arrival of a medical service request, the system must decide whether to assign local RUs from available vehicles or to offload the request to a remote cloud server, which guarantees service at the cost of higher delay and transmission overhead. Requests are clinically categorized as urgent or normal, with urgency levels encoded into the resource allocation policy to ensure appropriate quality of service (QoS). The system's behavior and decision-making are modeled using a semi-Markov decision process (SMDP), which captures the stochastic dynamics of vehicle mobility, request arrivals, and service completions. This architecture enables risk-aware, cost-efficient, and latency-sensitive medical response operations in highly dynamic environments.

Figure 1 illustrates the proposed vehicular mobile cloud VMC system where smart electric ambulances and medical drones form a dynamic computing infrastructure and operate within a mobile health ecosystem. Without loss of generality and to simplify the analysis, we assume that both types of vehicles have a basic computing resource (RU) in the VMC system. These vehicles provide these RUs to meet urgent medical needs, such as continuous monitoring of physiological parameters, automated injury assessment, and intelligent ambulance guidance via dynamic traffic algorithms.

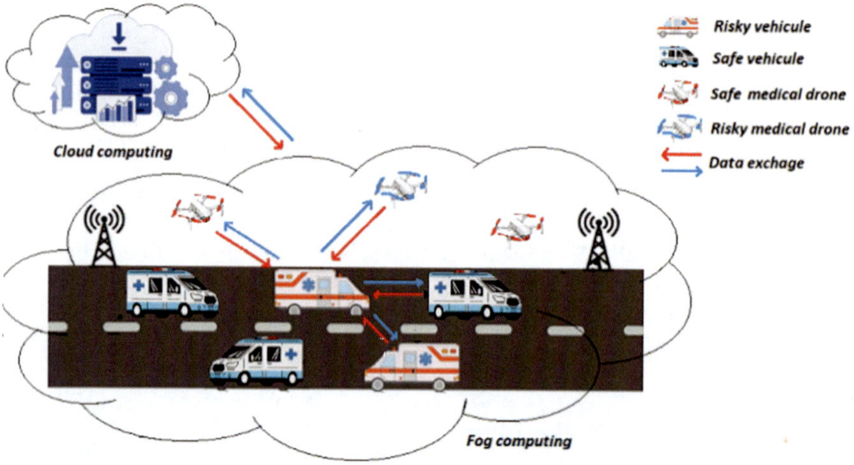

Fig. 1. VMC ecosystem integrating smart electric ambulances and medical drones for mobile health.

Risk-Aware Reliability Modeling. In VMC systems for emergency healthcare, node reliability is critical. Vehicles are classified as safe (V_s) or risky (V_r) based on a dynamic failure rate $\phi_v(t)$ that incorporates three risk factors: Mobility, Battery level and weather. Then the combined failure rate is defined by:

$$\phi_v(t) = \underbrace{c_m \left(1 + \frac{v(t)}{v_{\max}}\right)^p}_{\text{Mobility}} + \underbrace{c_e \left(1 - \frac{B(t)}{B_{\max}}\right)^q}_{\text{Battery}} + \underbrace{c_w W(t)}_{\text{Weather}} \tag{1}$$

where $W(t) \in 0, 1, 2$ represent the weather impact, c_m, c_e, c_w scale the sensitivity of each factor, and $p, q \geq 1$ control nonlinearity.

The total number of resource units available in the system for safe vehicles is noted by M_s, that of risky vehicles by M_r, and the sum of the two by M (i.e. $M = M_s + M_r$).

Operational Dynamics and Allocation Policy. The system processes requests according to the following policy, when a request arrives, the system must choose between: allocate i and j RUs from the class of available safe and risky vehicles, respectively (where $i \in \{1, \ldots, M_s\}$ and $j \in \{1, \ldots, M_r\}$), incurring a minor penalty for using the at-risk vehicle, or transfer the request to processing on a remote cloud RC with a major cost.

Moreover, according to standard triage protocols and clinical guidelines, in our VMC system, medical requests are classified into two distinct priority categories based on clinical urgency:

- Urgent requests (u): Time-critical cases (e.g., cardiac arrest, severe trauma) requiring immediate processing.
- Normal requests ($\bar{u}$): Routine monitoring of less critical cases (e.g., stable patient vitals) with flexible deadlines.

The system VMC operates under the following stochastic processes [?]:

- **Request arrivals**: Modeled as independent Poisson processes with rates λ_k ($k \in \{u, \bar{u}\}$) for every request type.
- **Request services**: Modeled as independent Poisson processes with rates μ in case of only one RU allocated, and $i\mu$ in case that i RUs are allocated.
- **Vehicles arrivals and departures**: Follow Poisson processes with rates λ_v (for arrivals) and μ_v (for departures).

3.2 State Space $\mathcal{S}$

As studied in [8], the state of our VMC system is affected by a set of stochastic events, denoted $\mathcal{E}$ and defined as:

$$\mathcal{E} = \{R, D_{ij}, V_{+s}, V_{-s}, V_{+r}, V_{-r}\}. \tag{2}$$

Here, R represents the arrival of a new medical request, D_{ij} denotes the service completion of a request that was allocated i safe and j risky RUs, V_{+s} and V_{-s} represent the arrival and departure of a safe vehicle, and V_{+r} and V_{-r} represent the arrival and departure of a risky one.

To capture the system's dynamics under these events, we extend the vehicular SMDP model from [8] by defining a generalized state space. A system state s is characterized by the tuple:

$$s = \big(e, M_s, M_r, (n)_s, (n)_r, k\big) \tag{3}$$

where, e is the current event type in the set $\mathcal{E}$, M_s/M_r is the number of resource units in the safe/risky class in the system, $(n)_s = (n_1, \ldots, n_i, \ldots, n_{M_s})$ is the safe allocation vector with n_i is the number of allocated service requests i RUs in the safe class, similarly $(n)_r$ is the risky allocation vector, and finally $k \in \{u, \bar{u}\}$ is the request type.

3.3 Action Space $\mathcal{A}$

For new service requests, the system choose between:

1. executes an allocation action $a = (i, j)$, where i represents resources from safe vehicles V_s and j denotes resources from risky vehicles V_r, provided that $(1 \leq i \leq M_s)$, $(1 \leq j \leq M_r)$ and $M_{\text{occupied}} + i + j \leq M$ (system capacity).
2. executes an allocation action $a = -1$, meaning transferring the request to the remote cloud RC, either strategically or in case of saturation.
3. during vehicle arrivals/departures or request completions, the system performs passive updates with $a = 0$.

The action space is formally defined as:

$$a = \begin{cases} (i, j) & \text{if } e = R,\ 1 \leq i \leq M_s,\ 1 \leq j \leq M_r,\ M_{\text{occupied}} + i + j \leq M \\ -1 & \text{if } e = R,\ \text{"Transfer to RC due to deliberate choice or saturation"} \\ 0 & \text{if } e \in \{V_{+s}, V_{-s}, V_{+r}, V_{-r}, D_{ij}\} \end{cases} \tag{4}$$

3.4 Transition Probabilities $p(s'|s; a)$

Building on the semi-Markov models in [8], we extend the transition formulation to incorporate the allocation policy $a = (i, j)$, clinical priorities (urgent, normal), and elastic resources (M_s, M_r). Transition rates and probabilities are now explicit functions of action a, capturing the allocation strategy and elastic capacity impact. This yields a state-action-dependent transition kernel that models interdependencies between allocation choices, priority scheduling, resource flexibility, and stochastic system evolution.

3.5 Reward Function $r(s, a)$

The reward function $r(s, a)$ for the Vehicular Mobile Cloud VMC system incorporates resource allocation efficiency, risk management, and criticality awareness. It is defined as:

$$r(s,a) = \begin{cases} \alpha_k \mu\,(i + \epsilon j) - \beta_r \left(j + \sum_{v \in V_r} \frac{\phi_v(t)}{\phi_{\text{risk}}} \right) & \text{if } a = (i, j) \\ -\alpha_k \beta_{rc} & \text{if } a = -1 \\ -\xi & \text{if } a = 0,\ M_{occupied} = M,\ e = V_{-r} \\ 0 & otherwise \end{cases} \tag{5}$$

As illustrated by the formula above, this reward function is composed of:

- **Allocation reward**: $\alpha_k \mu(i + \epsilon j)$ rewards processing with i safe and j risky RUs, weighted by clinical priority α_k and risk degradation factor ϵ

- **Risk penalty**: $\beta_r \left(j + \sum \frac{\phi_v(t)}{\phi_{\text{risk}}}\right)$ penalizes use of risky vehicles based on real-time failure probabilities
- **Cloud cost**: $-\alpha_k \beta_{rc}$ charges for remote cloud offloading
- **Failure penalty**: $-\xi$ applies when risky vehicle departure causes service failure during system saturation
- **Neutral events**: No reward for service departures or safe vehicle dynamics

3.6 Solution

According to Bellman's equation, the optimal reward satisfies:

$$\nu(s) = \max_{a \in \mathcal{A}} \left\{ r(s,a) + \gamma \sum_{s' \in \mathcal{S}} p(s'|s,a)\nu(s') \right\} \tag{6}$$

where $\gamma = e^{-\eta\tau(s,a)} \in (0,1]$ is the exponential discount factor for our SMDP [9], representing the present value of future rewards. Here, $\eta > 0$ is the discount rate determining how quickly future rewards are depreciated. $\tau(s,a) = \frac{1}{\sigma(s,a)}$ is the expected sojourn time in state s when taking action a.
Then, we propose the value iteration algorithm [9] [10] to solve the optimization problem given in (6).

Algorithm 1. Value Iteration Algorithm

Step 1 : Initialize $\nu(s)$ for all states s. Set $\epsilon > 0$ and $k = 0$. **Step 2** : For each state s, compute:

$$\nu_{k+1}(s) = \max_{a \in \mathcal{A}} \left\{ r(s,a) + \gamma \sum_{s' \in \mathcal{S}} p(s'|s,a)\nu_k(s') \right\}$$

Step 3 : If $\|\nu_{k+1} - \nu_k\| < \epsilon(1-\gamma)/2\gamma$, proceed to Step 4. Otherwise, increment k and return to Step 2. **Step 4** : Compute the stationary optimal policy:

$$d_\epsilon^* \in \operatorname*{argmax}_{a \in \mathcal{A}} \left[r(s,a) + \gamma \sum_{s' \in \mathcal{S}} p(s'|s,a)\nu_{k+1}(s') \right]$$

4 Numeral Results and Analysis

This section provides numerical results in order to evaluate the performances of the proposed computation resource allocation scheme. The main parameters used in our analysis are provided in Table 1.

4.1 Service Strategy Under the Availability of Resources

Case 0, Case 1, Case 2, and Case 3 correspond to different service strategies adopted by the VMC system. Specifically, **Case 0** represents cloud offloading, where the service request is entirely delegated to a remote cloud server without consuming any vehicular resources. **Case 1** indicates local allocation using a single safe RU, while **Case 2** involves the use of one risky RU. **Case 3** denotes a hybrid allocation with both one safe and one risky RU engaged.

Table 1. Summary of Simulation Parameters

Symbol	Value(s)	Symbol	Value(s)	Symbol	Value(s)
λ_u	2–14 req/min	$\lambda_{\bar{u}}$	2–14 req/min	λ_v	2–14 veh/min
μ	0.8 req/min	μ_v	0.1 veh/min	ϵ	0.5
ϕ_{risk}	0.3	c_m	0.1	c_e	0.2
c_w	0.1	p	1.0	q	1.5
α_u	2.0	$\alpha_{\bar{u}}$	0.5	β_r	0.1–5
β_{rc}	5–50	ξ	10.0	η	0.01
M_s	2	M_r	2		

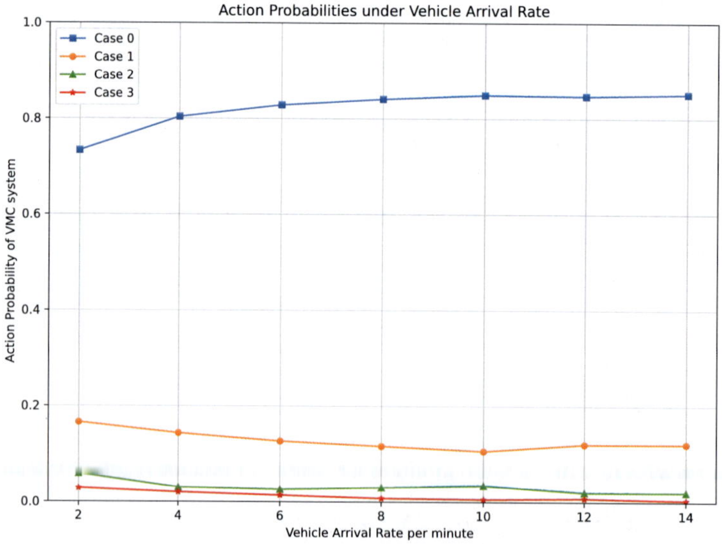

Fig. 2. Action probabilities under vehicle arrival rates λ_v

As illustrated in Fig. 2, the action probabilities for **Case 2** and **Case 3** exhibit a growing trend with the increase in the vehicle arrival rate λ_v. This suggests that the VMC system progressively favors allocating multi-RU resources (2 or 3) to service requests. This phenomenon occurs because a higher vehicle arrival rate generally leads to greater availability of computing resources within the vehicular cloud, thereby raising the probability of assigning more resources per task.

Conversely, the probabilities associated with **Case 0** (cloud offloading) and **Case 1** (single safe RU) demonstrate a gradual decrease. The reason is that an increased λ_v enhances the system's capability to process requests locally using a sufficient number of RUs, diminishing the necessity to offload to the remote cloud (**Case 0**) or to assign only a single RU (**Case 1**). In summary, these results validate that greater vehicle availability enhances local service capacity and promotes the adoption of scheduling strategies that utilize more resources.

4.2 SMDP vs GA and SA Policy Under the Availability of Resources

This section assesses the performance of the optimal SMDP policy against the Genetic Algorithm (GA) and Simulated Annealing (SA) heuristics under different levels of resource availability. The results, illustrated in Figs. 3, reveal a consistent and significant performance gap, highlighting the limitations of heuristic methods in this dynamically constrained environment.

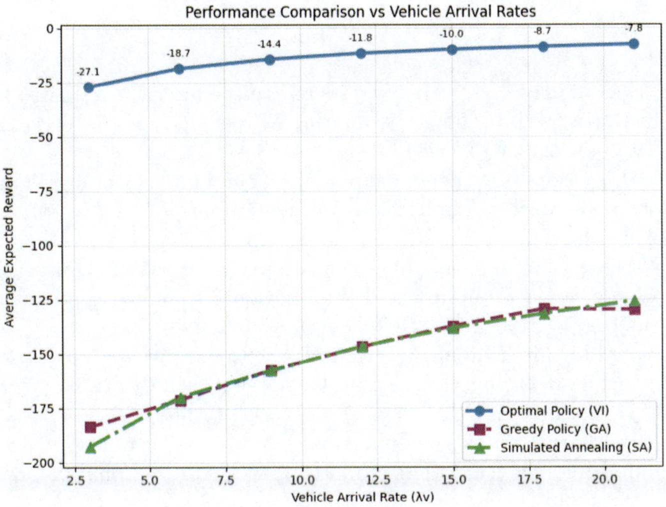

Fig. 3. SMDP policy vs. GA vs. SA under varying vehicle arrival rate (λ_v).

With an increasing vehicle arrival rate, all policies show an improvement in the average expected reward, as they benefit from enhanced resource availability. Nevertheless, as depicted in Fig. 3, the heuristic methods remain considerably suboptimal across the entire range of vehicle availability. At low λ_v values, the reward for GA and SA is approximately -900, compared to -126 for the SMDP policy. While this gap reduces under higher resource availability, it remains substantial (~ -600 for the heuristics versus ~ -37 for SMDP), demonstrating that merely increasing resources is insufficient to overcome the suboptimal decision-making inherent to GA and SA.

5 Conclusion and Future Work

This paper proposed a risk-aware resource allocation framework for emergency healthcare using a Vehicular Mobile Cloud (VMC). Formulated as a Semi-Markov Decision Process (SMDP), our model integrates a dynamic reliability scoring mechanism for mobile nodes (ambulances/drones) based on real-time mobility, battery, and weather conditions, alongside clinical priority-aware scheduling. A novel reward function optimizes the trade-off between performance, risk, and cost.

Results show the policy efficiently handles increasing workloads, favoring safer allocations and strategic cloud offloading as vehicle availability grows. Future work will explore non-Poisson traffic models, additional risk factors, and validation through realistic simulations. This framework provides a foundation for intelligent, reliable emergency healthcare platforms.

References

1. Bonomi, F. et al.: Fog computing and its role in the internet of things. In: Proceedings of 1st ACM Mobile Cloud Computing Workshop, pp. 13–16 (2012). https://doi.org/10.1145/2342509.2342513
2. Mohtadi, H.E., Hanini, M., Benmakhlouf, A., Haqiq, A.: A fault-tolerant queuing model for delay-sensitive applications in fog computing. IAENG Int. J. Comput. Sci. **52**(6) (2025)
3. Gia, T.N. et al.: Fog computing in healthcare IoT. In: Proceedings of IEEE International Conference on Computer and Information Technology, pp. 356–363 (2015). https://doi.org/10.1109/CIT/IUCC/DASC/PICOM.2015.51
4. Amzil, A., Abid, M., Hanini, M., Haqiq, A.: Stochastic analysis of fog computing and machine learning for scalable low-latency healthcare monitoring. Clust. Comput. **27**, 6097–6117 (2024). https://doi.org/10.1007/s10586-024-04285-x
5. Saidi, K., Bardou, D.: Task scheduling and VM placement to resource allocation in Cloud computing: challenges and opportunities. Cluster Comput. **26**, 3069–3087 (2023). https://doi.org/10.1007/s10586-023-04098-4
6. Mohtadi, H.E., Ouammou, A., Hanini, M., Haqiq, A.: Resilient vehicular fog computing networks: an analytical approach to system reliability under breakdown and vacation interruptions. Cluster Comput. 28(2), 108 (2025). https://doi.org/10.1007/s10586-025-05087-4
7. Lv, J., Chen, C.-M., Kumari, S., Li, K.: Resource allocation for AI-native healthcare systems in 6G dense networks using deep reinforcement learning. Digit. Commun. Netw. (2025). https://doi.org/10.1016/j.dcan.2025.01.004

8. Ezzidani, A., et al.: SMDP approach for vehicular cloud computing system. Math. Model. Eng. Probl. **8**(6), 928–936 (2021). https://doi.org/10.18280/mmep.080612
9. Puterman, M.L.: Markov Decision Processes: Discrete Stochastic Dynamic Programming. Wiley, New York (2014). https://doi.org/10.1002/9780470316887
10. Bertsekas, D.P.: Dynamic Programming and Optimal Control, vol. II, 4th edn. Athena Scientific, Belmont (2017)

A SWOT-Guided Hybrid Optimization Framework for Reducing EV Charging Time

Abdelghani Ezzahi(✉), Brahim Zraibi, Mohamed Mansouri, Adam Afadiss, and Omar Lamaamri

Hassan 1st University of Settat, National School of Applied Sciences of Berrechid LAMSAD Laboratory, Berrechid, Morocco
abdelghani.ezzahi.doc@uhp.ac.ma

Abstract. The rapid adoption of electric vehicles (EVs) increases pressure on charging infrastructures, where long charging times and uncoordinated patterns challenge both user convenience and grid stability. Several computational approaches—mathematical optimization, metaheuristics, and machine learning (ML) methods such as Deep Reinforcement Learning (DRL)—have been proposed to address these issues. Each family offers distinct strengths but also key limitations, and none alone can fully handle the complexity of EV charging optimization. This paper introduces a conceptual framework for reducing EV charging time, guided by SWOT (Strengths, Weaknesses, Opportunities, Threats) analysis. Building on a structured review of existing methods, SWOT is used to systematically identify complementarities among algorithmic families. The proposed three-layer workflow integrates short-term optimization for exact scheduling, metaheuristics for medium-term multi-objective planning, and DRL for long-term adaptive control. The framework targets the minimization of total sojourn time (from plug-in to target State-of-Charge (SoC)) across decision horizons of minutes, hours, and days, and is conceptually validated through a 10-EV, 3-station, 24-h scenario. It emphasizes robustness, scalability, and adaptability while acknowledging remaining challenges in computational cost, data dependency, and model interpretability. By employing SWOT as a methodological tool, the work provides a structured foundation for designing hybrid algorithms for sustainable EV-charging management.

Keywords: Electric vehicle charging · Charging optimization · SWOT analysis · Hybrid algorithms · Deep reinforcement learning

1 Introduction

The growing adoption of electric vehicles (EVs) [1] is transforming modern transportation, but it also brings new challenges for energy systems. Among these, charging time stands out as a central concern. Long charging sessions reduce user convenience and may discourage wider EV uptake, while uncoordinated charging risks stressing distribution networks by amplifying peak demand and causing voltage instability. Ensuring fast, reliable, and grid-friendly charging has therefore become a key challenge in the

A. E. Hassanien et al. (Eds.): AISDS 2025, LNDECT 286, pp. 164–173, 2027.
https://doi.org/10.1007/978-3-032-18758-1_16

transition toward sustainable mobility. A range of computational approaches has been proposed to address this challenge. Mathematical optimization methods, such as convex programming and Mixed-Integer Linear Programming (MILP) [2], deliver exact and robust solutions, yet their scalability is limited by computational complexity. Metaheuristic algorithms, including Genetic Algorithms (GA) [9], Particle Swarm Optimization (PSO) [8], and other bio-inspired strategies, provide greater flexibility in handling complex, multi-objective problems. However, they often converge slowly and are sensitive to parameter tuning. More recently, Deep Reinforcement Learning (DRL) [16] has shown promise for adaptive, real-time scheduling in uncertain environments, but it remains data-hungry, computationally expensive, and difficult to interpret in critical infrastructure applications. Taken together, these observations show that each family of methods contributes important strengths, but none is sufficient on its own to fully address the complex nature of EV charging optimization. What is missing is a systematic way to compare these approaches, identify their complementarities, and guide the design of hybrid solutions. To this end, we propose the use of SWOT (Strengths, Weaknesses, Opportunities, and Threats) analysis [21] as a comparative tool. Unlike traditional surveys, a SWOT perspective makes it possible not only to evaluate existing methods, but also to highlight how their strengths can be combined and their weaknesses mitigated. The main contributions of this paper are threefold. First, it provides a structured review of optimization, metaheuristic, and ML/DRL approaches for EV charging. Second it presents a SWOT-based comparative analysis highlighting the strengths, weaknesses, opportunities, and threats of these approaches. Finally, it proposes a conceptual framework that leverages SWOT insights to guide hybrid algorithmic integration aimed at reducing EV charging times. The remainder of this paper is organized as follows: Sect. 2 reviews the state of the art. Section 3 presents the SWOT-based analysis. Section 4 introduces the conceptual framework. Section 5 discusses advantages and limitations. Section 6 concludes and outlines future research directions.

2 Literature Review

The existing literature on electric vehicle (EV) charging optimization is typically divided into three main categories: mathematical optimization, metaheuristic algorithms, and ML approaches, particularly Deep Reinforcement Learning (DRL). Each of these families has its own strengths and weaknesses, which motivates the need for a comparative framework. Mathematical optimization approaches [1–6, 10] mainly rely on convex programming and Mixed-Integer Linear Programming (MILP). These methods provide exact and robust solutions, making them suitable for small- and medium-scale systems. However, they are limited by scalability, since computational complexity increases sharply when applied to large distribution networks. This restricts their applicability in real-world smart grid environments where uncertainty and variability are common. Metaheuristic approaches [7–13], including evolutionary algorithms such as Genetic Algorithms (GA), Particle Swarm Optimization (PSO), and more recent bio-inspired techniques such as the Red Deer Algorithm or Hiking Optimization, are often used to overcome the limits of exact optimization. Their flexibility and ability to handle multi-objective problems make them attractive for EV charging and discharging scheduling. Nevertheless, these methods typically suffer from slow convergence, risk of local minima, and high sensitivity to parameter tuning, resulting in inconsistent performance in different scenarios. Recent works have explored hybrid metaheuristics (e.g., GA combined with PSO or simulated annealing), showing potential for faster convergence and improved robustness. ML and DRL-based approaches [14–19] have recently gained significant attention. DRL has demonstrated strong adaptability in dynamic and uncertain environments, enabling real-time decision-making in EV charging management. Advanced models such as Graph Neural Networks (GNNs) have been employed to improve scalability and coordination across large fleets. Despite these advantages, such methods are data-hungry, computationally expensive, and often operate as "black boxes," posing interpretability and transparency challenges in critical infrastructure applications. In addition, fairness-oriented scheduling, decentralized frameworks, and Vehicle-to-Grid (V2G) integration are being explored as promising extensions, further broadening the scope of data-driven approaches. In summary, mathematical optimization ensures exactness but lacks scalability, metaheuristics provide flexibility at the cost of convergence speed, and DRL offers adaptability but depends heavily on data and computational resources. This diversity of strengths and weaknesses shows that no single approach can fully address the challenges of EV charging optimization. To capture these trade-offs and highlight complementarities, the next section introduces a SWOT analysis (Strengths, Weaknesses, Opportunities, and Threats). This review also aligns with canonical EV-charging formulations addressing queue-aware scheduling, rolling-horizon control, and grid-constrained optimization [3, 5, 11]. This structured comparison provides a foundation for the conceptual hybrid framework proposed in this paper.

3 SWOT-Based Comparative Analysis

To systematically compare the three families of approaches identified in the literature, we apply a SWOT analysis (Strengths, Weaknesses, Opportunities, Threats). This framework highlights not only the core characteristics of mathematical optimization, metaheuristics, and DRL-based methods, but also the potential complementarities that can guide the design of hybrid solutions. The results of this comparative assessment are summarized in Table 1, which presents the SWOT matrix for the three families of approaches.

The SWOT analysis shows that no single family of approaches is sufficient to tackle the complex challenges of EV charging optimization. Mathematical optimization methods excel in producing exact solutions with strong guarantees, yet their applicability is limited by scalability problems. Metaheuristics, on the other hand, provide flexibility and robustness in dealing with multi-objective and complex environments, but at the cost of slower convergence and inconsistent performance. DRL-based approaches are among the latest methods, offering adaptability and scalability through advanced models such as Graph Neural Networks. However, they remain computationally intensive and face interpretability challenges, which are important in smart grid applications. Taken together, each family of approaches offers unique strengths that can be used, while their weaknesses can potentially be reduced through hybridization. This motivates the conceptual framework presented in the next section, where SWOT serves as a guiding tool for designing integrated hybrid solutions that combine the strengths of these families of methods, as summarized in Table 1.

Table 1.

Approach	Strengths	Weaknesses	Opportunities	Threats
Mathematical Optimization	Exact and robust solutions; strong theoretical guarantees; well-suited for small/medium systems	Poor scalability; exponential computational growth; limited flexibility in dynamic contexts	Can be coupled with heuristics for decomposition; integration into short-term local scheduling	Combinatorial explosion in large-scale systems; sensitivity to uncertainty
Metaheuristics	Flexible and adaptable to complex, multi-objective problems; easy to extend to new scenarios	Slow convergence; risk of local minima; high sensitivity to parameter settings	Well-suited for large-scale and non-convex problems; potential to complement exact or learning-based methods	Highly variable performance; difficulty in ensuring reproducibility
ML/DRL	Adaptive learning in dynamic environments; scalable with advanced models (e.g., GNNs); potential for real-time scheduling	Data-hungry; computationally expensive; "black box" nature limits interpretability	Promising for smart grids, V2G, and demand-response integration; ability to generalize across scenarios	Computational overhead may limit adoption; trust and transparency issues in critical infrastructure

References supporting each SWOT category are included for traceability: Optimization [1, 3, 5], Metaheuristics [8, 10, 12, 13], DRL [16–20].

4 Conceptual Framework

Building on the SWOT analysis, we propose a conceptual framework for reducing EV charging time through hybrid algorithms. The framework uses the strengths of each family of approaches, addresses their weaknesses, and takes advantage of opportunities while reducing possible risks. By explicitly applying SWOT as a guiding tool, the framework offers a structured way to integrate mathematical optimization, metaheuristics, and DRL into an integrated hybrid solution. Importantly, the framework is conceptual rather than prescriptive, which makes it adaptable to different scenarios and future implementations.

4.1 Design Principles

The framework is guided by three overarching design principles:

Build on Strengths: Apply mathematical optimization for exact short-term scheduling and feasibility checks. Use metaheuristics to balance multi-objective trade-offs in complex environments, and leverage DRL for long-term adaptive decision-making under uncertainty.

Address Weaknesses: Combine DRL with heuristic techniques to accelerate convergence and avoid local minima, decompose large-scale optimization problems into smaller, tractable sub-problems, and develop hybrid approaches (e.g., optimization + learning or optimization + heuristics) to reduce computational burden.

Take Advantage of Opportunities and Manage Risks: Integrate smart grid features such as Vehicle-to-Grid (V2G), demand response, and renewable energy sources, handle computational challenges through distributed and cloud-edge infrastructures, and improve transparency by incorporating interpretable elements into data-driven models.

4.2 Hybrid Workflow

The framework is organized as a three-layer workflow corresponding to different planning horizons:

Short-Term Scheduling (Local Optimization): Mathematical optimization (e.g., MILP or convex programming) is applied to generate exact charging schedules that meet immediate grid and tariff constraints.

Medium-Term Planning (Metaheuristics): Metaheuristics such as PSO or GA are used to handle multi-objective trade-offs, providing flexible strategies for day-ahead or week-ahead planning.

Long-Term Policy Adaptation (DRL): DRL agents are trained to develop adaptive policies that work across different scenarios, enabling robust decision-making in dynamic contexts such as stochastic demand and renewable variability.

In this framework, "charging time" refers to the total sojourn duration from plug-in until the target state of charge (SoC) is reached, encompassing both waiting and active charging phases. Formally:

$$\min_{P_{i,t}} T_{sojourn} = T_{wait} + T_{charge} \tag{1}$$

subject to grid voltage limits

$$V_{min} \leq V_t \geq V_{max} \tag{2}$$

as well as feeder capacity and tariff-block constraints. The medium-term metaheuristic layer optimizes a multi-objective vector:

$$\left[F = T_{avg}, C_{energy}, R_{peak}\right] \tag{3}$$

which jointly captures mean charging time, energy cost, and peak-load ratio. The long-term DRL layer learns adaptive policies:

$$\pi(S_t) \rightarrow a_t \tag{4}$$

using state variables such as SoC distributions, queue lengths, and dynamic price signals. Inter-layer exchanges include feasible schedules, heuristic priors, and learned policy parameters to ensure consistency and scalability across horizons.

4.3 Conceptual Diagram

To illustrate the workflow described above, Fig. 1 presents a schematic diagram of the proposed framework. The diagram emphasizes the layered structure:

Input Layer: EV fleet demand, charging requirements, grid constraints, price signals, and renewable generation profiles.

SWOT-Guided Decision Layer: A central block indicating how SWOT categories inform the selection of algorithmic families (optimization, metaheuristics, DRL).

Hybrid Integration Layer: Three interconnected modules representing short-term optimization, medium-term planning, and long-term DRL policy learning.

Output Layer: Optimized charging schedules, reduced charging times, minimized costs, and enhanced grid stability.

The figure provides a visual representation of the hybrid workflow, linking the SWOT-based analysis (Sect. 3) to the proposed conceptual framework.

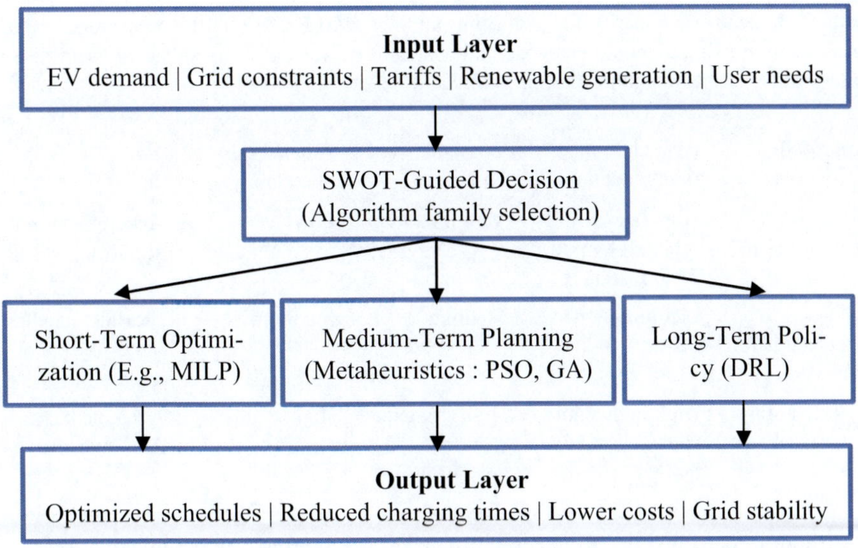

Fig. 1. Conceptual framework for EV charging optimization guided by SWOT

4.4 Advantages of the Proposed Framework

Compared to single-method approaches, the proposed framework offers several advantages: **robustness**, by combining exactness, flexibility, and adaptability; **scalability**, through decomposition and hybridization that enable deployment in large-scale systems; **adaptivity**, supported by DRL responsiveness to dynamic and uncertain environments; and **practical relevance**, as SWOT provides a structured foundation for evaluating approaches and guiding algorithmic integration.

5 Discussion

The proposed SWOT-guided conceptual framework stands out from existing approaches by proposing a structured way to evaluate and combine algorithmic families. Most of the reviewed studies adopt either a single class of methods, such as MILP formulations for short-term optimization or metaheuristics for multi-objective scheduling, or they present hybrid schemes without a clear rationale for algorithm selection. In contrast, SWOT provides a clear basis for comparing strengths and weaknesses, turning algorithm integration from an ad hoc process into a more systematic design strategy. An important advantage of the framework is that it is both flexible and robust. By explicitly linking mathematical optimization, metaheuristics, and DRL, it supports multiple levels of decision-making: exact feasibility checks at the short-term horizon, heuristic balancing of trade-offs at the medium-term horizon, and adaptive learning for long-term dynamics. This layered perspective provides greater resilience against uncertainties in EV demand, renewable variability, and grid constraints than any single method in isolation.

Table 2. Indicative comparison of algorithm families (MILP, PSO, DRL) by average sojourn-time reduction (%), computational cost, and adaptability. Values are hypothetical, derived from representative trends in published studies [3, 8, 14, 16].

Approach	Avg. charging time reduction	Computational cost	Adaptability to uncertainty	Notes
MILP (optimization)	High (exact schedules, up to **25% faster**)	Very high for >10 EVs	Low	Accurate but not scalable
PSO (metaheuristic)	Medium ($\approx$ **15% faster**)	Medium	Medium	Flexible, handles multi-objective trade-offs
DRL (learning)	High in dynamic contexts ($\approx$ **20% faster**)	High (training requires >10 k episodes)	High	Adaptive but data-intensive

In addition, using SWOT conceptually makes the approach applicable to other smart grid problems, such as demand response or renewable integration, where hybridization is also needed. To illustrate these points, a simplified illustrative scenario with 10 EVs (each with a 60 kWh battery and 7.2 kW chargers) distributed across 3 stations and scheduled within a 24-h horizon is considered. The three families of approaches are compared along three dimensions: charging time reduction, computational effort, and adaptability to uncertainty, serving as baseline references to highlight the complementary nature of the proposed hybrid framework. These comparative results are summarized in Table 2, which provides an indicative comparison of the representative methods (MILP, PSO, and DRL) from each family.

To visualize these results, Fig. 2 presents a bar chart comparing the three approaches. MILP excels at exactness but lacks scalability and adaptability, PSO provides balanced but moderate improvements, and DRL ensures adaptability at the cost of high computational requirements.

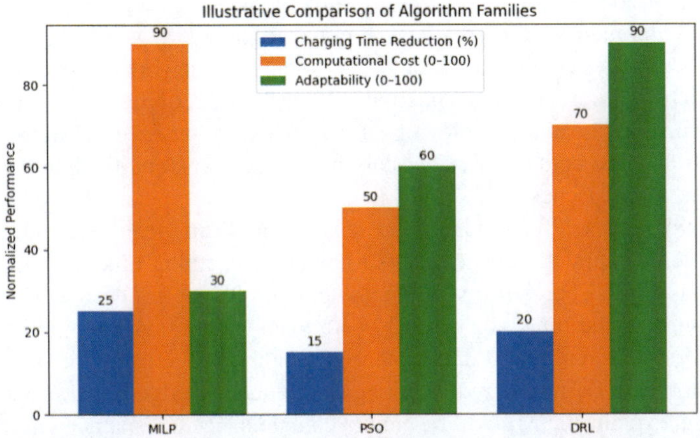

Fig. 2. Illustrative comparison of algorithm families (MILP, PSO, DRL) across charging-time reduction, computational cost, and adaptability. Bars normalized to baseline MILP = 100%; percentage values indicate relative reduction in average sojourn time. Values are hypothetical estimates inspired by typical performance ranges reported in the EV-charging optimization literature [3, 8, 14, 16].

This illustrative example reinforces the main message of the SWOT analysis: no single family of approaches is sufficient on its own. Instead, their complementarities motivate the design of hybrid solutions, where MILP can ensure feasibility, metaheuristics can balance trade-offs, and DRL can provide adaptability in dynamic contexts. At the same time, the framework has limitations. Being conceptual in nature, it lacks empirical validation at this stage. The practical deployment of such a hybrid system would face challenges related to computational complexity, data availability, and the interpretability of DRL models in critical infrastructure contexts. Overall, the SWOT-guided hybridization strategy offers a methodological contribution and a promising pathway for addressing EV charging challenges. Although the present work is conceptual, implementation and quantitative validation are currently under development. Future experiments will rely on benchmark datasets and realistic grid-simulation environments to evaluate performance across practical EV-charging scenarios. The absence of empirical results in this version reflects the ongoing nature of this research rather than a limitation of the proposed framework. These future efforts will ensure the framework's transition from conceptual design to real-world applicability, as outlined in the conclusion.

6 Conclusion and Future Work

This paper proposed a conceptual framework for reducing EV charging time based on hybrid algorithms guided by a SWOT (Strengths, Weaknesses, Opportunities, Threats) analysis. Through a systematic review of mathematical optimization, metaheuristics, and DRL-based approaches, the SWOT analysis revealed that no single family of approaches is sufficient to address the complex challenges of EV charging optimization. The proposed framework combines the complementary strengths of these methods while addressing their weaknesses, and offers a structured way to design hybrid solutions that are robust, scalable, and adaptive. While the framework makes a methodological contribution, it remains conceptual at this stage and requires further development before practical application. Deploying such a system in practice will raise additional challenges related to computational scalability, data quality, and the interpretability of learning-based approaches.

Future extensions will include full-scale validation using real EV-charging datasets and reproducible simulation benchmarks to quantify charging-time reductions against canonical baselines. These efforts will build upon the conceptual design outlined in this work and support the transition toward operational hybrid control systems. In parallel, integration with real-world data streams, including IoT sensors, smart meters, and renewable forecasts, will be explored, along with cloud and edge-based scalability tests. Furthermore, explainable ML will be employed to enhance transparency in DRL-based decision-making, while broader extensions will target smart grid applications such as demand response, renewable integration, and Vehicle-to-Grid (V2G) services, as well as application-focused concerns such as battery degradation [20]. In summary, this paper provides a methodological foundation for hybrid algorithm design in EV charging management. By using SWOT as a guiding tool, it opens new possibilities for the systematic integration of algorithm families and sets the stage for practical implementation and validation.

References

1. Wang, Y., Wang, H., Razzaghi, R., Jalili, M., Liebman, A.: Multi-objective coordinated EV charging strategy in distribution networks using an improved augmented epsilon-constrained method. Appl. Energy. **369**, 123547 (2024)
2. Ren, L., Yuan, M., Jiao, X.: Electric vehicle charging and discharging scheduling strategy based on dynamic electricity price. Eng. Appl. Artif. Intell. **123**, 106320 (2023)
3. Das, R., Wang, Y., Busawon, K., Putrus, G., Neaimeh, M.: Real-time multi-objective optimisation for electric vehicle charging management. J. Clean. Prod. **292**, 126066 (2021)
4. Singh, P.P., Das, S., Wen, F., Palu, I., Singh, A.K., Thakur, P.: Multi-objective planning of electric vehicles charging in distribution system considering priority-based vehicle-to-grid scheduling. Swarm Evol. Comput. **78**, 101234 (2023)
5. Zhang, L., Sun, C., Cai, G., Koh, L.H.: Charging and discharging optimization strategy for electric vehicles considering elasticity demand response. eTransportation. **18**, 100262 (2023)
6. Kaur, A.P., Singh, M.: Time-of-use tariff rates estimation for optimal demand-side management using electric vehicles. Energy. **273**, 127243 (2023)

7. Fang, B., Li, B., Li, X., Jia, Y., Xu, W., Liao, Y.: Multi-objective comprehensive charging/discharging scheduling strategy for electric vehicles based on the improved particle swarm optimization algorithm. Front. Energy Res. **9**, 811964 (2021)
8. Farhadi, F., Wang, S., Palacin, R., Blythe, P.: Data-driven multi-objective optimization for electric vehicle charging infrastructure. iScience. **26**, 107737 (2023)
9. Cao, G., Chang, L.: Optimal charging strategy of dynamic electricity price for electric vehicles based on improved particle swarm optimization. J. Electr. Eng. Technol. **20**, 1245–1254 (2025)
10. Huo, M., Pang, S., Zhao, H.: Improved PICEA-g-based multi-objective optimization scheduling method for distribution network with large-scale electric vehicles. Sci. Rep. **14**, 80184 (2024)
11. Pang, S., Fan, K., Huo, M.: Charge and discharge scheduling method for large-scale electric vehicles in V2G mode via MLGCSO. Sci. Rep. **15**, 00265 (2025)
12. Chen, Y., Zhao, H., Pang, S., Huo, M.: Optimal scheduling of solar powered EV charging stations in a radial distribution system using opposition-based competitive swarm optimization. Sci. Rep. **15**, 88758 (2025)
13. Boubaker, S., et al.: Multi-objective optimization framework for electric vehicle charging and discharging scheduling in distribution networks using the red deer algorithm. Sci. Rep. **15**, 97473 (2025)
14. Li, S., et al.: Electric vehicle charging management based on deep reinforcement learning. J. Mod. Power Syst. Clean Energy. **10**(3), 719–730 (2022)
15. Liu, D., Zeng, P., Cui, S., Song, C.: Deep reinforcement learning for charging scheduling of electric vehicles considering distribution network voltage stability. Sensors. **23**(3), 1618 (2023)
16. Yang, Y., Zhang, C., He, J., Li, Y., Liu, X.: Scalable reinforcement learning for large-scale coordination of electric vehicles using graph neural networks. Commun. Eng. **4**, 457 (2025)
17. Tushar, W., et al.: Reinforcement learning based EV charging management systems: a review. IEEE Access. **9**, 72478–72498 (2021)
18. Xiang, L., Zhang, Y., Xu, Z.: Advancements in EV charging optimization: a survey of reinforcement learning approaches. In: Proceedings of ICEMD 2024, pp. 87–94 (2024)
19. Lu, X., Wang, Y., Zhao, H., Song, W.: A deep reinforcement learning-based charging scheduling approach with augmented Lagrangian for electric vehicles. Appl. Energy. **376**, 124706 (2024)
20. Sun, W., Chen, Y., Zhang, G.: Electric vehicles charging management using deep reinforcement learning considering vehicle-to-grid operation and battery degradation. Energy Rep. **9**, 10031–10043 (2023)
21. Helms, M.M., Nixon, J.: Exploring SWOT analysis – where are we now? J. Strateg. Manag. **3**(3), 215–251 (2010)

Smart Material Selection for Electric Vehicle Chassis Using Artificial Intelligence: To Lighter, More Robust Chassis

Moussa Kalli Salissou[1,2(✉)], Yassine Zahidi[2,3], Mohamed El Moufid[2,4], Wafaa Dachry[1,2], Hassan Gziri[1,2], Hammadi Laghridat[2,3], and Hicham Medromi[2,5]

[1] Laboratory of Engineering and Innovation of Advanced Systems, Faculty of Sciences and Technologies (FST), Hassan I University, Settat, Morocco
salissou.moussakalli.doc@uhp.ac.ma
[2] Foundation for Research Development and Innovation in Science and Engineering, Mohammédia, Morocco
[3] Higher School of Water, Energy and Sustainable Development Technologies, Suptech Environnement Mohammedia, Mohammédia, Morocco
[4] Higher School of Biomedical Engineering and Health Techniques, Suptech Santé Mohammedia, Mohammédia, Morocco
[5] International Academy of Scientific Francophone (IASF), Rabat, Morocco

Abstract. The choice of materials for the chassis of electric vehicles is a critical step involving several criteria, including lightness, safety, environmental impact, and overall vehicle performance. Indeed, traditional approaches to material selection, based on engineering experience or customer feedback and the use of Ashby diagrams, are unable to encompass several criteria defined by current specifications. Our study therefore highlights an intelligent approach to optimizing the selection of materials for electric vehicle chassis through artificial neural networks (ANN). A structured database was constructed, based on reliable sources, containing several materials used in the manufacture of electric vehicles and incorporating several criteria, including physical properties (density), mechanical properties (toughness, elastic limit, modulus of elasticity), economic properties (cost), and environmental properties (recyclability). Thus, after rigorously pre-processing and standardizing the data, a multilayer neural network was trained to classify and select the most suitable materials. The remarkable results of our study demonstrate high accuracy and speed, as well as the ability to analyze complex relationships.

Keywords: Materials Database · Composite Materials · Machine Learning · Multi-objective Optimization · Artificial Intelligence · Electric Vehicle Chassis

1 Introduction

The transition to sustainable mobility is driving researchers to step up their efforts in the development and design of electric vehicles. This transition is part of the quest to find a compromise between performance and environmental impact. In this context, the

A. E. Hassanien et al. (Eds.): AISDS 2025, LNDECT 286, pp. 174–181, 2027.
https://doi.org/10.1007/978-3-032-18758-1_17

chassis plays a very important role, influencing the rigidity, lightness, safety, and energy efficiency of the vehicle [1, 2]. With this in mind, the major challenge that researchers are grappling with is to design a chassis that is both lightweight and strong. Indeed, even a minimal reduction in chassis weight could lead to a significant improvement in the vehicle's range and performance [3, 4]. However, the materials commonly used in chassis manufacturing, such as aluminum, steel, and composites, are unable to reconcile mechanical properties, ease of implementation, cost, and environmental impact [5]. In general, engineers typically base their choice of materials primarily on conventional approaches such as customer feedback, Ashby diagrams, and physical testing [6]. These subjective approaches are both weak in integrating multiple criteria and inadequate in meeting certain requirements such as carbon footprint, durability, and recyclability of materials [7, 8]. In this regard, the use of artificial intelligence (AI) in material selection is a promising development. Thanks to computer technology, it is now possible to exploit material databases and train predictive models that can automatically select the optimal materials based on the objectives to be achieved. These powerful models also make it possible to propose materials based on technical, economic, and environmental objectives [9, 10]. In this context, our study aims to propose an intelligent approach to selecting materials for electric vehicle chassis by highlighting machine learning applied to data comprising mechanical, physical, economic, and ecological properties. The objective of this work is therefore to demonstrate that the use of classification and regression algorithms makes it possible to reproduce decisions and discover innovative and economical alternatives. This study is part of an innovative data-driven eco-design approach, paving the way for more responsible, agile, and robust engineering. Like all scientific work, this study is divided into several parts. The first part concerns the collection, analysis, and interpretation of scientific documents in order to gain a better understanding of the advances and limitations in the use of algorithms for material selection; this part is commonly referred to as the "state of the art." The second part will outline the methodology that was adopted to carry out this work. The third part concerns the analysis and interpretation of the results, as well as the discussion, in order to conclude with a summary of our work.

2 State of the Art

The choice of materials has always been a critical step in the mechanical manufacturing process. Indeed, choosing materials for electric vehicle chassis is a major challenge because it requires rigidity, lightness, structural performance, and energy efficiency. For a long time, engineers have relied on traditional methods such as Ashby diagrams, which allow materials to be compared according to two criteria at a time [11]. Although effective, these traditional methods have their limitations when several criteria are considered, particularly mechanical, thermal, economic, and ecological constraints [12]. Some traditional multi-criteria methods, such as Analytic Hierarchy Process (AHP) and TOPSIS (Technique for Order of Preference by Similarity to Ideal Solution), have attempted to solve this problem, but without success, as they rely entirely on the subjectivity of the evaluator [13]. The increasing complexity of engineering means that data-driven methods based on artificial intelligence are gaining market share. These approaches have emerged thanks to the rise of materials databases and significant advances in machine learning algorithms. These innovative approaches now make it possible to model nonlinear correlations between material properties and performance [14]. Several studies have demonstrated the importance of using algorithms such as neural networks and Random Forest to find a compromise between new materials and mechanical or thermal constraints [15, 16]. With this in mind, several studies have attempted to merge materials and data science. Ramprasad's team has developed and designed an automated material selection pipeline based on machine learning algorithms. Their work demonstrates a significant improvement in design time [17]. Other techniques, such as unsupervised clustering, have led to the discovery of new materials with high potential in metal alloys [18]. It should be noted that these approaches also allow for the dynamic integration of important aspects in the development and sustainable design of electric vehicles, such as cost constraints, recyclability, and durability [19]. In addition, the use of optimization algorithms combined with knowledge graphs significantly increases the robustness of material selection [20]. In addition, certain platforms such as Citrination and Matminer facilitate access to large databases for training reliable artificial intelligence models [21]. At the same time, other efforts have been made, notably through techniques such as SHAP and LIME, to make these tools more interpretable, thereby increasing engineers' confidence in the recommendations generated [22, 23]. With regard to the development and design of electric vehicle chassis, much recent work has converged on the use of machine learning methods to replace traditional materials with innovative materials such as composites and aluminium alloys in order to evaluate their performance [24]. This is the case, for example, in the work proposed by the authors of reference [25], which aims to find a method for selecting chassis materials based on neural networks trained on data from multiphysics simulations. Other researchers, such as Kumar et al., have shown that combining experimental data with simulation data significantly increases the accuracy of artificial intelligence models in predicting sustainable automotive structures [26]. The integration of topological optimization coupled with intelligent material selection not only makes it possible to associate a material with an optimized structure, but also minimizes mass while ensuring rigidity [27]. In this regard, certain platforms such as Open Quantum Materials Database (OQMD) and OpenAI MatSci make it possible to integrate quantum physics into the pre-selection process [28]. However, all these innovations face several limitations, including the quality and standardization of the data used [29]. In addition, the implementation of these approaches by manufacturers requires design adaptation, which is still not widely automated in small and medium-sized enterprises (SMEs) [30].

In summary, although traditional methods have built a solid foundation for methodical materials engineering, it should be noted that intelligent methods such as data-driven and AI now expose a multi-scale relationship, introducing cost, sustainability, and performance. This study is also in line with this dynamic, offering a structured and comprehensive methodology for the intelligent selection of materials used in the design of electric vehicle chassis based on machine learning algorithms. The ultimate goal of this study is not only to demonstrate the feasibility of this approach, but also to provide a framework that is adaptable and reproducible for other components.

3 Methodology

Our study takes a rigorous approach combining materials engineering, machine learning, and data science to optimize material selection for electric vehicle chassis. First, a robust database was compiled from reliable sources such as CES EduPack, Granta EduPack MatWeb, etc., as well as scientific publications. The data includes several composite and polymer materials, metals, and aluminum alloys characterized by several properties, including mechanical, physical, economic, and environmental properties: elastic modulus, tensile strength, density, carbon footprint, recyclability, corrosion resistance, cost, etc. The suitability of each attribute is validated according to the structural requirements of the chassis. The data is pre-processed through rigorous management of missing values (targeted imputation) and normalization by Min-Max Scaling to ensure scale homogeneity between values in order to visualize and extract structural trends. The detection of interdependencies between properties (mechanical and economic) was made possible by an in-depth cross-correlation analysis, thus revealing the main criteria influencing the design of the chassis. The core of our methodology is based on the integration of multi-layer artificial neural networks (ANNs). These are chosen for their ability to model non-linear relationships between material properties and their performance in application. For better performance, the loss function used is the mean squared error. To optimize the hyperparameters, an adjustment by Grid Search was performed. Multi-criteria integration into the choice of materials was made possible by combining the ANN model with multi-criteria decision support. The use of AHP and TOPSIS not only allowed the criteria to be weighted subjectively according to requirements, but also enabled the materials to be ranked according to their distance from an ideal material. Cross-referencing these results with the neural network predictions produced a more interpretable and robust ranking. The analysis of gradients and Shapley Additive Explanations methods improved the interpretability of the neural network model in order to assess the relative impact of each property on the prediction. This not only guides engineers toward the most influential properties, but also strengthens confidence in decision-making. Figure 1 summarizes the methodology adopted.

4 Results and Discussion

Intensive training enabled the neural network-based material selection model to identify the best materials for designing and optimizing electric vehicle chassis while considering several criteria, displaying low convergence. The results of this study provide excellent recommendations on materials with specific performance properties. The recommended materials range from lightweight polymers for economical solutions to high-strength steel for structural integrity. This analysis is validated by the model's ability to handle multi-criteria processes while juggling economic, environmental, financial, and mechanical considerations in the design of sustainable electric vehicles.

The Fig. 1 shows the training results for ABS polymer. Indeed, a stable loss was observed around 3.615 and a gradient of 0.0010, confirming the successful convergence of the model after 1,000 epochs. This convergence demonstrates the effectiveness of the algorithm for polymer materials, with a precision adapted to the lightweight constraints of electric vehicle chassis.

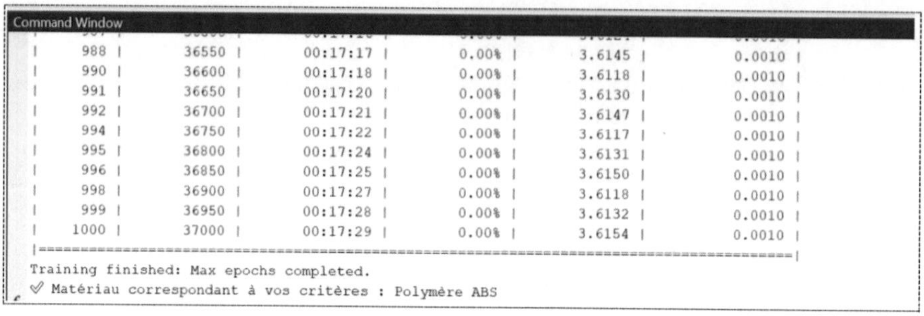

```
Command Window
|   988 |   36550 |   00:17:17 |   0.00% |   3.6145 |   0.0010 |
|   990 |   36600 |   00:17:18 |   0.00% |   3.6118 |   0.0010 |
|   991 |   36650 |   00:17:20 |   0.00% |   3.6130 |   0.0010 |
|   992 |   36700 |   00:17:21 |   0.00% |   3.6147 |   0.0010 |
|   994 |   36750 |   00:17:22 |   0.00% |   3.6117 |   0.0010 |
|   995 |   36800 |   00:17:24 |   0.00% |   3.6131 |   0.0010 |
|   996 |   36850 |   00:17:25 |   0.00% |   3.6150 |   0.0010 |
|   998 |   36900 |   00:17:27 |   0.00% |   3.6118 |   0.0010 |
|   999 |   36950 |   00:17:28 |   0.00% |   3.6132 |   0.0010 |
|  1000 |   37000 |   00:17:29 |   0.00% |   3.6154 |   0.0010 |
|==========================================================================|
Training finished: Max epochs completed.
✅ Matériau correspondant à vos critères : Polymère ABS
```

Fig. 1 Result of the first simulation

Table 1 Properties of ABS polymer

Material	Modulus of elasticity	Yield strength	Density	Recyclability	Carbon footprint	Cost
ABS polymer	2.3 GPa	45 MPa	1.05 g/cm^3	30%	6 Kg CO2/Kg	2 Euro/Kg

Table 1 shows that ABS polymer has a low modulus of elasticity and low yield strength. These properties make it unsuitable for parts subjected to high loads. However, its low carbon footprint and low density make it an excellent candidate for secondary parts where rigidity is less important. On the other hand, its low mechanical strength and poor recyclability limit its use in chassis applications where rigidity, robustness, and durability are essential. The chassis of electric vehicles has very stringent structural requirements, and metal materials demonstrate significantly superior properties, as shown in the following comparative data.

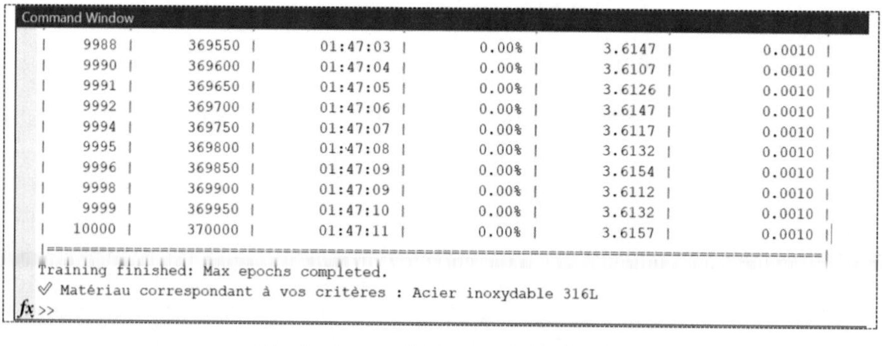

```
Command Window
|   9988 |   369550 |   01:47:03 |   0.00% |   3.6147 |   0.0010 |
|   9990 |   369600 |   01:47:04 |   0.00% |   3.6107 |   0.0010 |
|   9991 |   369650 |   01:47:05 |   0.00% |   3.6126 |   0.0010 |
|   9992 |   369700 |   01:47:06 |   0.00% |   3.6147 |   0.0010 |
|   9994 |   369750 |   01:47:07 |   0.00% |   3.6117 |   0.0010 |
|   9995 |   369800 |   01:47:08 |   0.00% |   3.6132 |   0.0010 |
|   9996 |   369850 |   01:47:09 |   0.00% |   3.6154 |   0.0010 |
|   9998 |   369900 |   01:47:09 |   0.00% |   3.6112 |   0.0010 |
|   9999 |   369950 |   01:47:10 |   0.00% |   3.6132 |   0.0010 |
|  10000 |   370000 |   01:47:11 |   0.00% |   3.6157 |   0.0010 |
|==========================================================================|
Training finished: Max epochs completed.
✅ Matériau correspondant à vos critères : Acier inoxydable 316L
fx >>
```

Fig. 2 Result of the second simulation

Figure 2 shows the results for 316L stainless steel, which required 10,000 epochs (370,000 iterations), with a final loss of 3.6157, similar to ABS but with a longer computation time (1 h 47 min). The stability of the gradient suggests robust optimization, which is important for chassis subjected to high mechanical stresses. These results highlight the major computational challenges associated with dense metallic materials, but validate the ability of artificial intelligence to handle complex mechanical properties.

Table 2 Properties of 316L stainless steel

Material	Modulus of elasticity	Yield strength	Density	Recyclability	Carbon footprint	Cost
ABS polymer	193 GPa	290 MPa	8 g/cm^3	95%	8 Kg CO2/Kg	3 Euro/Kg

Table 2 shows the properties of 316L stainless steel obtained after the second simulation. This material has a robust elastic limit and high elastic modulus, making it ideal for electric vehicle chassis where mechanical stresses are severe. Its high density is offset by its excellent recyclability and moderate carbon footprint. Furthermore, its competitive cost makes it attractive for applications requiring a balance between robustness, strength, durability, and environmental responsibility. Based on these results, it is clear that optimizing the performance-environment-economy trade-off is imperative when selecting materials for the electric automotive industry.

Neural networks provide a higher level of predictive accuracy, outperforming many of the approaches described in the literature using conventional methods. Compared with current industrial methods, which are based on empirical experience, this methodology offers a high degree of anticipation capacity and flexibility in terms of mechanical, economic, and environmental criteria.

5 Conclusion

Our study presented a methodology integrating both artificial intelligence and decision support for optimizing the selection of materials used in the design and optimization of electric vehicle chassis. In fact, it was possible to rank materials according to a set of physical, mechanical, economic, and environmental criteria by relying on the creation of a structured and representative database, advanced pre-processing and exploration techniques, and the application of machine learning models. The integration of approaches such as TOPSIS, although not very effective, made it possible to reconcile the search for robustness, lightness, and durability with a view to sustainable engineering. The results of this study confirm the relevance of this approach in guiding the design of high-performance, eco-friendly chassis. However, future work may be expanding the database, integrating criteria related to manufacturing processes, and thus coupling this methodology with multi-physics numerical simulations in order to enhance the reliability and accuracy of the choices made.

References

1. Jacob, S., Thiruvarasan, V., Surendhar, S., Senthamizh, R.: Design, analysis and optimization of all terrain vehicle chassis ensuring structural rigidity. Mater. Today Proc. **46**, 3786–3790 (2021)
2. Scurtu, I.L., Moldovanu, D.: Conceptual design of an electric vehicle chassis using topology optimization method. IOP Conf. Ser.: Mater. Sci. Eng. **1311**(1), 012022 (2024)
3. Kopp, G., Beeh, E., Schöll, R., Kobilke, A., Straßburger, P., Kriescher, M.: New lightweight structures for advanced automotive vehicles – safe and modular. Procedia. Soc. Behav. Sci. **48**, 350–362 (2012)
4. Burd, J. T. J.: Material substitution in electric vehicle manufacturing: comparing advanced high strength steel and aluminum. Master's Thesis, Massachusetts Institute of Technology (2019)
5. Ikpe, A.E., Onhorhoro, E.K., Gobir, A.: Engineering material selection for automotive exhaust systems using CES software. Int. J. Eng. Technol. **3**(2), 50–60 (2017)
6. Bhate, D., Penick, C.A., Ferry, L.A., Lee, C.: Classification and selection of cellular materials in mechanical design: engineering and biomimetic approaches. Designs. **3**(1), 19 (2019)
7. Chingo, C., Martinez-Gómez, J., Narváez, C., R. A.: Material selection using multi-criteria decision-making methods for geomembranes. Int. J. Math. Oper. Res. **16**(1), 24–52 (2020)
8. Roy, P., Tadele, D., Defersha, F., Misra, M., Mohanty, A.K.: Environmental and economic prospects of biomaterials in the automotive industry. Clean Techn. Environ. Policy. **21**, 1535–1555 (2019)
9. Jain, A., et al.: Commentary: the materials project: a materials genome approach to accelerating materials innovation. APL Mater. **1**(1), 011002 (2013)
10. Ramprasad, R., et al.: Machine learning in materials informatics. npj Comput. Mater. **3**(1), 54 (2017)
11. Ashby, M.F., Cebon, D.: Materials selection in mechanical design. J. Phys. IV. **3**(C7), C7-1–C7-9 (1993)
12. Rao, R.V.: A decision-making methodology for material selection using an improved compromise ranking method. Mater. Des. **29**(10), 1949–1954 (2008)
13. Jahan, A., Mustapha, F., Ismail, M.Y., Sapuan, S.M., Bahraminasab, M.: A comprehensive VIKOR method for material selection. Mater. Des. **32**(3), 1215–1221 (2011)
14. Ward, L., Agrawal, A., Choudhary, A., Wolverton, C.: A general-purpose machine learning framework for predicting properties of inorganic materials. npj Comput. Mater. **2**(1), 16028 (2016)
15. Xie, T., Grossman, J.C.: Crystal graph convolutional neural networks for an accurate and interpretable prediction of material properties. Phys. Rev. Lett. **120**(14), 145301 (2018)
16. Zhao, G. et al.: Machine-learning-assisted multiscale modeling strategy for predicting mechanical properties of carbon fiber reinforced polymers. Compos. Sci. Technol. **248**, 110455 (2024)
17. Pilania, G., Wang, C., Jiang, X., Rajasekaran, S., Ramprasad, R.: Accelerating materials property predictions using machine learning. Sci. Rep. **3**, 2810 (2013)
18. Isayev, O. et al.: Materials cartography: representing and mining materials space using structural and electronic fingerprints. Chem. Mater. **27**(3), 735–743 (2015)
19. Faraji, F., Faraji, H.: Optimizing lightweight building construction with artificial intelligence: a focus on efficiency, sustainability, and practical design solutions. In: Conference Paper presented at the International Conference on Urban Planning, Architecture, Civil Engineering and Environment (2025)
20. Raccuglia, P. et al.: Machine-learning-assisted materials discovery using failed experiments. Nature. **533**(7601), 73–76 (2016)

21. Saal, J.E., Kirklin, S., Aykol, M., Meredig, B., Wolverton, C.: Materials design and discovery with high-throughput density functional theory: the open quantum materials database (OQMD). JOM. **65**(11), 1501–1509 (2013)
22. Lundberg, S.M., Lee, S.-I.: A unified approach to interpreting model predictions. Adv. Neural Inf. Proces. Syst. **30**, 4765–4774 (2017)
23. Ribeiro, M.T., Singh, S., Guestrin, C.: "Why should I trust you?": explaining the predictions of any classifier. In: Proceedings of the 22nd ACM SIGKDD International Conference on Knowledge Discovery and Data Mining, pp. 1135–1144. ACM (2016)
24. Zhou, X., Jiang, J., Hu, Z., Hua, L.: Lightweight materials in electric vehicles. Int. J. Automot. Manuf. Mater. **1**, 3 (2022)
25. Cabrera, M., Ninic, J., Tizani, W.: Fusion of experimental and synthetic data for reliable prediction of steel connection behaviour using machine learning. Eng. Comput. **39**(6), 3993–4011 (2023)
26. Chen, C., et al.: ML-based prediction of mechanical performance of hybrid composites in automotive. Compos. B. **250** (2023)
27. Brown, N.K., Garland, A.P., Fadel, G.M., Li, G.: Deep reinforcement learning for engineering design through topology optimization of elementally discretized design domains. Mater. Des. **218**, 110672 (2022)
28. Tran, R. et al.: Surface energies of elemental crystals. Sci. Data. **3**, 160080 (2016)
29. Himenen, L., Geurts, A., Foster, A.S., Rinke, P.: Data-driven materials science: status, challenges, and perspectives. Adv. Sci. **6**(21), 1900808 (2019)
30. Pollice, R. et al.: Data-driven strategies for accelerated materials design. Acc. Chem. Res. **54**(4), 849–860 (2021)

Energy-Efficient Power Allocation for 6G Massive Communications: A Mean-Field Game Analysis

Salma Tika[1(✉)], Essaid Sabir[2], and Abdelkrim Haqiq[1]

[1] Hassan First University of Settat, Faculty of Sciences and Techniques, Computer, Networks, Mobility and Modeling Laboratory: IR2M, 26000 Settat, Morocco
{salma.tika.doc,abdelkrim.haqiq}@uhp.ac.ma

[2] Department of Science and Technology, TÉLUQ, University of Quebec, Montreal H2S 3L4, Canada
Essaid.Sabir@teluq.ca

Abstract. Sixth-generation (6G) wireless networks aim to provide ultra-reliable, low-latency, and energy-efficient communications for massively dense networks. The increasing number of connected devices, however, introduces significant challenges in balancing energy efficiency with the maintenance of Quality-of-Service (QoS). In this paper, we focus on energy-efficient power control in ultra-dense 6G networks using a mean field game (MFG) framework. By modeling a representative IoT device interacting with the average network behavior, we address the transmit power control challenge by formulating it as a differential game and solve it through a finite difference method. Numerical analysis and results confirm that the proposed approach effectively balances energy consumption and transmit performance, providing a scalable solution for massive networks. This work highlights the potential of mean-field game theory as a powerful tool for designing adaptive and energy-aware 6G systems.

Keywords: 6G · Energy-Efficiency · Ultra-Dense Networks (UDNs) · Power Allocation · Mean Field Game (MFG)

1 Introduction

The evolution toward sixth-generation (6G) networks introduces major challenges in balancing high throughput, ultra-low latency, and massive connectivity with sustainability and Energy Efficiency (EE)[1–3]. In ultra-dense networks with massive MIMO and millimeter-wave communication, energy consumption has become a central constraint. Energy Efficiency, expressed as the number of bits transmitted per joule, is now a key performance metric guiding the design of greener architectures, protocols, and hardware solutions.

Over the years, energy efficiency has been improved through network planning, base station sleeping, energy harvesting (e.g., SWIPT), and efficient transmission via beamforming and resource allocation[4,5]. More recently, Artificial

A. E. Hassanien et al. (Eds.): AISDS 2025, LNDECT 286, pp. 182–191, 2027.
https://doi.org/10.1007/978-3-032-18758-1_18

Intelligence (AI) as well as Machine Learning (ML) have been increasingly applied to support adaptive resource allocation, traffic prediction, and real-time decision-making, with approaches such as supervised learning, reinforcement learning, and federated learning showing promising results in power control and computation offloading [6]. However, these methods face challenges in ultra-dense environments due to their computational complexity and limited scalability [7,8].

In this context, Mean-Field Game (MFG) theory has emerged as a promising framework for scalable modeling and optimization of energy-efficient 6G networks [9]. By approximating interactions among a massive number of agents through aggregate behavior, MFG enables tractable yet effective resource management in complex wireless environments [10]. In this study, a MFG model is proposed to address the EE problem in dense small cell deployments. By modeling the interactions among Small Base Stations (SBSs) through the mean-field approximation, we capture the aggregate effect of interference and energy consumption in large-scale deployments.

The rest of this paper is structured as follows: Sect. 2 outlines the system and network models and presents the cost function, the state representation, the action space, and the control policy. Section 3 introduces the mean-field formulation and derives the associated equations. Section 4 provides the numerical analysis using the finite difference method, followed by the numerical results. Section 5 presents the key conclusions of this study.

2 Problem Formulation

2.1 Network Model

The present work addresses an ultra-dense cellular network where SBSs are deployed according to a homogeneous PPP $\Phi_{\text{SBS}} = \{y_1, y_2, \ldots\}$ with intensity λ_{SBS}. Each SBS is connected to multiple IoT devices, with user association determined by proximity to the nearest SBS. The system primarily operates in the uplink domain, where IoT devices belong to closed subscriber groups [11].

Our study focuses on uplink power control for active IoT devices that transmit simultaneously over the same frequency band within a circular region of radius R_{IoT}. In each SBS coverage area, the group of associated active IoT devices is random. To capture this spatial randomness, we adopt a Matérn Cluster Process (MCP), where each cluster center represents an SBS. For a given cluster center y, the number of connected devices M^y, uniformly distributed within a circle of radius R_{SBS}, is assumed to follow a Poisson distribution with mean λ_{m}.

The device-to-SBS distance is characterized by the following density function:

$$f(d) = \frac{2d}{R_{\text{SBS}}^2}, \qquad d \leq R_{\text{SBS}}. \tag{1}$$

The resulting MCP, expressed as $\mathcal{N} = \bigcup_{y \in \Phi_{\text{SBS}}} M^y$, results in a stationary point process with density $\lambda_{\text{SBS}} \lambda_{\text{m}}$. Consequently, the total number of active devices within a circular region of radius R_{IoT}, denoted by $\mathcal{N}_T$, is a random variable with mean $\lambda_{\mathrm{SBS}}\lambda_{\mathrm{m}}\pi R_{\mathrm{IoT}}^2$.

The channels between transmitters and their associated SBSs are assumed to experience i.i.d. Rayleigh fading, and a path loss $l(d) = \min(1, d^{-\alpha})$ where the exponent satisfies $\alpha > 2$. Accordingly, the SINR of IoT device k at time t at its SBS is defined as:

$$\gamma_k(t) = \frac{p_k(t)h_{k,k}(t)l_{k,k}(r)}{I_k(t) + \sigma_0}, \tag{2}$$

with $p_k(t)$ refers to the transmit power of device k, $h_{n,k}(t)$ represents the Rayleigh fading channel gain and $l_{n,k}(r)$ denotes the corresponding path loss function from transmitter n to the SBS associated k, σ_0 is the noise power, while $I_k(t) = \sum_{n=1,n\neq k}^{\mathcal{N}_T} p_n(t)h_{n,k}(t)l_{n,k}(r)$ represents the aggregate interference power plus noise perceived by receiver node k brought by all other transmit nodes in $\mathcal{N}_T$.

2.2 Cost Function

Each IoT device k is powered by a finite battery with an initial energy budget, denoted by e_{max}, to be used within a given period. To avoid energy waste caused by unnecessary retransmissions, device k must guarantee that its $\gamma_k(t)$, at any time t, satisfies the condition:

$$\gamma_k(t) \geq \gamma_{th}, \tag{3}$$

where γ_{th} is the minimum SINR threshold required for successful transmission. In a large-scale IoT deployments, enforcing this strict constraint often proves infeasible, as it is generally impossible for all devices to simultaneously satisfy the SINR requirement due to interference. To address this, we replace the hard constraint (3) by defining a utility function $g_k(\gamma_k(t))$, which characterizes the Quality of Service (QoS) satisfaction of k.

The cost function corresponding to device k at time t is defined as a combination of two functions, given by:

Utility Function $g_k(\gamma_k(t))$ reflecting its satisfaction with the achieved QoS:

$$g_k(\gamma_k(t)) = \log(1 + \gamma_k(t)), \tag{4}$$

$g_k(\gamma_k)$ is modeled as a logarithmic function of the SINR, reflecting its proportionality to the Shannon capacity. The Shannon capacity represents the highest reliable data rate attainable over a communication channel, with a certain level of noise.

Cost Function $C_k(p_k)$ to reflect the energy consumption:

$$C_k(p_k) = c_k p_k(t), \tag{5}$$

where c_k is the pricing coefficient.

For every IoT device k, the objective is to optimize its individual cost function F_k over T, given by:

$$F_k(t) := F_k(t, p_k, p_{-k}) = C_k\left(p_k(t)\right) - g_k\left(\gamma_k(t)\right), \tag{6}$$

Here, p_{-k} denotes the vector of transmit powers corresponding to all devices excluding k.

2.3 Control Policy

For an IoT device k, the state at time t is characterized by its available energy level, denoted as $e_k(t)$. Accordingly, the state space of device k is defined as:

$$\mathscr{E}_k = [0, e_k(0)]$$

with $e_k(0) = e_{max}$ represents its initial energy at $t = 0$. The overall system state at time t is represented by: $e(t) = [e_k(t)\forall k]^T$. The action for the device k corresponds to its transmit power $\mathscr{P}_k = [0, p_{max}]$, constrained by the maximum power p_{max}. We also define the *state equation* of the system as follows.

Definition 1. The evolution of the state of device k, represented by $e_k(t) \in [0, e_{max}]$, is described by the following equation:

$$de_k(t) = -p_k(t)dt, \quad 0 \le t \le T \tag{7}$$

This expression implies that the decrease in energy during an infinitesimal interval dt is proportional to the power consumed for transmission. The *control policy* specifies how the state of a device is mapped to its corresponding action within the finite time horizon T. For player k, the control policy over the interval $[0, T]$ is denoted by $p_k(0 \to T)$. The corresponding optimal power control policy, $p_k^*(0 \to T)$ minimizes the average cost of device k while ensuring efficient energy usage and reliable communication.

3 Mean-Field Formulation

3.1 Power Allocation as a Differential Game

The DG is defined over a time interval, denoted by $\in [0, T]$, where each player selects a transmit power at time t based on both its own state dynamics and the actions of the others, with the objective of minimizing its average cost over the time:

$$\mathscr{U}_k(P(0 \to T)) = \mathbb{E}\left[\int_0^T F_k(t, p_k, p_{-k})dt\right], \tag{8}$$

where $P(0 \to T) := [p_k(0 \to T), p_{-k}(0 \to T)]$ represents the power control policy profile.

To characterize the Nash Equilibrium (NE) in this differential game, we consider the value function u_k given by:

$$u_k(t) = \min_{p_k(t \to T)} \mathbb{E}\left[\int_0^T F_k(s, p_k, p_{-k})\, \mathrm{d}s\right]. \tag{9}$$

This value function is characterized by the HamiltonJacobiBellman (HJB) (10), which must be satisfied for each IoT device $k \in \mathcal{N}_T$:

$$\frac{\partial u_k(t)}{\partial t} + H_k\left(e_k(t), \frac{\partial u_k(t)}{\partial e_k}\right) = 0, \tag{10}$$

where $H_k(x, y) = \min_{p_k(t)}\left[F_k(t, x) - p_k(t)\frac{\partial u_k(t)}{\partial e_k}\right]$ is the Hamiltonian.

Definition 2. We define $P^*(0 \to T) := \left[p_k^*(0 \to T), p_{-k}^*(0 \to T)\right]$ as a NE of the DG under the condition that, for every IoT $\forall k \in \mathcal{N}_T$:

$$p_k^*(0 \to T) = \min_{p_k(0 \to T)} \mathscr{U}_k\left(p_k, p_{-k}^*\right) \tag{11}$$

However, in massive IoT networks, solving the large number of coupled HJB equations to find the Nash equilibrium is too complex. Therefore, a MFG formulation reduces the problem to just two coupled equations.

3.2 Mean-Field Regime

The mean-field framework relies on the following set of assumptions:

- **Large population assumption:** It is assumed that the device population in the network is extremely large, enabling the use of the mean-field approximation.
- **Exchangeability:** This property is guaranteed under the following conditions:
  1. Each IoT device has knowledge only of its own local state (e.g., remaining energy).
  2. Homogeneous power transmission:
  $$p_k(t) = p(t, e_k).$$
  3. $\mathbb{E}\left[\int_0^T p(t, e_k(t)\, dt)\right]^2 < \infty.$

First, We begin by introducing the mean-field distribution $m(t, s)$ as:

$$m(t, e) = \lim_{\mathcal{N}_T \to \infty} \frac{1}{\mathcal{N}_T} \sum_{k=1}^{\mathcal{N}_T} \mathbf{1}_{\{e_k(t) = e\}}, \tag{12}$$

where $\mathbf{1}_{\{.\}}$ indicates the indicator function, equal to 1 when $e_k(t) = e$ is satisfied, and 0 otherwise.

Proposition 1. *In ultra-dense IoT networks, the finite mean-field interference can be expressed as [9]:*

$$I^{MF}(t) = 2\pi\lambda_{SBS}\left[\frac{1}{2} + \frac{1 - R_{IoT}^{2-\alpha}}{\alpha - 2}\right] P^{MF}(t), \tag{13}$$

where $P^{MF}(t) = \int_0^{e_{\max}} p(t,e)m(t,e)\,de$.

Next, we define the mean-field SINR and the associated utility function, both relying on the transmit power of a representative IoT device considered jointly with the mean-field interference.

Mean-Field SINR: is introduced and expressed as:

$$\gamma^{MF}(t) = \frac{p(t,e)R_0}{I^{MF} + N_0}, \tag{14}$$

where $R_0 = \frac{1}{R_{SBS}^2} + \frac{2(1-R_{SBS}^{2-\alpha})}{R_{SBS}^2(\alpha-2)}$.

Mean-Field Utility Function: captures the utility achieved by a representative device in the mean-field regime, it is expressed as:

$$\mathscr{U}^{MF} = \int_0^{e_{max}} \int_0^{T} F^{MF}(t,E)\, m(t,e)\, \mathrm{d}e\, \mathrm{d}t \tag{15}$$

Mean-Field Cost Function: The cost function of a representative IoT device is:

$$F^{MF}(t,e) = cp(t,e) - g\left(\gamma^{MF}(t)\right) \tag{16}$$

3.3 Mean-Field Equations

The mean-field power control problem is characterized by a system of two coupled equations: (HJB) equation and the (FPK) equation:

The HJB Equation characterizes the backward evolution of the value function, which represents the optimal cost-to-go for a representative IoT device, expressed as:

$$\frac{\partial u}{\partial t} + H\left(e, \frac{\partial u(t,e)}{\partial e}\right) = 0 \tag{17}$$

The FPK Equation describes the forward-time evolution of the states of all players. Specifically, it describes how $m(t,e)$ evolves based on the collective transmission strategies of all devices:

$$\frac{\partial m(t,e)}{\partial t} + \frac{\partial}{\partial e}(m(t,e)p(t,e)) = 0 \tag{18}$$

4 Numerical Analysis

4.1 Finite Difference Method

To achieve the Mean-Field Equilibrium (MFE), the FPK equation (18) must be solved jointly with the HamiltonJacobiBellman (HJB) equation (17). The resulting system of coupled PDEs is:

$$\begin{cases} \partial_t u + H\left(e, \partial_e u\right) = 0 \text{ in} & [0,T]\times[0,e_{\max}] \\ \partial_t m - \partial_e(mp) = 0 \text{ in} & [0,T]\times[0,e_{\max}] \\ u(T,.) = 0, m(0,.) = m_0, \quad \displaystyle\int_0^{E_{\max}} m(e,.)\mathrm{d}e = 1. & \end{cases} \tag{19}$$

A finite difference method is adopted to solve this system. The algorithm is guaranteed to converge to the MFE only if the F^{MF} is convex.
In this approach, the time horizon $[0,T]$ and the state space $[0,e_{\max}]$ are discretized into a grid of $X\times Y$ points, where $\Delta t = \frac{T}{X}$, $t_i = i\Delta t$ for $i = 0,1,\ldots,X$, and $\Delta e = \frac{e_{\max}}{Y}$, $e_j = j\Delta e$ for $j = 0,1,\ldots,Y$.

Forward Equation Solution: This is achieved through an iterative first-order finite difference scheme, given by:

$$m_j^{i+1} = m_j^i + \frac{\Delta t}{\Delta e}\left(p_{j+1}^i m_{j+1}^i - p_j^i m_j^i\right). \tag{20}$$

Backward Equation Solution: Solving the HJB equation is achieved by reformulating it as an optimal control problem, with the forward equation as a constraint. The associated optimization problem can be formulated as follows:

$$\begin{cases} \displaystyle\min_{p(t,e),m(t,e)} \int_0^T \int_0^{e_{\max}} F^{MF}(t,e)m(t,e)\,\mathrm{d}e\,\mathrm{d}t \\ \text{subject to} \\ \dfrac{\partial m(t,e)}{\partial t} - \dfrac{\partial}{\partial e}(m(t,e)p(t,e)) = 0 \quad [0,T]\times[0,e_{\max}] \\ m(0,.) = m^0, \quad \displaystyle\int_0^{e_{\max}} m(e,.)\mathrm{d}e = 1. \end{cases} \tag{21}$$

For the considered problem, the Lagrangian $L(p(t,E), m(t,E), \nu(t,E))$ is formulated using the Lagrange multiplier $\nu(t,E)_{\forall t,e}$, and can be expressed as:

$$L(p,m,\nu) = \int_0^{e_{\max}} \int_0^T F^{MF} m(t,e)\,\mathrm{d}e\,\mathrm{d}t + \int_0^{e_{\max}} \int_0^T \nu\left[\partial_t m - \partial_e(mp)\,\mathrm{d}e\,\mathrm{d}t\right] \tag{22}$$

The discretized version of the Lagrangian L_{dis} is expressed as:

$$L_{dis}(p,m,\nu) = \sum_{i,j} m_j^i F_j^i + \sum_{i,j} \nu_j^i \left[\frac{m_j^{i+1} - m_j^i}{\Delta t} - \frac{p_{j+1}^i m_{j+1}^i - p_j^i m_j^i}{\Delta e}\right], \tag{23}$$

where ν_j^i represents the Lagrange multiplier, and F_j^i denotes the cost function evaluated at the discretized grid point (i,j).

Optimality of the decision variables (p^*, m^*, ν^*) is obtained by satisfying the KKT conditions. At each discretized grid point (i, j), the optimal solution should satisfy:

- $\frac{\partial L_{\text{disc}}}{\partial m_j^i} = 0,$

$$\nu_j^{i-1} = \nu_j^i - \frac{\Delta t}{\Delta e} p_j^i \left[\nu_j^i - \nu_{j-1}^i\right] - F_j^i \Delta t \tag{24}$$

- $\frac{\partial L_{disc}}{\partial p_j^i} = 0,$

$$m_j^i \frac{\partial F_j^i}{\partial p_j^i} + \frac{m_j^i}{\Delta e} \left[\nu_j^i - \nu_{j-1}^i\right] = 0 \tag{25}$$

Equations (20), (24), and (25) can be addressed through an iterative procedure, repeated until convergence is achieved. The step-by-step procedure for seeking the mean-field equilibrium is summarized in Algorithm 1:

Algorithm 1. Iterative Seeking of Mean-Field Equilibrium

Initialization:

1. Start with a randomly generated initial power allocation P_0;

Learning pattern:

2. Compute the distribution m using equation (20), subject to m_0;
3. Evaluate the MF interference I^{MF} according to Proposition 1;
4. Find ν using (24) with the final condition;
5. Adjust power allocation P using (25);
6. Repeat steps 25 until the iterative process converges to the MFE;

4.2 Numerical Results

In this section, We evaluate the performance of (1) using numerical analysis. For this purpose, the discretized grid is set to X = 100 and Y = 40 forming the discretized space X × Y, with the following parameters: α = 3, σ_0= 10^{-14}, c= 0.4, p_{max} =0.1 W, T = 1 s, R_{SBS}= 40, λ_{SBS}= 500 SBS/Km^2 and e_{max}= $0.2J$. In ultra-dense IoT scenarios, numerous devices attempt to access the network and transmit data simultaneously, and the initial energy distribution m_0 is considered uniform. Figure 1 illustrates the distribution of the mean field over time and energy levels with a uniform initial state. As time progresses, the proportion of IoT devices characterized by high energy budgets gradually decreases, while the proportion of energy-depleted devices increases. Due to the cost function, devices with larger initial reserves conserve part of their energy instead of transmitting at full power. This approach allows devices to implicitly coordinate their energy usage by responding to the average network behavior.

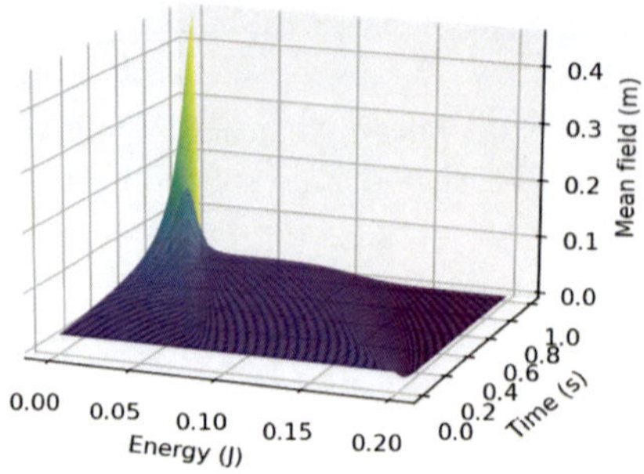

Fig. 1. Mean-field behavior at equilibrium with a uniform m_0

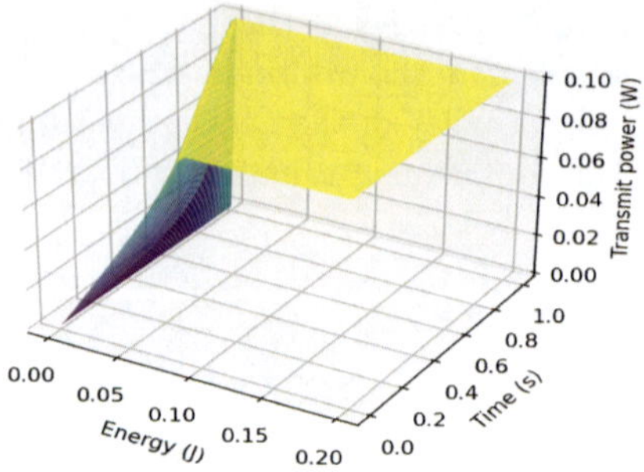

Fig. 2. Time and energy dependent optimal power control

Figure 2 illustrates the optimal power control over time and energy. It can be observed that devices progressively converge toward a common transmission power. Devices with limited remaining energy transmit conservatively at lower power levels, while those with higher initial energy transmit at higher power, constrained by the maximum allowable limit. This adaptive behavior is a direct result of the mean-field interaction. Each device reacts to the aggregate interference level rather than to individual actions, leading the overall system toward a stable and scalable equilibrium. Without the mean-field approach, devices act selfishly, causing rapid energy consumption, unstable power allocation, and higher interference, whereas the mean-field framework ensures coordinated, fair, and energy-efficient transmission.

5 Conclusion

This paper investigated energy-efficient power control in ultra-dense 6G networks using a MFG framework. By modeling a representative IoT device interacting with the average network behavior, we derived the coupled HJBFokkerPlanck system and solved it using a finite difference approach. Numerical results demonstrate that the proposed method effectively balances transmit power and energy consumption while ensuring network-wide performance. The mean-field formulation provides a scalable solution for massive networks, highlighting its potential for practical 6G deployment. Future work may extend this framework to multicell scenarios and incorporate learning-based strategies for real-time adaptation.

References

1. Kitanov, S., Petrov, I., Nikolić, V., Janevski, T.: Energy efficiency in 6G mobile networks. J. Electr. Eng. Inf. Technol. **9**, 33–41 (2024). https://doi.org/10.51466/JEEIT2491217033k
2. Li, L.: Research on future 6G Green Wireless Networks. Green Technol. Sustain. **3**(2), 100156 (2025). https://doi.org/10.1016/j.grets.2024.100156

3. Alhussien, N., Aaron Gulliver, T.: Toward AI-enabled green 6G networks: a resource management perspective. IEEE Access **12** (2024). https://doi.org/10.1109/ACCESS.2024.3460656
4. Gandotra, P., Jha, R.K., Jain, S.: Green communication in next generation cellular networks: a survey. IEEE Access **5**, 11727–11758 (2017). https://doi.org/10.1109/ACCESS.2017.2711784
5. Mughees, A., Tahir, M., Sheikh, M.A., Ahad, A.: Energy-efficient ultra-dense 5G networks: recent advances, taxonomy and future research directions. IEEE Access **9** (2021). https://doi.org/10.1109/ACCESS.2021.3123577
6. Yang, H., Alphones, A., Xiong, Z., Niyato, D., Zhao, J., Wu, K.: Artificial intelligence-enabled intelligent 6G networks. IEEE Netw. **34**(6), 272–280 (2020). https://doi.org/10.1109/MNET.011.2000195.1912.05744 [cs]
7. Noman, H.M.F., et al.: Machine learning empowered emerging wireless networks in 6G: recent advancements, challenges and future trends. IEEE Access **11**, 83017–83051. https://doi.org/10.1109/ACCESS.2023.3302250
8. Mabrouk, A., Kobbane, A., Sabir, E., Ben-Othman, J., El Koutbi, M.: Meeting always-best-connected paradigm in heterogeneous vehicular networks: a graph theory and a signaling game analysis. Veh. Commun. **5**, 1–8 (2016). https://doi.org/10.1016/j.vehcom.2016.04.002
9. Nadif, S., Sabir, E., Elbiaze, H., Haqiq, A.: Green grant-free power allocation for ultra-dense Internet of Things: a mean-field perspective. J. Netw. Comput. Appl. **229**, 103908 (2024) https://doi.org/10.1016/j.jnca.2024.103908
10. Lasry, J.-M., Lions, P.-L.: Mean field games. Jpn. J. Math. **2**, 229–260 (2007). https://doi.org/10.1007/s11537-007-0657-8
11. Nadif, S., Sabir, E., Elbiaze, H., Haqiq, A.: Traffic-aware mean-field power allocation for ultradense NB-IoT networks. IEEE Internet Things J. **9**(21), 21811–21824 (2022). https://doi.org/10.1109/JIOT.2022.3182854

Scalable Identification of Influential Nodes: XGBoost-Based Closeness Centrality Approximation

Hicham El Bousty(✉) and Issam Qaffou

Cadi Ayyad University, Marrakech, Morocco
{h.elbousty.ced,qaffou}@uca.ac.ma

Abstract. We propose a machine learning methodology that can identify the most central nodes in networks without computing full closeness centrality. We only used 20% of the nodes for training, and based our XGBoost classifier on PageRank, Two Hop density and Louvain communities' membership to identify potential high-centrality nodes. We validated our model using five networks, and found excellent results (mean F1-score = 0.88, balanced accuracy = 0.94) suggest that we can precisely rank influential nodes without needing the expensive shortest-path computations characteristic of classical measures. The high balanced accuracy suggests the classifier had discriminative performance across the two classes, while the mean F1-score indicates an overall robustness between precision and recall. We believe that our approach provides two performance benefits: (1) scalability, since we used only 20% of node-based topology, (2) generalizability, since the classifier performed comparably across disparate networks. Our approach will be particularly beneficial in large-scale or dynamic networks where traditional measures, such as centrality, cannot be feasibly computed while achieving an acceptable trade-off with 94% accuracy and computational cost.

Keywords: Closeness Centrality · Influential Nodes · Machine learning · Network Topology · Computational Efficiency · Scalable Network Analysis

1 Introduction

The identification of influential nodes can be beneficial in many scientific and applied fields, allowing researchers to target interventions where they will have the most overall effect on the network with the least amount of resource expenditure. Recent advancements in network science have shown that systematic identification of influential nodes can increase information diffusion in social networks by 88.6% compared to random targeting [14]. Closeness centrality is one of methods that can identify efficiently influential nodes. It quantifies the ease with which a node can connect to others via shortest-possible paths. Targeting vertices that have higher closeness centrality can expedite the flow of information in social networks [15] and decrease the length of time delays in transportation systems [6]. The core idea of closeness centrality in tracking social contagions, identified by epidemiologists, as highly central nodes, that will ultimately accelerate

A. E. Hassanien et al. (Eds.): AISDS 2025, LNDECT 286, pp. 192–201, 2027.
https://doi.org/10.1007/978-3-032-18758-1_19

an epidemic because of super-connection [4]. Closeness centrality is also helpful to organizations, allowing management to find employees who can most effectively communicate and share information [11]. Centrality measures have also been used to cluster documents in information retrieval and question-answering systems [7].

Closeness centrality provides interesting information about how influential a node can be, but there are important limitations in applying it to today's modern large-scale networks. The central challenge is computational complexity. Calculating the exact closeness centrality of nodes requires calculating the shortest paths between all pairs of nodes, which results in O(nm) time complexity for implementations using continued iterations of the Breadth-First Search (BFS) algorithm, or O(n(m+n log n)) when using Dijkstra's algorithm. This quadratic-to-cubic scaling makes it computationally unviable (at least, not without a super-computer) to calculate closeness centrality for networks that contain millions or billions of nodes, such as social media networks or internet graphs as seen today [2].

Approaches based on approximating centrality, such as landmark-based sampling [5] or parallel algorithm-based approaches [12], offer partial solutions to the computational challenge, but these approaches introduce varying accuracy drawback in comparison to using exact computations [13]. In light of these challenges, interest has developed in alternative centrality metrics and machine learning methods that may yield similar interpretations on networks at web-scale while exhibiting better computational characteristics; Thus, our goal is to develop a Machine Learning approach, balancing speed and precision, for efficiently identifying influential nodes through the prediction of Closeness Centrality scores, using topological features.

To this end, we propose a machine learning framework that addresses these challenges through three specific contributions: (1) an XGBoost classifier that predicts nodes as high-closeness using only 20% of the nodes for training, (2) a feature engineering approach that incorporates candidate local topological properties (two-hope density) and global centrality properties (PageRank) to encapsulate influence patterns across scales, and (3) validation of the model with numerous real-world and synthetic datasets, showing that the framework consistently performs with a 0.8 F1-score across different network sizes and community structures.

This paper thoroughly examined our machine learning framework to predict high closeness centrality. Section 2 sets the context for the reader, initially placing the problem in context, then discusses our feature engineering and finally our pipeline. In Sect. 3, we defined the characteristics of the datasets and the hyperparameters of the model. In Section 4, we summarized and explained our experimental results. Overall, these sections have demonstrated the predictive potential of our model and that our methods are practical and scalable methods for practical real-world use.

2 Methodology Design

2.1 Problem Formulation

We converted the computationally expensive closeness centrality problem (O(nm) complexity) to a binary classification problem, by labeling nodes as high-Closeness centrality nodes/Influents (top 9%) or low-Closeness centrality nodes/non-Influents (bottom 91%)

utilizing exact closeness values. We set the classification threshold at the 91st percentile P_{91}. Formally, for node v:

$$y(v) = \begin{cases} 1 \ if \ C(v) \geq P_{91} \\ 0 \quad \text{otherwise} \end{cases}$$

The 9% threshold was selected because it provides sufficient positive examples (9:91 class ratio) for effective model training while maintaining real-world relevance

This approach has three benefits: (1) no need to compute shortest-path distance every time you deploy the model, (2) high-centrality nodes are interpretable providing clear insight into their potential influence within the network (3) cutoff value for various precision/recall requirements can be easily created.

We used stratified sampling to ensure the 9:91 class distributions were maintained in all data splits and that the training and validation set contained a representative sample of the data.

2.2 Features Engineering

Closeness centrality is a basic metric used in the analysis of networks that measures the ability of a node to spread information to all other nodes in a network [9]. Predicting this measure requires careful selection of features, which should encompass many aspects of the position and power of a node in relation to other nodes in the network. The following six features were considered for selection due to their theoretical and empirical significance in indicating node properties that relate directly to closeness centrality.

Topological Features

Degree Centrality. Degree centrality quantifies direct connections, and generally degree centrality will rank high-degree nodes near a network's center. Because closeness depends on the highest number of short path lengths, well connected nodes will tend to have shorter short path lengths to other nodes, thereby increasing closeness centrality.

Louvain Community. Is an algorithm extensively used for detecting communities within networks, specifically large networks by maximizing the modularity measure. Although community detection does not directly reflect whether a node can efficiently reach another node, the characteristic of a node's community is fundamentally related to its Closeness Centrality.

*Local Density***.** Local density assesses connection density of a node's neighborhood. If density is high, it may imply tight clusters and short internal pathways with greater likelihood of improving closeness centrality, provided the cluster is broadly central.

Triangles counts. indicate local clustering, which often correlates with high closeness centrality. Nodes embedded in many triangles typically have shorter paths to neighbors, accelerating information spread. This local connectivity mirrors closeness centrality's reliance on shortest paths, making triangles a useful proxy.

Two-Hop Density. Two-hop density is how many nodes can be reached by traversing the network in two steps from a node. Bigger Two-Hop Density value means that a node will be able to reach a large part of the network quickly over a very short distance, which contributes directly to closeness overall.

PageRank. Is a centrality measure that was originally created to rank web pages; however, it evaluates the relative importance of a node within its network in terms of the number and quality of incoming links [8]. In terms of Closeness Centrality, a node with high PageRank implies an influential access point in the network that could quickly reach others.

Features Correlation. To allow for a thorough investigation of potential multicollinearity amongst the final selection of the network features (i.e., degree centrality, local density, and other relevant measures), we began with an overall Pearson correlation analysis. This statistical test assesses the size and direction of linear relationships between variables. This information can help identify interdependencies between features which could compromise the predictive ability of the classification model.

The correlation analysis found a strong linear relationship between PageRank, Degree Centrality and Triangles counts ($r > 0.85$), demonstrating considerable redundancy. To further optimize feature space, two would have to be eliminated, typically Degree Centrality and Triangles counts, because PageRank includes influence of the neighbors (Fig. 1).

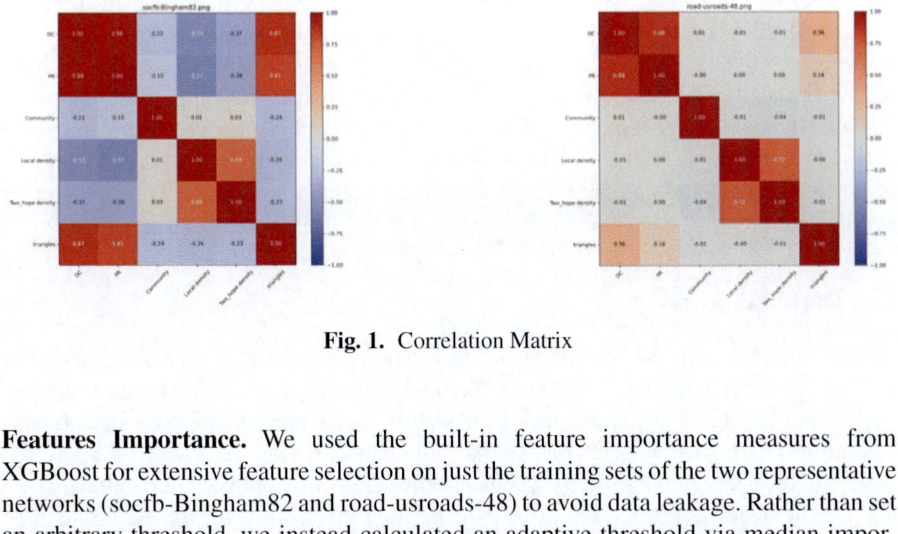

Fig. 1. Correlation Matrix

Features Importance. We used the built-in feature importance measures from XGBoost for extensive feature selection on just the training sets of the two representative networks (socfb-Bingham82 and road-usroads-48) to avoid data leakage. Rather than set an arbitrary threshold, we instead calculated an adaptive threshold via median importance value across all features. In this case, we were able to identify three consistent features: PageRank (a global measure of influence), Louvain community membership (a meso-scale structural role), and 2-hop density (connectivity in the local neighborhood) Fig. 2.

2.3 Algorithm Selection

XGBoost (Extreme Gradient Boosting) is a well-known machine learning algorithm, with proven usefulness for classification problems. XGBoost was developed by Chen and Guestrin [3] as a scalable and optimized version of the gradient boosting framework.

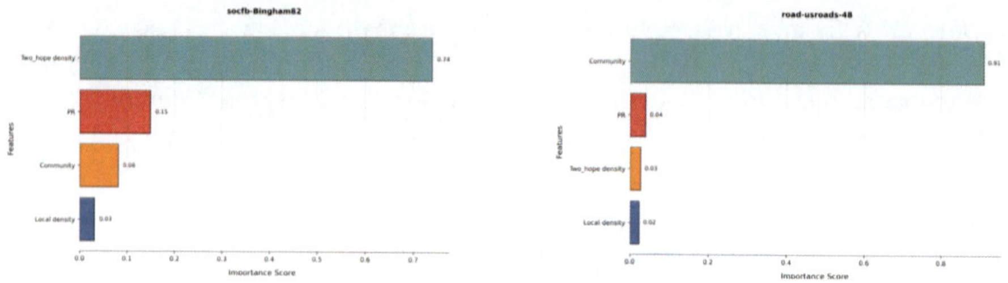

Fig. 2. Features importance

The algorithm builds an ensemble of decision trees by adding trees sequentially; each new tree is trained on the residuals of the prediction errors made by the ensemble of previous trees.

We chose to work with XGBoost for these graph-based classification tasks for many reasons. Its algorithm captures complex interactions between features since the behavior of a node might not stem from a single network property, but from a combination of multiple properties in the network. In practice, XGBoost frequently achieves state-of-the-art results on tabular datasets which is encouraging when graph metrics serve as features [1]. XGBoost is also efficient and highly scalable, especially when you have to process a huge graph with a multitude of nodes where many metrics could have been calculated. XGBoost offers a strong high-performance baseline or alternative to GNNs, especially when a number of sufficiently rich and informative node features are available. In practice, XGBoost has subjectively demonstrated success across a number of applications, including classifying nodes in blockchain transaction graphs based on engineered features [10].

2.4 Overall Pipeline

We developed an efficient machine learning pipeline that reframed closeness centrality estimation as a binary classification problem where nodes were labeled high-centrality (top 9%) or low-centrality (bottom 91%) based on a set of pre-computed exact closeness centralities. Our starting point was six possible features: PageRank, degree centrality, triangles, local density, two-hop density, and Louvain community. After following the feature engineering process, we arrived at the final three most discriminative features by using correlation analysis and an importance-based ranking by XGBoost. The final model consisted of PageRank, two-hop density, and Louvain community. After feature subset selection, we used our three features in conjunction to create a binary classifier for high-centrality nodes, using XGBoost on 20% of nodes (stratified sample) which we could then apply to the other 80% (Fig. 3).

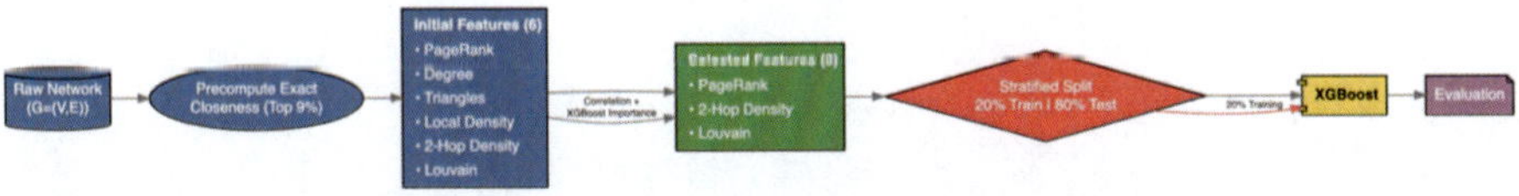

Fig. 3. Overall pipeline of the proposed framework

3 Experimental Setup

3.1 Datasets

To thoroughly evaluate model scalability and generalizability, we performed a set of experiments on five benchmark graphs: 3 real networks (social, collaboration, and infrastructural) and two synthetic graphs under systematic variation of graph structure. We identified our datasets to include:

Table 1. Network Datasets: Number of nodes, Density and Average Degree

Network	Nodes	Density	Average degree (k)	Description
socfb-Bingham82	10001	7,26E-03	128	*Facebook social network* (university-scale).
ca-AstroPh	17903	1,23E-03	78	*Collaboration network* (arXiv astrophysics).
road-usroads-48	126146	2,04E-05	3	*US highway network.*
30k_ graph	30000	7,00E-03	211	Extreme-degree synthetic graph
250k_graph	250000	2,03E-05	6	Large synthetic sparse graph

3.2 Model Configuration and Hyper Parameters

The model architecture was specifically configured to handle our binary classification problem of identifying high-closeness centrality nodes. The model was trained with the following optimized hyperparameters Fig. 4. The 0.3 learning rate and maxdepth = 5 with 200 trees is a very efficient combination of hyperparameters which trains fairly quickly without taking up too much memory, which allows obtaining reasonable output without costing a significant number of computational resources compared with deeper trees or more estimators.

```
params = {
        'colsample_bytree': 0.8,
        'gamma': 0.1,
        'learning_rate': 0.3,
        'max_depth': 5,
        'n_estimators': 200,
        'reg_alpha': 0.1,
        'reg_lambda': 10.0,
        'subsample': 1.0,
        'scale_pos_weight': ratio,
        'random_state': random_state,
        'eval_metric': 'logloss'
```

Fig. 4. XGboost Hyper parameters

3.3 Evaluation Metrics

Since the influent nodes is 9% of the total nodes in each dataset, we have created a class imbalance. As a result, we focused our evaluation on F1-score (1) and balanced accuracy (2) in order to fairly evaluate model performance. In general, both metrics convey complementary properties of robustness when classifying. The F1-score is particularly useful for highly imbalanced datasets, and it penalizes a model that is overly focused on maximizing precision at the expense of recall (or vice versa). In other words, balanced accuracy is the arithmetic mean of a model's per-class recall scores. It helps resolve class imbalance, as it treats all classes equally While F1-score is preferred when false negatives and positives are of similar consequence (e.g. spam detection), balanced accuracy is preferred when we care equally about all classes regardless of sample size (e.g. disease sub type classification).

During the training phase, we prioritize maximizing the F1 score while maintaining strong balanced accuracy.

$$F1 = 2 * \frac{Precision * Recall}{Precision + Recall} \tag{1}$$

$$Balanced\ Accuracy = \frac{1}{C}\sum_{c=1}^{C} Recall_c \tag{2}$$

4 Results and Discussion

4.1 Results

All five networks Table.1, displayed consistent performance across the model with a balanced accuracy ranging from 0.9295 (250k_graph) to 0.9452 (30k_graph) and F1-scores between 0.8273 (road-usroads-48) and 0.9042 (30kgraph). The two highest scores for F1 were obtained for both 30kgraph (0.9042) and socfb-Bingham82 (0.8919), with the two networks possessing the highest density values (>7.00E-03) and the largest average degrees (211 and 128, respectively). We can see that the performance for ca-AstroPh (F1 = 0.8893, average degree = 78) and 250kgraph (F1 = 0.8660) was slightly lower but remained a strong performance compared to road-usroads-48, which had the worst performance (F1 = 0.8273). Notably, road-usroads-48 and 250kgraph had nearly identical network densities (2.04E-05 and 2.03E-05, respectively), yet 250k_graph outperformed road-usroads-48 (F1 = 0.8660 vs. 0.8273). This difference can be attributed to their average degrees: 250k_graph had a higher average degree (6 vs. 3), suggesting that node connectivity was as influential as density in model performance. Overall, precision had the most variation across networks (range = 0.1633), balanced accuracy had the least variation across networks (range = 0.0157). Recall values across all networks remained high (>0.86) with road-usroads-48 (0.9174) achieving the highest recalled value even it had the lowest F1-score (Fig. 5 and Table 2).

Table 2. Model Performance Across Five Graph Datasets

Graph	Balanced_Accuracy	precision	recall	F1
socfb-Bingham82	0,9301	0,9167	0,8684	0,8919
ca-AstroPh	0,9392	0,8886	0,8900	0,8893
road-usroads-48	0,9430	0,7534	0,9174	0,8273
30k_graph	**0,9452**	0,9084	0,9000	**0,9042**
250000_graph	0,9295	0,8578	0,8743	0,8660

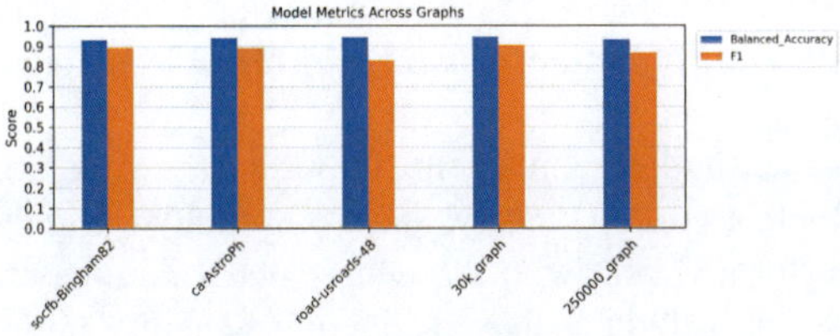

Fig. 5. Performance Across Five Graph Datasets

4.2 Discussion

The performance patterns uncover multiple important aspects of the model's performance across various types of networks:

Generalizability Across Network Types. The model shows an impressive ability to generalize across a variety of structures, from social networks (socfb-Bingham82) to infrastructural networks (road-usroads-48) and synthetic graphs. The consistent, high level of performance (Balanced Accuracy: 0.9295–0.9452) indicates that our feature set captures fundamental topological characteristics. It appears as though the centrality characteristics are useful across domains, regardless of the network type. Specifically, it seems that both scale-free networks (which are typical for social systems) and more uniform degree distributions (such as road networks), are well-captured by the same approach.

Scalability Information. The 250k_graph performing better than road-usroads-48 (despite both have similar density) indicates that size is a substitute for sparsity. The sheer number of nodes within 250k_graph realizes more topological information which allowed the model to learn a more useful pattern, than it did for road-usroads-48.

Effectiveness of Our Features. The strong performance on the social networks and collaboration networks (socfb-Bingham82 and ca-AstroPh) allows us to have confidence in our feature engineering, considering these two networks should characteristically show mixing of global and local centrality patterns that we were hoping to capture with PageRank, two-hop density and Louvain community features.

Overall, these results indicate that our machine learning framework is successfully approximating closeness centrality across many types of networks, and reveals interesting advantages.

5 Conclusion

In this research, we have presented a new machine learning approach for identifying important nodes in complex networks by treating centrality as a holistic phenomenon. We developed a method that reframed the closeness centrality calculation to a binary classification problem together with topological features. The framework was able to identify important nodes across a wide range of networks, from relatively dense social networks and a sparse infrastructure network, with stable predictions (F1 score and Balanced Accuracy).

Part of the model's strength is representing important network properties in three critical features: PageRank, Two-hop density, and Community using the Louvain method. The new features performed best with the denser networks, but still provided results with the sparse networks, as long as they used a representative number of nodes.

Future work may include expanding the feature space to include many other metrics or measures from networks to increase predictive power, particularly with sparse networks. Our experience with this study should be coupled with consideration to time complexity, and a comparison to the costs that are associated with traditional methods. Given the successes in the framework suggest it might be extended to other network analytical processes that researchers/practitioners may require performant and interpretable results.

This work also adds to the field of network science, as it demonstrates an example of how machine learning can produce meaningful structural insight while still being domain-related and capable of accepting the need for interpretability. For this reason, the model and method could be a useful tool across any type of network analysis, including, but not limited to, examining social influence, and identification of critical aspects of infrastructure networks, even when available network data is limited.

References

1. Bentéjac, C., Csörgő, A., Martínez-Muñoz, G.: A comparative analysis of XGBoost. arXiv preprint. https://arxiv.org/abs/1911.01914 (2019)
2. Bisenius, P., Bergamini, E., Angriman, E., Meyerhenke, H.: Computing top-k closeness centrality in fully dynamic graphs. arXiv preprint. arXiv:1710.01143v1 (2017)
3. Chen, T., Guestrin, C.: XGBoost: a scalable tree boosting system. In: Proceedings of the 22nd ACM SIGKDD International Conference on Knowledge Discovery and Data Mining, pp. 785–794. ACM, New York (2016). https://doi.org/10.1145/2939672.2939785
4. Christley, R.M. et al.: Infection in social networks: using network analysis to identify high-risk individuals. Am. J. Epidemiol. **162**(10), 1024–1031 (2005). https://doi.org/10.1093/aje/kwi308
5. Cohen, E., Delling, D., Pajor, T., Werneck, R.F.: Computing classic closeness centrality at scale. SIAM J. Comput. **43**(2), 217–238 (2014) https://doi.org/10.48550/arXiv.1409.003
6. Crucitti, P., Latora, V., Porta, S.: Centrality measures in spatial networks of urban streets. Phys. Rev. E. **73**(3), 036125 (2006). https://doi.org/10.1103/PhysRevE.73.036125
7. Edress, Z., Al-Zoubi, A.: A novel centrality-driven clustering approach for information retrieval and question answering. IEEE Access. **12**, 12345–12360 (2024). https://doi.org/10.1109/IDAP64064.2024.10711092
8. Evans, T.S., Chen, B.: Linking the network centrality measures closeness and degree. Commun. Phys. **5**(1), 172 (2022). https://doi.org/10.1038/s42005-022-00949-5
9. Freeman, L.C.: Centrality in social networks: conceptual clarification. Soc. Networks. **1**(3), 215–239 (1978). https://doi.org/10.1016/0378-8733(78)90021-7
10. Fu, B., Wang, Y., Feng, T.: CT-GCN+: a high-performance cryptocurrency transaction graph convolutional model for phishing node classification. Cybersecurity. **7**(1), 3 (2024). https://doi.org/10.1186/s42400-023-00194-5
11. Hansen, D.L., Shneiderman, B., Smith, M.A.: Calculating and visualizing network metrics. In: Analyzing Social Media Networks with NodeXL, pp. 69–78. Morgan Kaufmann, Burlington (2011). https://doi.org/10.1016/B978-0-12-382229-1.00005-3
12. Kang, U., Lee, J.-Y., Koutra, D., Faloutsos, C.: Net-Ray: visualizing and mining billion-scale graphs. IEEE Trans. Parallel Distrib. Syst. **30**(1), 1–14 (2019)
13. Matta, J., Spielman, S., Tuccillo, J.: Comparing the speed and accuracy of approaches to calculating betweenness centrality in large networks. Comput. Soc. Netw. **6**(1), 2 (2019). https://doi.org/10.1186/s40649-019-0062-5
14. Olanrewaju, A.A.-S., Ahmad, R., Mahmudin, M.: Influence maximization towards target users on social networks for information diffusion. In: Proceedings of the International Conference Reliable Information and Communication Technology, pp. 1–10. Springer, Cham (2017). https://doi.org/10.1007/978-3-319-59427-9_87
15. Valente, T.W.: Network interventions. Science. **337**(6090), 49–53 (2012). https://doi.org/10.1126/science.1217330

Comparison of LSTM and CNN Models for SOC Estimation in Lithium Polymer Batteries

O. Lammamri(✉), B. Zraibi, M. Mansouri, A. Ezzahi, and A. Afadiss

Hassan First University of Settat, National School of Applied Sciences of Berrechid, Laboratory Lamsad, Berrechid, Morocco
omar.lammamri.doc@uhp.ac.ma

Abstract. State of Charge (SOC) estimation is a key function of Battery Management Systems (BMS) in electric vehicles, ensuring safe operation, optimal performance, and prolonged battery life. This study investigates the application of deep learning models specifically Long Short-Term Memory networks and Convolutional Neural Networks to estimate the SOC of lithium polymer batteries. The two models are evaluated based on architectural design, training complexity, and predictive accuracy under varying operating conditions. LSTM models are designed to capture long-term temporal dependencies in battery behavior, while CNN models are optimized for local feature extraction and noise reduction in time-series data. The dataset used for training and testing includes real sensor measurements of voltage, current, and temperature. The analysis relies on three metrics: Mean Absolute Error (MAE), Root Mean Square Error (RMSE), and the coefficient of determination (R^2) to measure performance. Results indicate that CNN achieves slightly better prediction accuracy and generalization capabilities compared to LSTM, particularly during rapid transitions and in low-SOC regions. This comparative analysis provides insights for selecting appropriate neural network architectures for SOC estimation applications in electric vehicle battery management systems.

Keywords: LSTM · CNN · State of Charge · Lithium Polymer Battery · Battery Management System

1 Introduction

Transport emissions are being increasingly targeted by global mitigation policies; the IEA Global EV Outlook 2023 assigns approximately 22% of CO_2 to the sector, and urban exposure to $PM_{2.5}$, O_3, NO_n, and CO is causing health and policy pressure [1].

In order to meet air-quality and climate goals, Governments around the world have created policies that favor electric mobility, such as purchase subsidies, tax relief, and zero-emission mandates. However, the uptake of large-scale EVs remains uneven. The most significant obstacles are the total cost of ownership, reliable access to public and depot charging, range sufficiency throughout duty cycles, and confidence in battery durability [2]. Within available storage options, lithium-polymer cells are often selected

A. E. Hassanien et al. (Eds.): AISDS 2025, LNDECT 286, pp. 202–211, 2027.
https://doi.org/10.1007/978-3-032-18758-1_20

for their high gravimetric/volumetric energy density relative to lead–acid. Despite this, there are still practical constraints such as cycle and calendar aging that cause service life to be shortened, thermal control can be challenging, fast-charge capability is limited, model uncertainty complicates state estimation, and range confidence remains a concern [3].

In modern BMS architectures, determining SOC is one of the primary control functions, as it directly impacts vehicle performance, safety, and battery longevity. Accurate SOC estimation enables optimal energy management, prevents overcharging and deep discharge conditions, and provides reliable range predictions for drivers [4]. Conventional approaches to SOC estimation fall into three broad classes: conventional methods (such as coulomb counting and open-circuit voltage), model-based methods such as equivalent-circuit and electrochemical formulations, and data-driven approaches built on machine-learning algorithms [5].

Data-driven techniques have demonstrated superior performance in SOC prediction due to their ability to learn complex non-linear relationships from battery operational data without requiring detailed knowledge of battery chemistry or physics [6]. Among these approaches, deep learning methods have gained significant attention, with Long Short-Term Memory (LSTM) networks and Convolutional Neural Networks (CNN) emerging as prominent solutions for SOC estimation tasks.

LSTM networks are designed to retain information across long sequences, making them well suited to represent battery dynamics that evolve over extended time spans. Their gated architecture allows selective retention and forgetting of information, enabling effective modeling of complex battery charging and discharging patterns [7]. Conversely, CNN models demonstrate exceptional capability in local feature extraction and pattern recognition, making them effective for identifying short-term variations and noise reduction in battery sensor data [8].

This paper presents a comparison between LSTM and CNN approaches for SOC estimation in lithium polymer batteries. The analysis examines architectural differences, computational requirements, accuracy performance, and practical implementation considerations. Additionally, the study explores hybrid approaches that combine both methods to leverage their complementary strengths. The findings provide valuable insights for researchers and engineers developing advanced BMS solutions for electric vehicle applications.

2 Model Design

2.1 LSTM Modeling

LSTMs extend recurrent neural networks with gating and memory cells, enabling stable training by reducing vanishing-gradient effects seen in standard RNNs. The LSTM architecture incorporates a sophisticated gating mechanism that enables selective information retention and forgetting, making it particularly effective for modeling long-term temporal dependencies in sequential data [9].

In each timestep, the LSTM manages memory state via three gating functions: forget, input, and output, which respectively control retention, write-in, and exposure of information. The forget gate computes a gating vector that attenuates the previous memory before adding the candidate update [10] Formally:

Forget gate:

$$f_t = \sigma\left(W_f\left[h_{t-1}, x_t\right] + b_f\right. \tag{1}$$

The input gate filters new information before storing it in the cell state; there are two components of this mechanism:

Input gate:

$$i_t = \sigma\left(W_i\left[h_{t-1}, x_t\right] + b_j\right. \tag{2}$$

Intermediate Memory Cell:

$$\overline{c}_t = \tanh\left(W_c\left[h_{t-1}, x_t\right] + b_c\right. \tag{3}$$

The cell's memory is refreshed by fusing the retention term from the forget gate with the update from the input gate:

Memory cell:

$$C_t = f_t * C_{t-1} + i_t * \overline{C}_t \tag{4}$$

At the end of the update, the output gate extracts the usable part of the cell state to form the output signal:

Output gate:

$$o_t = \sigma\left(W_o\left[h_{t-1}, x_t\right] + b_o\right. \tag{5}$$

Hidden state:

$$h_t = o_t{}^* \tanh\left(\overline{C}_t\right) \tag{6}$$

For SOC estimation applications, LSTM networks process sequential battery data including voltage, current, temperature, and power measurements. The network learns to identify temporal patterns and dependencies that correlate with SOC variations, enabling accurate predictions based on historical battery behavior.

This output is taken as an input by next time step as a hidden state and the whole process starts again. Complete LSTM network with different layers is shown in figure below (see Fig. 1).

2.2 CNN Modeling

Convolutional Neural Networks (CNN) have shown great effectiveness in recognizing patterns and extracting features. Although they have mainly been used for image tasks, 1D CNNs have also been very successful in analyzing time-series data and processing

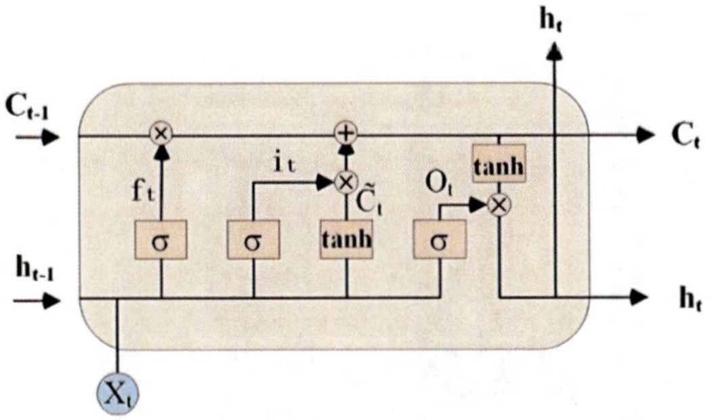

Fig. 1. Architecture of LSTM model.

signals, such as estimating battery state of charge (SOC) [11]. The key operation in a CNN is convolution, which uses learnable filters on input data to capture local features. For 1D time-series data, convolution is defined as:

$$y[n] = \sum_{m=0}^{M-1} x[n+m] * w[m] \tag{7}$$

where y[n] represents the feature map element, $x[n]$ is he corresponding input sample, $w[m]$ is the filter, and M is the size of the kernel.

CNN models for SOC estimation usually include several convolutional layers followed by pooling layers to reduce the dimensions. The convolutional layers extract local temporal features from battery sensor data, while pooling layers decrease computational load and ensure translation invariance [12]. The mathematical form of a convolutional layer with ReLU activation is:

$$a^{(l)} = ReLU\left(W^{(l)} * a^{(l-1)} + b^{(l)}\right. \tag{8}$$

Using $a^{(l)}$ as the output (activation) of layer l , with parameters $W^{(l)}$ and $b^{(l)}$ for its weights and bias, respectively. Max pooling performs windowed reduction, replacing each patch with its maximum value to reduce dimensionality and retain the strongest features:

$$P_{i,j} = \max_{(m,n)\epsilon R_{i,j}} a_{m,n} \tag{9}$$

where $P_{i,j}$ is the pooled output and $R_{i,j}$ is the pooling area.

For SOC estimation, CNN models are skilled at spotting local patterns in battery data, such as voltage changes, current spikes, and temperature shifts that relate to specific SOC levels. The ability to extract features hierarchically allows the network to learn more complex representations from raw sensor data [13] (Fig. 2).

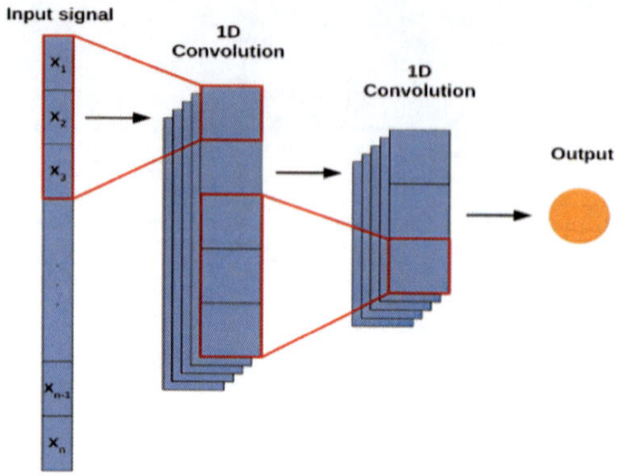

Fig. 2. Architecture of CNN model.

3 Dataset Creation and Model Building

3.1 Dataset

Data is essential to any data-driven model. Generally, there are two main approaches to obtaining data: (i) generating measurements through controlled experiments on the target cell, or (ii) utilizing datasets published by reliable third parties. The first method is typically preferred, as datasets that precisely match the required operating conditions are often unavailable. In this study, however, the dataset used is a private one provided by the author of reference [14].

The models for SOC estimation are built for a lithium-polymer cell whose key specifications are summarized in Table 1.

Table 1. Li-polymer cell: key specifications.

Attribute	Details
Manufacturer	Power-Xtra
Nominal capacity	350 mAh
Charging voltage	4.2 V ± 0.05 V
Nominal voltage	3.7 V ± 0.05 V
Discharge cut-off voltage	*3.0 V*
Weight	*10.5 g ± 0.5 g*

SOC and usable capacity were obtained with the Coulomb counting method: we integrate the measured current over time and normalize by the rated capacity. We set the initial SOC from a calibrated reference cycle, then update SOC forward using the measured current (discharge current reduces SOC; charge current increases SOC). The resulting SOC trajectory is used as the ground-truth label for training the deep-learning models [15]. The computation is given in Eq. (10):

$$SOC(t) = SOC(t_0) + \frac{\int_{t_0}^{t_0+\tau} I_c * d\tau}{Q_{rated}} * 100\% \tag{10}$$

From Eq. (10), $SOC(t_0)$ *is initial state of charge at* t_0 , Q_{rated} nominal capacity, and I_c current. Coulomb counting integrating current over time and normalizing by capacity, provides a low-cost SOC estimate [15]. We used it to generate reference SOC for a lithium-polymer cell, then trained and tested a deep-learning model to predict SOC from current, voltage, and temperature.

3.2 Experimental Procedure

This section describes the steps for training and testing a deep learning model using a dataset with three inputs: temperature, voltage, and current. The data are first cleaned and arranged in a suitable format for model training. After preprocessing, we partition the data into training and test sets. Both sets contain paired inputs and targets recorded from laboratory measurements of the cell. We then select the model hyperparameters and fit the network; training time depends on the chosen configuration and task. During training, the model iteratively adjusts its weights to match the target outputs. Performance is quantified on the held-out test set to measure accuracy on unseen data. Several performance indicators are applied to evaluate the accuracy of SOC estimates. After testing, the model can be deployed in real-world conditions and its performance analyzed. As outlined in the figure (see Fig. 3), the workflow proceeds from dataset preprocessing to model training and final testing. Before building the neural network model, its hyperparameters must be specified [16].

Tables 2 and 3 list the hyperparameters used for each model. Values were chosen to balance accuracy and comparability, and to make the CNN's capacity and training dynamics broadly comparable to the LSTM for a fair evaluation. All models were implemented and trained in Google Colab with high performance GPU.

LSTM training dataset:

- **Input:** 3 × 20560 data points (voltage, current, temperature)
- **Output:** 1 × 20560 data points (SOC)

CNN training dataset:

- **Input:** 5 × 20560 data points (voltage, current, temperature)
- **Output:** 1 × 20560 data points (SOC)

The two additional inputs (mean voltage and mean current) help capture the time-varying behavior of voltage, current, and SOC. We also vary the number of time steps to study the effect of sequence length on model performance.

The following summarizes the input–output configurations used to test the LSTM and CNN models.

LSTM model (testing):

- **Input:** 3 × 5141 data points (voltage, current, temperature, mean voltage, mean current)

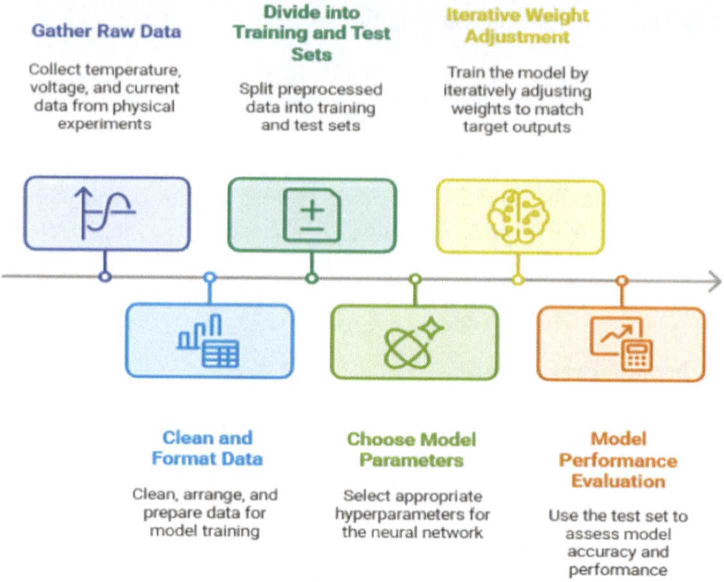

Fig. 3. Steps in data acquisition and preprocessing.

Table 2. Hyperparameters of the LSTM model.

Hyperparameters	Value
Layers	4
Hidden Units	128
Epochs	500
Learning rate	0,001
Optimizer	AdamW

Table 3. Hyperparameters of the CNN model.

Hyperparameters	Value
Layers	2
Hidden Units	128
Epochs	500
Learning rate	0,001
Optimizer	AdamW

- **Output:** 1×5141 data points (SOC)

CNN Model (Testing):

- **Input:** 5 × 5141 data points (voltage, current, temperature, mean voltage, mean current)
- **Output:** 1 × 5141 data points (SOC)

The two models are evaluated using MAE, RMSE, and R^2 to measure their performance in predicting the SOC. The metrics are defined as follows:

$$MAE = \frac{1}{K}\sum_{k=1}^{k}|y_k - \hat{y_k}| \tag{11}$$

$$RMSE = \sqrt{\frac{1}{N}\sum_{i=1}^{N}(y_k - \hat{y_k})^2} \tag{12}$$

$$R^2 = 1 - \frac{\sum_{i=1}^{n}(y_i - \hat{y}_i)^2}{\sum_{i=1}^{n}(y_i - \overline{y})^2} \tag{13}$$

where y_k and $\hat{y_k}$ denote the true and predicted SOC for sample k, $\overline{y}$ is the mean of the true value, and K is the number of samples.

4 Results and Discussion

This section reports and interprets the results of our comparative study on lithium-polymer battery state-of-charge (SoC) estimation using two deep-learning models: a CNN and an LSTM. Both models were trained on 20560 data points and evaluated on 5141 unseen points. presenting prediction SOC (see Fig. 4) against real SOC, where the blue curve represents the true SoC and the orange curve denotes the model prediction, followed by quantitative metrics including MAE, RMSE, and R^2 across operating conditions. The comparative evaluation reveals distinct performance characteristics and computational trade-offs: overall, the CNN achieves better generalization than the LSTM showing tighter alignment during rapid transients and in low-SoC regions, while also offering favorable inference efficiency (Table 4).

In our evaluation, the one-dimensional CNN consistently outperformed the LSTM for state-of-charge (SOC) estimation from voltage, current, and temperature time series. Quantitatively, the CNN achieved $R^2 = 0.9945$, MAE =0.0174, and RMSE =0.0224, whereas the LSTM obtained $R^2 = 0.9906$, MAE =0.0204, and RMSE =0.0272. This advantage is attributable to the CNN's learned convolutional filters, which act as data-driven finite-impulse-response kernels that efficiently extract local motifs; current steps, IR drop and recovery, and brief rest periods; while batch normalization and global averaging attenuate sensor noise without blurring sharp transients. By contrast, the LSTM must propagate information through a recurrent state; over long sequences, gate saturation and hidden-state lag can soften rapid load changes, yielding short bursts of error

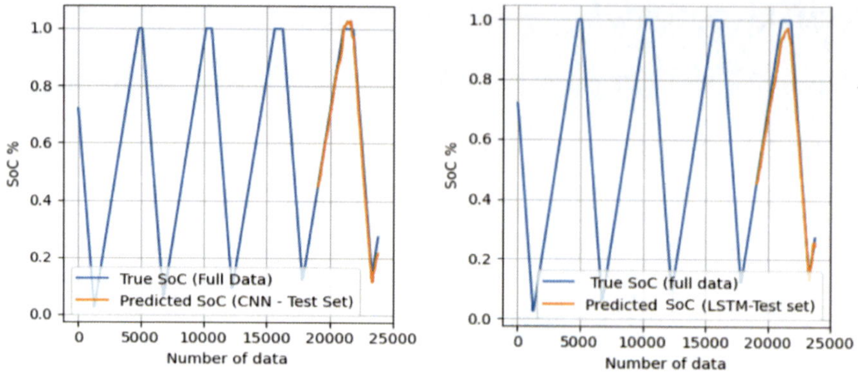

Fig. 4. presents the SOC estimation results by CNN and LSTM model.

Table 4. Comparison of error metrics for LSTM and CNN model.

Methods	MAE	RMSE	R^2%
LSTM	0.0204	0.0272	99.06
CNN	0.0174	0.0224	99.45

during high-load events and in low-SOC regions where voltage–SOC nonlinearity is pronounced. Notwithstanding its strengths, the CNN may exhibit slow drift over extended, low-dynamic intervals or under pronounced hysteresis if its receptive field is too short, whereas the LSTM can be comparatively stable in such long-horizon contexts. These complementary behaviors motivate a compact hybrid design in which a CNN front end performs local feature extraction and denoising, followed by a recurrent or attention module to capture slower trends and hysteresis, thereby combining the advantages of both approaches.

5 Conclusions

This study presented a comparative analysis of Long Short-Term Memory (LSTM) and Convolutional Neural Network (CNN) models for estimating the State of Charge in lithium polymer batteries, a critical function in Battery Management Systems for electric vehicles. Both models demonstrated strong predictive capabilities, leveraging deep learning to process complex temporal and sensor data.

The experiments indicate that while LSTM are effective in capturing long-term temporal dependencies, CNN models offered better performance in accuracy terms, generalization, and computational efficiency. CNNs were particularly well-suited for capturing local variations and reducing noise in battery data, leading to more precise SOC predictions.

Future work will focus on developing hybrid architectures that combine the strengths of LSTM and CNN models. These hybrid models will be tested across a wide range of temperatures to evaluate their robustness and adaptability in varying environmental conditions.

References

1. International Energy Agency: Global EV outlook 2023. IEA Publications (2023)
2. International Energy Agency: Transport – energy system (2024)
3. Pelletier, S., Jabali, O., Laporte, G.: Battery degradation and behaviour for electric vehicle battery-swapping systems. Transp. Res. C. **85**, 206–233 (2017)
4. Chen, H., et al.: State of charge estimation for lithium-ion battery using long short-term memory networks. J. Phys. **2890**(1) (2024)
5. Zhang, Y., Zhao, Z., Zhou, J., Zhang, Z., Zhou, H.: State of charge prediction algorithm of lithium battery based on multi-step time window and time attention mechanism. IEEE (2024)
6. Naguib, M., Kollmeyer, P. J., Emadi, A.: State of charge estimation of lithium-ion batteries: comparison of GRU, LSTM, and temporal convolutional deep neural networks. IEEE (2023)
7. Zraibi, B., Mansouri, M., Okar, C.: Comparing single and hybrid methods of deep learning for remaining RUL of lithium-ion batteries. E3S Web Conf. **297**, 01043 (2021)
8. Shahriar, S.M., Bhuiyan, E.A., Nahiduzzaman, M., Ahsan, M.: State of charge estimation for electric vehicle battery management systems using hybrid recurrent learning. Energies. **15**(21) (2022)
9. Hao, T., Ding, J., Tu, T.: A hybrid Kalman filter for SOC estimation of lithium-ion batteries. In: 41st Chinese Control Conference (CCC), Hefei, China. IEEE (2022)
10. Xiao, F., et al.: A CNN-LSTM method based on voltage deviation for predicting the state of health of lithium-ion batteries. Batter. Energy. **4**(3), e20240036 (2024)
11. Dong, Z., Zhou, X., Liu, Y.: Comparison of neural network models in lithium-ion batteries' state of charge estimation. In: Proceedings of the IEEE (2023)
12. Mondal, P., Bhavsar, D.N., Mittal, K., Mittal, M.: Estimating state-of-charge in lithium-ion batteries through deep learning techniques: a comparative evaluation. IEEE Access. **12**, 78773–78786 (2024)
13. Taş, G., Bal, C., Uysal, A.: Performance comparison of lithium polymer battery SOC estimation using GWO-BiLSTM and cutting-edge deep learning methods. Electr. Eng. **105**(5), 3383–3397 (2023)
14. Taş, G., Uysal, A., Bal, C.: A new lithium polymer battery dataset with different discharge levels: SOC estimation of lithium polymer batteries with different convolutional neural network models. Arab. J. Sci. Eng. **48**(5), 6873–6888 (2023)
15. Hoque, M. A., et al.: Prognosis of lithium-ion battery health with hybrid EKF-CNN+LSTM model using differential capacity. https://arxiv.org/abs/2504.13956 (2025)
16. Kandru, V.K., Kuthuri, N.R.: Comparison of FNN and LSTM models for SOC estimation of EV battery. Adv. Dengineering Sci. (2022)
17. Zraibi, B., Mansouri, M., Lammamri, O., Loukili, S.E., Said, M.A.: Advanced hybrid deep learning model for accurate RUL prediction of lithium-ion batteries. In: Advances in Intelligent Systems and Digital Applications, vol. 1485, pp. 190–198. Springer, Cham (2025)

AI for Security and Trustworthy Systems

A Hybrid AI Model for Enhanced Malware Detection and Classification

Wajdi Elhamzi[1,2](✉) and Mohamed Atri[3]

[1] College of Computer Engineering and Sciences, Prince Sattam bin Abdulaziz University City, Al-Kharj, Saudi Arabia
wajdi.elhamzi@essths.rnu.tn

[2] The Higher School of Science and Technology of Hammam Sousse, University of Sousse, Hammam Sousse, Tunisia

[3] Laboratory of Condensed Matter and Nanosciences, Faculty of Sciences of Monastir, University of Monastir, Monastir, Tunisia

Abstract. Malware is becoming increasingly sophisticated, particularly in its use of obfuscation and zero-day vulnerabilities. This implies that traditional threat-detection methods based on signatures are becoming less valuable. This research presents a robust malware detection solution utilizing a hybrid artificial intelligence (AI) model, specifically designed for the static analysis of Windows Portable Executable (PE) files. The proposed technique operates without necessitating file execution, ensuring both safety and efficiency. We obtained a comprehensive set of static properties from multiple ST-WinMal datasets. These features include binary size, entropy, byte and opcode n-grams, and API call patterns. We tested several machine learning models, including Random Forest and Gradient Boosting, as well as deep learning models such as Convolutional Neural Networks. We then combined them into a weighted hybrid ensemble. Tests indicate that the hybrid approach outperforms all other methods, achieving 99.66% accuracy in multi-class classification. This research demonstrates that a hybrid AI technique utilizing static features offers a highly accurate, scalable, and practical solution for modern malware detection and classification, providing an attractive alternative to more resource-intensive strategies.

Keywords: Malware Detection · Static Analysis · Artificial Intelligence · Cybersecurity · PE Files

1 Introduction

Cyberattacks are the biggest concern in contemporary technology. Malware is any software or package of instructions that is meant to damage a computer, user, company, or computer system [1]. Fred Cohen created the first known computer virus program in 1983 that could replicate itself and spread [2]. Subsequently, various kinds of malware emerged, including viruses, Trojan horses, ransomware, spyware, adware, rogue software, wipers, and scareware. Malicious software is any code that runs without the user's knowledge or permission [3].

A. E. Hassanien et al. (Eds.): AISDS 2025, LNDECT 286, pp. 215–224, 2027.
https://doi.org/10.1007/978-3-032-18758-1_21

In fact, Malware Cybercriminals who operate under the hacker moniker create intrusive programs known as malware to steal data and cause system damage and disruption. The various objectives of malware include disrupting normal computer operations, stealing confidential data, gaining unauthorized access to networks, and displaying unwanted advertisements. The ability of malware attacks to break weak passwords enables them to penetrate deeply into systems and spread throughout networks, resulting in significant disruptions to businesses and organizations during their daily operations. Malware exists in various forms that can lead to the encryption of important files, as well as the flooding of advertisements, a decline in system performance, and the redirection of websites to dangerous sites. The primary tool used by cyber attackers to execute their operations is malicious software, which enables major data breaches that trigger extensive identity theft and financial fraud incidents. Malware is the primary factor behind ransomware incidents, which annually result in millions of dollars in economic losses. These attacks aim at both personal users, businesses, and governmental organizations.

Malware analysis is the process of examining malicious software to understand its functionality, origin, impact, and the techniques used for infection and evasion. This discipline plays a crucial role in cybersecurity, enabling organizations and researchers to detect, respond to, and mitigate malware attacks. Malware analysis can be employed both reactively and proactively to identify threats and strengthen defenses. Indeed, malware analysis is a structured process designed to uncover the functionality, objectives, and potential impacts of malicious software on targeted systems. Understanding the various phases of malware analysis is crucial for effective threat detection, mitigation, and response.

The analysis of malicious software has traditionally relied on a combination of static, dynamic, and hybrid techniques. These conventional approaches involve gathering multiple features from malware samples to facilitate detection and classification. However, this task is still limited to modern and zero-day malware. In contrast, artificial intelligence makes detection and analysis more efficient and straightforward. Based on a large dataset, the AI model can distinguish between different types of malware attacks and their variants. Furthermore, deep learning enhances classification accuracy and comprehensiveness by extracting a broader range of features than traditional machine learning methods. This is achieved via multiple layers of feature abstraction, which allow deep learning models to continuously learn and recognize emerging malware patterns beyond their initial training phase.

In this context, we focus on this paper, which applies various machine learning algorithms and deep learning models to detect malicious files and extract features. The mixing of machine learning and deep learning methods leads to an efficient hybrid technique for binary and multi-classification

This paper is organized as follows: Sect. 1 introduces AI techniques for malware detection and analysis. Section 2 focuses on the literature review. Section 3 describes the proposed methodology. After that, the experimental environment is presented in Sect. 4. The results are presented and discussed in Sect. 5. Finally, the conclusion in Sect. 6 summarizes the study results and future work.

2 Related Work

It's worth noting the traditional approaches to malware detection before explaining how artificial intelligence is used. Malware detection and analysis have used static, dynamic, and hybrid methods for years. These methods analyze file formats, code activity, and system interactions and work well in many cases. Traditional approaches have many advantages, but they may struggle to keep up with new threats. This section analyzes major studies on the merits and cons of conventional malware detection systems.

In 2018, Alzahraniet al. [4] conducted a systematic review to identify the literature and describe the merits, shortcomings, accuracy, and limitations of each malware analysis methodology. However, effective and easy to use, static analysis did not work against "zero-day" malware. Dynamic analysis revealed complex, previously undisclosed malware, but significantly increased resource consumption and execution time. A hybrid method combining static and dynamic analysis yielded the highest accuracy, reaching 98.7% in some cases.

Modern virus types can threaten human life, safety, and business continuity in cyberspace. Malware has been identified on supported operating systems in several tests. Jerlin et al. [5] created the first machine learning-based malware detection system utilizing the naïve Bayesian approach. The authors examined PE file character strings, APIs, and byte sequences.

AI-based malware detection has grown in popularity because it can better predict infestations. Since there are various kinds of malware, extracting features is difficult. This makes AI-based malware detection difficult. Li et al. [6] developed a graph convolutional network-based malware classifier to accommodate variability in malware properties. First, extract the malware code's API call sequence to generate a directed cycle graph. Markov chains and principal component analysis are used to create the graph's feature map. The following analyzes and compares the performance of a graph convolutional network-based classifier. The methodology outperforms other algorithms across most detections, achieving 98.32% accuracy.

Moujoud et al. [7] propose three ensemble model scenarios to test ensemble learning algorithms for malware detection. All malware types and behaviors are tested using a thorough sample set. Extensive testing and validation show that ensemble learning models outperform single-model methods at detecting advanced malware.

Chen et al. [8] propose a simple neural network for malware classification using RGB memory dump images and attention. Compared to previous deep learning-based systems, the proposed system decreases trainable parameters while preserving classification accuracy. This model gets 98.1% recognition accuracy on Dumpware10.

Dhalaria et al. [9] explore the impact of various combinations of learning models on malware detection in Android systems. They evaluated five classifiers (SVM, DT, NB, K-NN, and RF) based on their accuracy. The results demonstrate that KNN_RF overcomes other combinations, with an accuracy of 98.02%.

To detect malware in IoT devices, Khan et al. [10] proposed a system called Deep Squeezed-Boosted and Ensemble Learning. This boosting-based architecture, integrating three classifiers—SVM, MLP, and AdaBoost—achieved an accuracy of 98.50%.

Gupta et al. [11] proposed a methodology using ensemble learning and big data to enhance malware detection efficiency. The suggested method included five distinct classifiers: Naive Bayes (NB), K-Nearest Neighbors (KNN), Decision Tree (DT), Support Vector Machine (SVM), and Random Forest (RF), attaining an accuracy of 99.2%.

3 Methodology

This section describes the approach taken for designing, implementing, and evaluating a static-analysis-based malware detection system for Windows. The method integrates conventional machine learning (ML) and deep learning (DL) to enhance accuracy, robustness, and generalization across various malware families.

The system described in Fig. 1 is a modular, organized pipeline consisting of four phases: the first phase loads PE files, followed by the feature extraction phase, the preprocessing stage, and finally the classification phase. The architecture is intended to be efficient, scalable, and secure, without requiring dynamic analysis or execution of the dangerous file. The system comprises five key components: input ingestion, preprocessing, static feature extraction, classification, and result interpretation. The system output indicates the malware family when it is malicious.

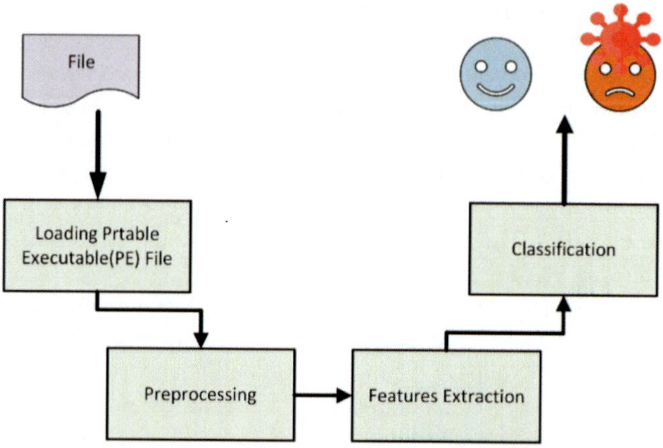

Fig. 1. High-Level Overview of the System

3.1 Loading PE Files

The proposed system is designed to analyze Windows Portable Executable (PE) files, the standard file format for executables, object code, and DLLs in Windows operating systems. The model's input consists of a dataset comprising both benign software and malicious PE samples.

The primary dataset used in this study is the ST-WinMal dataset, a comprehensive collection of thousands of reclassified media files. This dataset encompasses a diverse range of malware families, including Trojans, Worms, and Backdoors, providing a broad spectrum of malicious behaviors for the model to learn. The files are loaded in batches and parsed statically, meaning the analysis is performed by inspecting the file's structure and content without executing any code. This approach ensures the safety and efficiency of the detection process, mitigating the risks associated with handling live malware.

3.2 Preprocessing

To ensure data quality and integrity, a comprehensive pre-processing pipeline was implemented. The raw data was first cleansed by identifying and removing missing or non-meaningful values. Subsequently, all numerical features were normalized to a standard variance scale using StandardScaler. Categorical labels, specifically malware family names, were converted to integers via label encoding to make them suitable for supervised learning algorithms. Finally, to address the challenge of imbalanced class distributions, techniques such as class weighting or oversampling were applied. This critical step ensures the model is trained on a more representative dataset, preventing bias toward the majority classes.

3.3 Feature Extraction

The first step is to extract malware features, which will serve as input to our machine learning and deep learning algorithms and will be preserved for all subsequent tests. Each instance in the dataset is represented by a vector of manually crafted features, forming an $N \times M$ matrix, where N is the cardinality of the set of instances and M the number of extracted features. The matrix's dimensionality varies depending on the extracted malware attributes. For illustration, extracting the size of binary files generates a one-dimensional matrix (one vector per column), while extracting the frequency of opcodes on it generates an $S \times N$ matrix, where S represents the number of samples and N the number of API function calls recorded. An.npz file was used to save the extracted features. The NumPy library produced these files to facilitate the simultaneous storage of multiple NumPy arrays within these archives. The processing time for extracted features depends on the type of attribute considered.

The proposed system relies on a comprehensive static analysis pipeline for feature extraction, which meticulously examines PE files without executing them. The extracted feature set encompasses:

- **Binary Size:** The total size of the executable file.
- **Entropy:** A measure of the file's randomness, used to detect encrypted or packed code.
- **Opcode Frequency:** A count of the low-level instructions present in the code.
- **N-gram Patterns:** Bytes 3-g and Opcode 4-g are analyzed to identify significant sequential and structural patterns within the binary.
- **API Calls:** The frequency and sequential order of Windows API invocations are tracked to understand the program's intended behavior.

A custom feature engineering function aggregates all extracted features into a consolidated matrix. This unified representation serves as input to downstream machine learning and deep learning classifiers. The selected static features offer a combination of structural, statistical, and behavioral indicators for differentiating malware from benign files without requiring execution. Binary size is a rapid heuristic for determining file type. Entropy helps detect obfuscated or packed files. Low-level instruction-level and instruction-sequence patterns are learned and represented using opcode frequency and an n-gram model, which are commonly observed in malware. Patterns in API call frequency and ordering disclose typical misbehaviors on system contact. Collectively, these properties create a strong, execution less foundation for effective malware detection.

3.4 Classification

To ensure both the accuracy and stability of the novel system, three families of models were constructed, and their performance was tested before combining them. Detection of malware remains a significant challenge in cybersecurity. We initiated our work in the first phase by employing classical machine learning algorithms for malware classification. We used Random Forest (RF) [12], Gradient Boosting (GB) [13], and a convolutional neural network (CNN) [14] to classify malware.

To improve the model's performance and generalization, a Hybrid Model was also designed. A combination of Random Forest, Gradient Boosting, and CNN is suggested to form a single classification pipeline. The design is based on the idea that different malware types exhibit distinct behavioral patterns that can be identified with complementary analytical tools. CNN is the deep learning backbone that extracts hierarchical characteristics from consecutive API request patterns. Additionally, the Random Forest component provides robust, ensemble-based decision-making capabilities that excel at handling the heterogeneous feature types in the ST-WinMal dataset, including binary indicators, categorical variables, and numerical measurements of system interactions. Gradient Boosting's sequential error correction and boosting techniques enhance the model's understanding of complex decision boundaries between malware variants and benign software.

4 Experimental Environment

4.1 Dataset

In this section, we detail the datasets evaluated for this study and outline the final selections. Our process involved reviewing several prominent datasets commonly used in malware research. A significant part of this effort was dedicated to examining ST-WinMal [15], a dataset developed by a fellow thesis student at Politecnico di Torino. Curating an appropriate dataset proved challenging, as our goal was to identify collections that not only provided labeled samples and a broad set of pre-extracted features but also included the original binary files.

Table 1 presents the Malware families in the ST-WinMal Dataset, along with the number of samples per family.

Table 1. Table Malware families in ST-Winmal dataset

Family Name	Type	Number of Samples	Class Number
Ramnit	Trojan	1,499	0
Zbot	Trojan	1,159	1
IceId	Trojan	860	2
Trickbot	Trojan	725	3
Virlock	Ransomware	744	4
Stop	Ransomware	1,000	5
Magniber	Ransomware	804	6
Wannacry	Ransomware	1,444	7
Benign Samples	Benign File	955	8

4.2 Performance Metrics

Multiple standard metrics enabled the assessment of the proposed model's effectiveness. Evaluation metrics consist of Accuracy as well as Precision, Recall, F1 Score, and Receiver Operating Characteristic - Area Under the Curve (ROC AUC). These metrics offer individual insights into the model's performance, which becomes essential when analyzing binary or multi-class model behavior:

- **Accuracy** reveals the total ratio of instances that models identify correctly. The ratio of verified positive outcomes, together with verified negative outcomes, to the total number of samples is used to calculate this metric (Eq. 1).

$$Accuracy = \frac{TP + TN}{TP + TN + FP + FN} \tag{1}$$

- **Prediction** measures the validity of positive predictions. Accuracy represents the ratio of true positives divided by the sum of true positives and false positives. A high accuracy value indicates that the system produces few false positives (Eq. 2).

$$\text{Prediction} = \frac{TP}{\text{TP} + \text{FP}} \tag{2}$$

- **Recall**: The detection capability of models against all pertinent cases is determined by recall, also known as sensitivity and true positive rate. The ratio calculates true positive results by dividing them by the sum of true positives and false negatives (Eq. 3).

$$\text{RecalTl} = \frac{TP}{\text{TP} + \text{FN}} \tag{3}$$

- **F1 Score**: This score represents a harmonic average between precision and recall values. The measure offers a well-balanced evaluation, as it performs best on datasets with uneven class distributions (Eq. 4).

$$\text{F1 Score} = \times 2\,\frac{\text{Prediction} \times \text{Recall}}{\text{Prediction} + \text{Recall}} \quad (4)$$

- **ROC AUC**: The ROC AUC evaluation method assesses a model's ability to distinguish between positive and negative class assignments. The ROC curve illustrates the relationship between the true positive rate (TPR) and the false positive rate (FPR) across various threshold settings. The calculation of TPR and FPR follows Eqs. 5 and 6.

$$\text{FPR} = \frac{FP}{\text{FP} + \text{TN}} \quad (5)$$

$$\text{TPR} = \frac{TP}{\text{TP} + \text{FN}} \quad (6)$$

5 Results and Discussions

This section provides a detailed evaluation of the performance of the Random Forest-Gradient Boosting-Convolutional Neural Networks (CNN) hybrid model for detecting and classifying malware. The evaluation analyzes the multi-class Classification of multiple malware types, using the ST-WinMal dataset. To evaluate model effectiveness. In Table 2, the system utilized accuracy, precision, recall, and F1 score. The hybrid model exhibits greater predictive accuracy and robustness than single classifiers by combining traditional machine learning with deep learning components, according to the evaluation results.

Table 2. Performance results for the different models

Class	Accuracy	Precision	Recall	F1 Score
Random Forest	0.9945	0.9941	1.0000	0.9970
Gradient Boosting	0.9954	0.9963	1.0000	0.9982
CNN	0.9798	0.9954	0.9968	0.9961
Hybrid	0.9966	0.9963	1.0000	0.9982

Label encoding turned malware family names into numerical representations for machine learning models. ML algorithms cannot process textual data; hence, they require numerical data. RF and GB standard models yielded consistent findings, while the Hybrid model combined predictions to improve dependability and reduce prediction variation. CNN excelled in class 2 but was uneven overall. The Hybrid model has the highest accuracy and nearly flawless classification. The Class 4 representation in the dataset requires adjustments to the CNN model parameters or improvements to the training data.

Figure 2 compares the performance of Gradient Boosting, Convolutional Neural Network, Random Forest, and Hybrid classification using four confusion matrices. In classes 1, 2, and 3, the Random Forest model correctly classifies nearly all occurrences, achieving the highest classification accuracy. Gradient Boosting works well in classes 2 and 1, but some data from classes 3 and 7 are misclassified. Although effective, the CNN model misclassifies more objects in class 1, labeling numerous items as classes 0, 2, 6, 7, and 8, demonstrating that it fails to distinguish between closely related classes. CNNs misclassify most classes, resulting in lower precision than tree-based models.

The Hybrid-CNN-RF-GB combination outperforms its basis models in performance and robustness, as indicated by the confusion matrix. The simple Convolutional Neural Network showed significant inter-class confusion, especially for Classes 1 and 7, which had many dispersed False Negatives across the matrix. CNNs' robust deep feature extraction and RF and GB's classification stability via soft voting produced a near-perfect diagonal matrix for the Hybrid model. This result shows that hybrid architectures can mitigate the CNN's classification weaknesses by leveraging ensemble methods to establish cleaner, more distinct decision boundaries, thereby reducing False Positives and False Negatives across all classes.

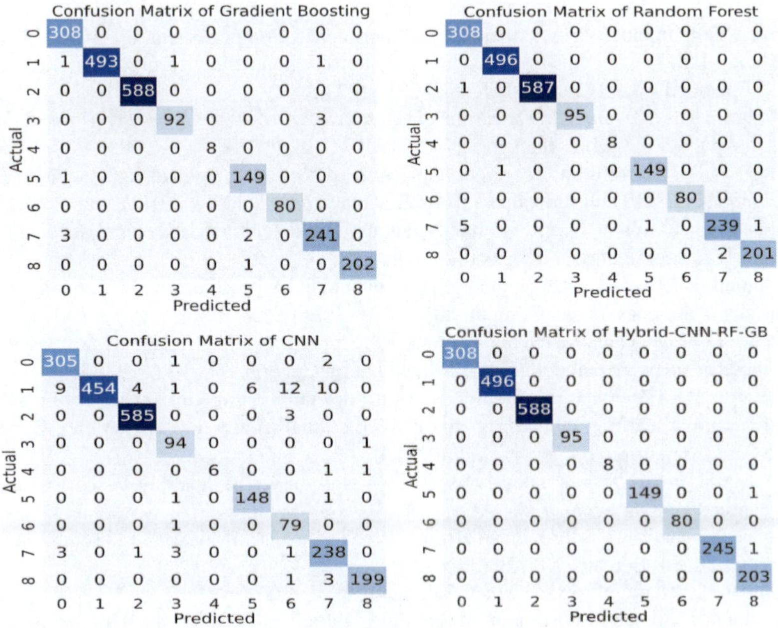

Fig. 2. Confusion Matrix for GB, RF, CNN, and Hybrid techniques

6 Conclusion

The final analysis of the confusion matrices unequivocally concludes that the Hybrid-CNN-RF-GB ensemble model provides the superior classification performance and the highest stability for this multi-class problem. The individual models, particularly the simple CNN, suffered from significant intra-class confusion, leading to a high number of False Negatives and dispersed prediction errors. By contrast, the Hybrid-CNN-RF-GB architecture effectively combined the deep feature-extraction power of the CNN with the robustness to errors of the tree-based classifiers (Random Forest and Gradient Boosting). This fusion strategy yielded a near-perfect diagonal confusion matrix, demonstrating the near-complete elimination of prediction errors and establishing the most precise decision boundaries among the 9 classes. However, all the models are strong in their respective areas, and Random Forest demonstrates the best stability and accuracy. At the same time, CNNs require enhancements to achieve improved performance on more complex classification tasks. The Hybrid model achieved its highest accuracy and almost perfect classification results. Consequently, the ensemble approach is validated as the most reliable and accurate technique for this classification task.

References

1. Nikam, U.V., Deshmuh, V.M.: Performance evaluation of machine learning classifiers in malware detection. In: Proceedings of the 2022 IEEE International Conference on Distributed Computing and Electrical Circuits and Electronics, pp. 1–5. IEEE, India (2022)
2. Cohen, F.: Computer viruses: theory and experiments. Comput. Secur. **6**, 22–35 (1984)
3. Akhtar, M.S., Feng, T.: IOTA based anomaly detection machine learning in mobile sensing. EAI Endorsed Trans. Create. Tech. **9**, 172814 (2022)
4. Alzahrani, A.S., Awan, I.U.: Detection and classification of malware based on hybrid analysis. Int. J. Adv. Sci. Eng. Inf. Technol. **8**(4–2), 1630–1636 (2018)
5. Jerlin, M.A., Marimuthu, K.: A new malware detection system using machine learning techniques for API call sequences. J. Appl. Secur. Res. **13**, 45–62 (2018)
6. Li, S., Zhou, Q., Zhou, R., Lv, Q.: Intelligent malware detection based on graph convolutional network. J. Supercomput. **78**(3), 4182–4198 (2022)
7. Moujoud, L., Ayache, M., Belmekki, A.: Enhancing malware detection through ensemble learning techniques. Clust. Comput. **28**(7), 477 (2025)
8. Chen, J., Wu, M., Gan, W., Huang, H., Lau, T.S.C.: A lightweight CNN malware classification method for software detection. Int. J. Mach. Learn. Cybern. **16**(10), 1–19 (2025)
9. Dhalaria, M., Gandotra, E.: Android malware detection using chi-square feature selection and ensemble learning method. In: Sixth International Conference on Parallel, Distributed and Grid Computing, pp. 36–41. India (2020)
10. Khan, S.H. et al.: A new deep boosted CNN and ensemble learning based IoT malware detection. Comput. Secur. **133**, 103385 (2023)
11. Gupta, D., Rani, R.: Improving malware detection using big data and ensemble learning. Comput. Electr. Eng. **86**, 106729 (2020)
12. Breiman, L.: Random forests. Mach. Learn. **45**, 5–32 (2001)
13. Friedman, J.: Greedy boosting approximation: a gradient boosting machine. Ann. Stat. **29**, 1189–1232 (2001)
14. Ayadi, W., Elhamzi, W., Charfi, I., Atri, M.: Deep CNN for brain tumor classification. Neural. Process. Lett. **53**(1), 671–700 (2021)
15. Marco S.: Machine Learning for malware characterization and identification. https://github.com/saci98/ML_Malware_Classification (2025). Accessed 21 Sept 2025

Towards Secure and Auditable Agentic AI Using Blockchain: A Review and Taxonomy

Kamal Bouhassoune[1](✉), Sam Goundar[2], and Abdelkrim Haqiq[1]

[1] Hassan First University, Faculty of Sciences and Techniques, Computer, Networks, Mobility and Modeling Laboratory:IR2M, Casablanca Street, Box 577, 26000 Settat, Morocco
k.bouhassoune@uhp.ac.ma

[2] University of Central Asia, 310 Lenin Street, Naryn 722918, Kyrgyzstan

Abstract. Agent-based artificial intelligence (AI) and blockchain are two rapidly advancing paradigms that have deep transformative effects on both technical and societal domains. Agentic AI introduces autonomous systems capable of adaptive decision-making, goal-directed execution, and self-organization in open-ended conditions. Blockchain, as a distributed ledger technology, provides trustless coordination, immutability of transactions, and incentive-compatible mechanisms for decentralized cooperation. Independently, these technologies address distinct challenges: agentic AI advances intelligent autonomy, while blockchain ensures verifiable security, transparency, and accountability. Their convergence creates a powerful synergy, with blockchain reinforcing the integrity, provenance, and auditability of autonomous AI agents and their decisions. This paper maps the landscape of blockchain-enabled, secure, and auditable autonomous AI agents. It proposes a taxonomy of security and auditability dimensions. It examines how blockchain enforces safeguards for identity, integrity, and interaction, while supporting auditability through provenance tracing, decision accountability, and agent transactions. Finally, it identifies research directions to explore the opportunities and address the limitations of blockchain integration with agentic AI.

Keywords: Agentic AI · Autonomous AI agents · Auditability · Blockchain · Security · Taxonomy

1 Introduction

Trust and accountability are fundamental requirements for the adoption of autonomous systems, particularly as many forms of Artificial Intelligence (AI) evolve toward greater independence. Agentic AI, which operates with a significant degree of autonomy in decision-making and transaction execution, emerges as a promising field of research on the limitation of human and AI in-the-loop models [1]. Moreover, the deployment of such systems in critical sectors like healthcare, education and supply chain requires robust frameworks to ensure the security, auditability, and compliance of AI-driven outcomes.

Blockchain technology provides a distributed infrastructure to meet security and auditability requirements. Its intrinsic trustlessness, immutability [2] and transactional transparency, provide a reliable infrastructure for agentic AI operations.

A. E. Hassanien et al. (Eds.): AISDS 2025, LNDECT 286, pp. 225–234, 2027.
https://doi.org/10.1007/978-3-032-18758-1_22

First, its secure and tamper-proof ledger creates an immutable audit trail for decisions and transactions; second, its consensus-based verification process guarantees the integrity and verifiability of AI-driven actions. The integration of AI and blockchain technology can enhance corporate trust by mitigating challenges associated with data centralization and the opacity of AI systems. Moreover, it enables AI systems to comply with ethical, social, and legal requirements, it is also providing a foundation for converging ethical frameworks [3].

Our research explores blockchain enabling security and auditability in autonomous AI agent systems. For accountability, compliance, and retrospective verification, we prioritize auditability while keeping transparency. Auditability involves immutable timestamped traces, while transparency provides immediate exposure and explainability. In the remainder of this paper, we first provide, in the next section, a background on agentic AI and blockchain technology; Next, we present our methodology and taxonomy. Then, we discuss implications and limitations. The paper concludes by entailing open perspectives and directions for further research.

2 Background

2.1 Agentic Artificial Intelligence

Agentic AI, or autonomous AI agents, refers to a class of intelligent systems characterized by context awareness and adaptive behavior. Unlike AI models constrained by predefined tasks, agentic AI exhibits purposive, goal-directed action, enabling autonomous operation across diverse and dynamically evolving contexts. By exhibiting initiative traditionally associated with human cognition, these systems extend beyond reactive computation to engage in deliberate decision-making and proactive interaction. This autonomy allows AI agents not only to adapt but also to influence and shape outcomes in complex, open-ended conditions [4].

Recent progress in large language models (LLMs) further reinforces this evolution by extending agents' reasoning and communicative capacities. The open-ended capacity of such systems refers to the ability to take initiative by asking questions and dynamically uncovering information, this capacity makes its evaluation more complex than closed-ended systems [5]. Core algorithms that support goal-directed behavior, contextual adaptation, and autonomous decision-making are fundamental to the development of such autonomous systems. Agentic AI can achieve higher levels of resilience through the integration of three key components: (i) reinforcement learning, which allows agents to acquire knowledge through iterative trial-and-error interactions and optimize cumulative rewards over time; (ii) goal-oriented architectures, which enable agents to prioritize and pursue multiple objectives simultaneously; and (iii) adaptive control systems, which facilitate adjustment to dynamic environments and sustain robust, efficient performance under varying conditions [6].

2.2 Blockchain Technology

Blockchain has emerged as a transformative proposition for decentralized trust and tamper-resistant record-keeping. As a distributed ledger maintained across peer-to-peer

(P2P) network, blockchain has fundamental properties of tamper resistance, transparency, and verifiability through consensus protocols and cryptographic primitives [7]. Beyond its applications in cryptocurrencies, Blockchain technology now underpins the secure exchange of data, tracking of provenance, and decentralized coordination in finance, supply chains, and healthcare, where the integrity of decisions and parties is crucial. Blockchain technology eliminates the need for a centralized governing organ, establishes trustless and distributed coordination, and provides a relevant and appropriate infrastructure for systems that require security and auditability.

Based on a consistent corpus scope and size, comprising well-cited surveys and papers predominantly published in first-quartile venues, Table 1 summarizes the common themes and highlights the main research focuses in the blockchain literature.

Table 1. A body of relevant surveys on blockchain technology

Theme	Focus	Highlights	Ref.
General	Applications, challenges	Functional and technical concerns	[8–10]
	Decentralization	Governance, consensus	[11, 12]
	Application layer	Layer-Two protocols, blockchain-like system	[13, 14]
	Infrastructure and network	Decentralized Physical Infrastructure Networks (DePIN)	[15, 16]
Architecture	Design decision	Software, storage	[17]
	Consensus	Resource, IoT	[18–20]
Interoperability	Standards, exchange	Oracles, cross-chain	[21–23]
Scalability	Performance	Consensus, throughput	[24–26]
Security	Privacy, immutability, incentive	Identity, access control, proof	[27–30]

2.3 Research Motivation

Most existing surveys focus on blockchain architecture and design, addressing challenges related to decentralization, security, scalability and interoperability, as well as blockchain applications in domains such as the Internet of Things [31], edge computing [32] or artificial intelligence in general [33]. Yet, to the best of our knowledge, very few works have systematically examined blockchain within the specific context of agentic artificial intelligence.

Advancing this direction requires a multidisciplinary approach, drawing on expertise in distributed programming, cryptography, and artificial intelligence. Such approach is essential to understand the challenges, requirements, and opportunities associated with leveraging blockchain to strengthen the security, auditability, and trustworthiness of AI-powered agent systems. The combination of agentic AI and blockchains marks a major change in autonomous systems. To make sure that critical applications are trustworthy and accountable, agentic AI systems need strong security protocols and auditable transactional frameworks.

We assume blockchain-based AI agents require rigorous accountability mechanisms, especially in domains where irreversible transactions carry irreversible consequences. Existing safeguards, including validation systems and human verification, are insufficient in the absence of frameworks for recording decision logic and maintaining immutable audit evidence. Advancing blockchain-native audit systems that capture both actions and underlying reasoning is essential to ensure regulatory compliance [34]. Our taxonomy addresses this gap by establishing security and auditability concerns as the cornerstones of accountable agent autonomy.

3 Methodology

3.1 Literature Review

The present research is informed by a corpus of studies and surveys, predominantly published after 2019, which have been sourced from leading indexing systems such as Scopus and Google Scholar, as well as scientific repositories including the ACM Digital Library, IEEE Xplore, ScienceDirect, SpringerLink, and preprints from the online repository archive and index (arXiv).

The relevant literature was identified using targeted queries, incorporating the keywords *blockchain*, *security*, *auditability*, *verifiability*, and *transparency*. A close examination of this extensive body of work reveals several significant research trends, including but not limited to: blockchain-driven *privacy* and *traceability*, *self-sovereign identity*, *multi-agent systems* and the rise of *generative AI*.

3.2 Security and Auditability Taxonomy

To analyze the mechanisms that underpin blockchain-enabled secure and auditable agentic AI, we propose a taxonomy with two dimensions (Fig. 1).

Within the security dimension, identity encompasses on-chain recognition and decentralized identifiers (DID), access control and verifiable credentials (VC) [35, 36], ensuring agents are authenticated and accountable. Interaction security covers adversarial defense and coordinated verification [37] providing robustness against malicious behavior and promoting trustworthy multi-agent exchanges. Integrity mechanisms, including hashing, on-chain verification of AI models, and distributed storage guarantee that agents operate only with untampered resources.

The auditability dimension addresses three criteria. First is provenance through immutable task assignment, dispute mitigation, and versioning [38]. Second, decision logic with alignment, proofs and explainability [37], ensure transparency and accountability of AI agent actions. Third, transactions encompass incentive mechanisms that support accountability and traceable exchanges across autonomous AI agents in peer-to-peer networks.

Blockchain-enabled autonomous AI agents

- **Security**
  - Identity : Decentralized authentication (DID) | Access control (VC)
  - Interaction : Communication protocols | Multi-agent consensus
  - Integrity : Hashing | On-chain verification and storage
- **Auditability**
  - Provenance : Assignment | Mitigation | Versioning
  - Decision : Proofs | Alignment | Explainability
  - Transaction : Incentive-compatible | Traceability and accountability

Fig. 1. A taxonomy of blockchain-enabled security and auditability for agentic AI.

4 Discussion

Blockchain reinforces agentic AI along two dimensions: security and auditability. Together, these mechanisms provide the foundation of trust that enables autonomous agents to operate transparently and in accordance with governance requirements.

4.1 Verifiable Security

Identity: Assessing Verifiable Autonomy
A core challenge for agentic AI is the assurance of authentic and non-replicable identities. Blockchain enables cryptographically anchored identifiers, such as decentralized identifiers (DIDs), that provide immutable and verifiable credentials (VC) for autonomous agents. In marketplaces for AI services, blockchain-backed identity ensures that only authorized and accountable agents can participate, extending identity standards originally developed for Internet-of-Things (IoT) communication into the field of intelligent multi-agent autonomy [37, 39].

Interaction: Validation Agent Coordination
Secure interaction among autonomous agents requires mechanisms that validate communication and confirm accountability. Blockchains provide tamper-proof coordination where smart contracts may govern negotiations and settlements, while immutable traces prevent agents from repudiating their prior commitments. Such verifiable interaction histories reduce collusion risks and provide auditable evidence for issue mitigation. Emerging frameworks [37] and blockchain-assisted secure communication architectures demonstrate how distributed ledgers function as neutral infrastructures for multi-agent negotiation and coordination.

Integrity: Ensuring Models and Decision Trustworthiness
The integrity of models, decisions and datasets is essential for trustworthy autonomous systems. Blockchain supports this by recording certified hashes of resources, enabling provenance tracking and verification against tampering This ensures that agents only operate with approved and auditable components, thereby reducing exposure to malicious model injections. For example, in cybersecurity contexts, the integration of

blockchain with AI has been shown to securely log decision metadata and immutably record anomaly detection outputs, ensuring that both datasets and model-driven alerts remain tamper-resistant and verifiable [40].

4.2 Auditable Transparency

Provenance: Tracking Agent's Actions
Blockchain enables fine-grained provenance by recording which agent performed which task, under what conditions, and with what outcome. This ensures transparency and accountability in contexts where the traceability of decisions is essential for compliance such as supply chains [41] and intellectual property rights [42].

Decision: Tracing Cognitive Reasoning
Beyond actions, blockchain can capture the decision logic of AI-based agents [43], linking inputs, decision, and outcomes within immutable records. This transforms autonomous decisions from an opaque process into verifiable and accountable trails. In fields such as healthcare [44], finance and cybersecurity [45], the integration of explainable AI with distributed ledgers enables decisions to be accurate, interpretable and tamper-resistant. It also allows the verification of post hoc justification, alignment, and reliability of autonomous agents.

Transaction: Verifying Multi-agent Exchanges
Auditability also extends to transactions between agents. Blockchain technology secures and tracks exchanges across peer-to-peer networks. It enables verifiable cooperation among AI agents operating within its record system. In supply chain management, for example, AI agents can autonomously handle tasks such as control, optimization, and prediction [46]. As peers in a blockchain, agents' exchanges can be immutably recorded to provide trails that support accountability and mitigation concerns.

4.3 Conditions and Limitations

Security and auditability position blockchain technology as a promising trust infrastructure for AI-based multi-agent systems. Unlike secure databases or audited cloud solutions, blockchains enable distributed agentic intelligence, consensus, and on-chain mechanisms for incentive and governance, thereby reducing dependence on centralized trust infrastructures and mitigating single points of failure [47]. Numerous empirical studies have explored the integration of blockchain and artificial intelligence in domains such as security [48], and autonomous agents in decentralized finance [49]. However, fewer works have examined blockchain, proof-of-thought with LLM-based and cooperative multi-agent systems [50].

This combination also warrants analysis of its success factors and practical limitations. The success of blockchain and AI integration in multi-agent systems depends on at least three conditions. First, bounded autonomy ensures that agent behavior remains aligned with predefined rules and resource constraints, which can be enforced through smart contracts. Second, blockchain coherence maintains consistency between off-chain

AI learning models, on-chain state and oracles anchoring among agents. Third, explainability and accountability require that agent decisions and model updates need to be verifiable on-chain. In terms of limitations, the scalability of evaluation metrics, role and resource allocation among AI agents remains a challenge that often necessitates simplifying assumptions [50].

5 Conclusion

This paper examined how blockchain technology reinforces agentic AI through two complementary dimensions: verifiable security and auditability. The taxonomy emphasizes security in terms of identity assurance, validation of interactions, and model integrity. It also highlights auditability in terms of provenance, decision logic, and transaction traceability.

Further research is needed to build on our analysis. One promising direction is integrating game-theoretic approaches with blockchain-enabled agentic AI, especially cooperative and coopetitive solution concepts. Game theory provides rigorous basis to analyze strategic interactions among rational and autonomous players. Another promising technical area of research is the scalability of blockchain-enabled multi-agent systems, especially for the real-time evaluation and allocation of resources.

References

1. Natarajan, S., Mathur, S., Sidheekh, S., Stammer, W., Kersting, K.: Human-in-the-loop or AI-in-the-loop? Automate or collaborate? Proc. AAAI Conf. Artif. Intell. **39**(27), 28594–28600 (2025)
2. Politou, E., Casino, F., Alepis, E., Patsakis, C.: Blockchain mutability: challenges and proposed solutions. IEEE Trans. Emerg. Top. Comput. **9**, 1972–1986 (2019). https://doi.org/10.1109/TETC.2019.2949510
3. Nehme, E., El Sibai, R., Bou Abdo, J., Taylor, A.R., Demerjian, J.: Converged AI, IoT, and blockchain technologies: a conceptual ethics framework. AI Ethics. **2**, 129–143 (2022). https://doi.org/10.1007/s43681-021-00079-8
4. Huang, K. (ed.): Agentic AI: Theories and Practices. Springer Nature Switzerland, Cham (2025). https://doi.org/10.1007/978-3-031-90026-6
5. Wang, C., Shi, Y., Xiao, X.: A framework for evaluating AI agents in open-ended conversations via scripted simulation. In: Proceedings of the 31st ACM SIGKDD Conference on Knowledge Discovery and Data Mining V.2, pp. 5810–5818. ACM, Toronto ON Canada (2025). https://doi.org/10.1145/3711896.3737390
6. Acharya, D.B., Kuppan, K., Divya, B.: Agentic AI: autonomous intelligence for complex goals—a comprehensive survey. IEEE Access. **13**, 18912–18936 (2025). https://doi.org/10.1109/ACCESS.2025.3532853
7. Abdelhamid, M., Sliman, L., Ben Djemaa, R., Perboli, G.: A review on blockchain technology, current challenges, and AI-driven solutions. ACM Comput. Surv. **57**, 1–39 (2024)
8. Zheng, Z., Xie, S., Dai, H.N., Chen, X., Wang, H.: Blockchain challenges and opportunities: a survey. Int. J. Web Grid Serv. **14**, 352 (2018). https://doi.org/10.1504/IJWGS.2018.095647
9. Casino, F., Dasaklis, T.K., Patsakis, C.: A systematic literature review of blockchain-based applications: current status, classification and open issues. Telematics Inform. **36**, 55–81 (2019). https://doi.org/10.1016/j.tele.2018.11.006

10. Tasca, P., Tessone, C.J.: A taxonomy of blockchain technologies: principles of identification and classification. ledger. **4** (2019). https://doi.org/10.5195/ledger.2019.140
11. Sai, A.R., Buckley, J., Fitzgerald, B., Gear, A.L.: Taxonomy of centralization in public blockchain systems: a systematic literature review. Inf. Process. Manag. **58**, 102584 (2021). https://doi.org/10.1016/j.ipm.2021.102584
12. Zarrin, J., Wen Phang, H., Babu Saheer, L., Zarrin, B.: Blockchain for decentralization of internet: prospects, trends, and challenges. Clust. Comput. **24**, 2841–2866 (2021). https://doi.org/10.1007/s10586-021-03301-8
13. Gudgeon, L., Moreno-Sanchez, P., Roos, S., McCorry, P., Gervais, A.: SoK: layer-two blockchain protocols. In: International Conference on Financial Cryptography and Data Security, pp. 201–226 (2020). https://doi.org/10.1007/978-3-030-51280-4_12
14. Bellaj, B., Ouaddah, A., Bertin, E., Crespi, N., Mezrioui, A.: Drawing the boundaries between blockchain and blockchain-like systems: a comprehensive survey on distributed ledger technologies. Proc. IEEE. **112**, 247–299 (2024). https://doi.org/10.1109/JPROC.2024.3386257
15. Ballandies, M.C., Wang, H., Chee Law, A.C., Yang, J.C., Gösken, C., Andrew, M.: A taxonomy for blockchain-based decentralized physical infrastructure networks (DePIN). In: 2023 IEEE 9th World Forum on Internet of Things (WF-IoT), pp. 1–6. IEEE, Aveiro, Portugal (2023). https://doi.org/10.1109/WF-IoT58464.2023.10539514
16. Lin, Z., Wang, T., Shi, L., Zhang, S., Cao, B.: Decentralized physical infrastructure networks (DePIN): challenges and opportunities. IEEE Netw. **39**, 91–99 (2025). https://doi.org/10.1109/MNET.2024.3487924
17. Xu, X. et al.: A taxonomy of blockchain-based systems for architecture design. In: 2017 IEEE International Conference on Software Architecture (ICSA), pp. 243–252. IEEE, Gothenburg, Sweden (2017). https://doi.org/10.1109/ICSA.2017.33
18. Lashkari, B., Musilek, P.: A comprehensive review of blockchain consensus mechanisms. IEEE Access. **9**, 43620–43652 (2021). https://doi.org/10.1109/ACCESS.2021.3065880
19. Bouraga, S.: A taxonomy of blockchain consensus protocols: a survey and classification framework. Expert Syst. Appl. **168**, 114384 (2021). https://doi.org/10.1016/j.eswa.2020.114384
20. Hussein, Z., Salama, M.A., El-Rahman, S.A.: Evolution of blockchain consensus algorithms: a review on the latest milestones of blockchain consensus algorithms. Cybersecurity. **6**, 30 (2023). https://doi.org/10.1186/s42400-023-00163-y
21. Belchior, R., Vasconcelos, A., Guerreiro, S., Correia, M.: A survey on blockchain interoperability: past, present, and future trends. ACM Comput. Surv. **54**, 1–41 (2022). https://doi.org/10.1145/3471140
22. Pasdar, A., Lee, Y.C., Dong, Z.: Connect API with blockchain: a survey on blockchain oracle implementation. ACM Comput. Surv. **55**, 1–39 (2023). https://doi.org/10.1145/3567582
23. Li, W., Liu, Z., Chen, J., Liu, Z., He, Q.: Towards blockchain interoperability: a comprehensive survey on cross-chain solutions. Blockchain: Res. Appl. **100286** (2025). https://doi.org/10.1016/j.bcra.2025.100286
24. Khan, D., Jung, L.T., Hashmani, M.A.: Systematic literature review of challenges in blockchain scalability. Appl. Sci. **11**, 9372 (2021). https://doi.org/10.3390/app11209372
25. Rebello, G.A.F. et al.: A survey on blockchain scalability: from hardware to layer-two protocols. IEEE Commun Surv Tutor. **26**, 2411–2458 (2024). https://doi.org/10.1109/COMST.2024.3376252
26. Yang, C., Chin, K.-W., Wang, J., Wang, X., Liu, Y., Zheng, Z.: Scaling blockchains with error correction codes: a survey on coded blockchains. ACM Comput. Surv. **56**, 1–33 (2024). https://doi.org/10.1145/3637224

27. Li, X., Jiang, P., Chen, T., Luo, X., Wen, Q.: A survey on the security of blockchain systems. Futur. Gener. Comput. Syst. **107**, 841–853 (2020). https://doi.org/10.1016/j.future.2017.08.020
28. Liu, Y., He, D., Obaidat, M.S., Kumar, N., Khan, M.K., Raymond Choo, K.-K.: Blockchain-based identity management systems: a review. J. Netw. Comput. Appl. **166**, 102731 (2020). https://doi.org/10.1016/j.jnca.2020.102731
29. Peng, L., Feng, W., Yan, Z., Li, Y., Zhou, X., Shimizu, S.: Privacy preservation in permissionless blockchain: a survey. Digit Commun Netw. **7**, 295–307 (2021). https://doi.org/10.1016/j.dcan.2020.05.008
30. Han, R., Yan, Z., Liang, X., Yang, L.T.: How can incentive mechanisms and blockchain benefit with each other? A survey. ACM Comput. Surv. **55**, 1–38 (2022). https://doi.org/10.1145/3539604
31. Uddin, M.A., Stranieri, A., Gondal, I., Balasubramanian, V.: A survey on the adoption of blockchain in IoT: challenges and solutions. Blockchain: Res. Appl. **2**, 100006 (2021). https://doi.org/10.1016/j.bcra.2021.100006
32. Nguyen, T., Nguyen, H., Nguyen Gia, T.: Exploring the integration of edge computing and blockchain IoT: principles, architectures, security, and applications. J. Netw. Comput. Appl. **226**, 103884 (2024). https://doi.org/10.1016/j.jnca.2024.103884
33. Hussain, A.A., Al-Turjman, F.: Artificial intelligence and blockchain: a review. Trans. Emerg. Telecommun. Technol. **32**, e4268 (2021). https://doi.org/10.1002/ett.4268
34. Romandini, N., Mazzocca, C., Otsuki, K., Montanari, R.: SoK: security and privacy of AI agents forblockchain. In: 2025 7th International Conference on Blockchain Computing and Applications (BCCA). pp. 708–720. IEEE, Durbovnic, Croatia (2025). https://doi.org/10.1109/BCCA66705.2025.11229689
35. Mazzocca, C., Acar, A., Uluagac, S., Montanari, R., Bellavista, P., Conti, M.: A survey on decentralized identifiers and verifiable credentials. IEEE Commun Surv Tutor. (2025). https://doi.org/10.1109/COMST.2025.3543197
36. Di Francesco Maesa, D., Lisi, A., Mori, P., Ricci, L., Boschi, G.: Self sovereign and blockchain based access control: supporting attributes privacy with zero knowledge. J. Netw. Comput. Appl. **212**, 103577 (2023). https://doi.org/10.1016/j.jnca.2022.103577
37. Zou, Z., Liu, Z., Zhao, L., Zhan, Q.: Blocka2a: towards secure and verifiable agent-to-agent interoperability. arXiv preprint. https://arxiv.org/abs/2508.01332 (2025)
38. Lüthi, P., Gagnaux, T., Gygli, M.: Distributed ledger for provenance tracking of artificial intelligence assets. In: Friedewald, M., Önen, M., Lievens, E., Krenn, S., and Fricker, S. (eds.) Privacy and Identity Management. Data for Better Living: AI and Privacy. pp. 411–426. Springer International Publishing, Cham (2020). https://doi.org/10.1007/978-3-030-42504-3_26
39. Wang, Y., Pan, Y., Guo, S., Su, Z.: Security of internet of agents: attacks and countermeasures. IEEE Open J. Comput. Soc., 1–12 (2025). https://doi.org/10.1109/OJCS.2025.3589638
40. Goundar, S., Gondal, I.: AI-blockchain integration for real-time cybersecurity: system design and evaluation. J. Cybersecur. Priv. **5**, 59 (2025). https://doi.org/10.3390/jcp5030059
41. Malik, S., Dedeoglu, V., Kanhere, S.S., Jurdak, R.: PrivChain: provenance and privacy preservation in blockchain enabled supply chains. In: 2022 IEEE International Conference on Blockchain (Blockchain), pp. 157–166. IEEE, Espoo, Finland (2022). https://doi.org/10.1109/Blockchain55522.2022.00030
42. Wang, Q., Yu, G., Sai, Y., Bandara, H.M.N.D., Chen, S.: Is your AI truly yours? Leveraging B lockchain for copyright s, provenance, and lineage. IEEE Trans. Serv. Comput., 1–19 (2025). https://doi.org/10.1109/TSC.2025.3586091
43. Singh, J., Cobbe, J., Norval, C.: Decision provenance: harnessing data flow for accountable systems. IEEE Access. **7**, 6562–6574 (2019). https://doi.org/10.1109/ACCESS.2018.2887201

44. Bhardwaj, T., Sumangali, K.: An explainable federated blockchain framework with privacy-preserving AI optimization for securing healthcare data. Sci. Rep. **15**, 21799 (2025). https://doi.org/10.1038/s41598-025-04083-4
45. Kumar, P., Javeed, D., Kumar, R., Islam, A.K.M.N.: Blockchain and explainable AI for enhanced decision making in cyber threat detection. Softw. Pract. Experience. **54**, 1337–1360 (2024). https://doi.org/10.1002/spe.3319
46. Karim, M.M., Van, D.H., Khan, S., Qu, Q., Kholodov, Y.: AI agents meet blockchain: a survey on secure and scalable collaboration for multi-agents. Future Internet. **17**, 57 (2025). https://doi.org/10.3390/fi17020057
47. Shamsan Saleh, A.M.: Blockchain for secure and decentralized artificial intelligence in cybersecurity: a comprehensive review. Blockchain: Res. Appl. **5**, 100193 (2024). https://doi.org/10.1016/j.bcra.2024.100193
48. Kuznetsov, O., Sernani, P., Romeo, L., Frontoni, E., Mancini, A.: On the integration of artificial intelligence and blockchain technology: a perspective about security. IEEE Access. **12**, 3881–3897 (2024). https://doi.org/10.1109/ACCESS.2023.3349019
49. Gularso, S.N., Wiputra, R.: Are autonomous agents the inevitable future of decentralized finance? A review and forward outlook. In: 2025 17th International Conference on Intelligent Human-Machine Systems and Cybernetics (IHMSC), pp. 140–143. IEEE, Hangzhou, China (2025). https://doi.org/10.1109/IHMSC66529.2025.00038
50. Chen, B., Li, G., Lin, X., Wang, Z., Li, J.: BlockAgents: towards byzantine-robust LLM-based multi-agent coordination via blockchain. In: ACM Turing Award Celebration Conference 2024, pp. 187–192. ACM, Changsha China (2024). https://doi.org/10.1145/3674399.3674445

Supervised Classification for Intelligent Decision-Making Systems: Critical Review, Applications, and Challenges

Mouataz Idrissi Khaldi[1](✉), Allae Erraissi[2], Mustapha Hain[1], and Mouad Banane[1]

[1] ENSAM, Hassan II University, Casablanca, Morocco
mo3tazik@gmail.com
[2] Chouaib Doukkali University, El Jadida, Morocco

Abstract. At a time when digital technologies are transforming practices in critical areas such as healthcare, supervised classification algorithms are essential tools for automating decision-making and ensuring its reliability. This paper presents an in-depth comparative study of the most widely used models through a multi-criteria evaluation that integrates aspects of performance, interpretability, and operational constraints.

The methodological approach is based on the analysis of recent publications and concrete examples of medical applications, such as the prediction of chronic diseases, the analysis of biomedical signals, and the automation of medical coding. The results highlight the superiority of ensemble models and margin classifiers for raw performance, while explainable models retain a decisive advantage in regulated or low-resource contexts.

The article provides a visual and tabular summary of the strengths and weaknesses of each algorithm, accompanied by a discussion of the trade-offs between accuracy, transparency, and technical constraints. In conclusion, concrete recommendations are made to guide the choice and implementation of models, highlighting the growing interest in hybrid, responsible, and adaptable solutions in the context of reliable and sustainable artificial intelligence applied to healthcare.

Keywords: supervised classification · machine learning · artificial intelligence · intelligent systems · digital health

1 Introduction

In a data-driven world, the explosion in volumes of digital information in sectors as varied as healthcare, finance and education reinforces the need for intelligent systems capable of extracting relevant knowledge and effectively assisting decision-making [1]. Artificial intelligence (AI), and machine learning in particular, is emerging as a strategic response to these challenges, with supervised classification occupying a central place in intelligent decision-making systems [2]. This involves learning from labeled data to predict a class or outcome (e.g. medical diagnosis, fraud detection, risk assessment). This ability to automate decisions makes classification indispensable in critical environments, notably in healthcare, where it contributes to screening, diagnosis and prevention [1, 3].

A. E. Hassanien et al. (Eds.): AISDS 2025, LNDECT 286, pp. 235–245, 2027.
https://doi.org/10.1007/978-3-032-18758-1_23

Despite the rise of advanced approaches such as deep neural networks or reinforcement learning, classical supervised classification algorithms - such as Naive Bayes, k-Nearest Neighbors (k-NN), Support Vector Machines (SVM), decision trees or random forests - retain a marked relevance in real systems. Their simplicity of implementation, robustness against small datasets and interpretability make them the tools of choice, particularly in contexts where computational resources are scarce [2, 4]. These qualities are particularly crucial in connected healthcare systems, where transparency of decisions and confidence in models are fundamental requirements [1].

In this context, it is essential to reassess the performance, limitations and applicability of conventional algorithms in the face of the growing demands of modern decision-making systems. Can they still be trusted in dynamic, noisy or regulated environments? Which methods remain the most relevant depending on the use case, particularly in the healthcare sector, where interpretability and reliability are paramount?

This study adopts a structured literature review approach, primarily focused on recent scientific publications (2023–2025) to capture the most up-to-date advances in supervised classification. However, earlier foundational works (2000–2022) were also included when they provided essential theoretical or methodological background. The rapid evolution of artificial intelligence, particularly in healthcare, justifies this temporal focus on the last three years. The works are analyzed according to a common framework: the algorithm's operating principle, advantages, limitations, complexity, and fields of application. Particular emphasis is placed on healthcare use cases, in order to assess their relevance to critical decision-making systems.

A thematic review makes it possible to synthesize contributions scattered throughout the literature, offering a panoramic and up-to-date view. It thus provides a useful analytical framework for researchers, practitioners and decision-makers, in a context where the choice of an algorithm depends not only on its raw performance, but also on criteria such as interpretability, complexity or robustness [3].

The review will identify contexts in which classical algorithms are still relevant, or even preferable to more complex approaches. It will also propose recommendations for their integration into intelligent decision-making systems, with a particular focus on healthcare applications.

This study defends the idea that classical supervised classification algorithms, far from being obsolete, continue to meet real needs, particularly in constrained or highly regulated contexts. Their role in the interpretability, reliability and transparency of algorithmic decisions remains fundamental. Moreover, their hybridization with modern approaches represents a promising avenue for future research.

2 Methodology

This study adopts a structured comparative review approach to analyze supervised classification algorithms applied to intelligent decision-making systems, with a particular focus on healthcare applications. Although not a systematic PRISMA-type review, it follows a rigorous methodology for source selection, analysis, and synthesis, in line with best practices for comparative narrative reviews. The objective is to assess the performance, robustness, complexity, and applicability of six widely used supervised algorithms—Naive Bayes, k-NN, Decision Tree, Random Forest, Support Vector Machine

(SVM), and Logistic Regression—in real-world, high-demand environments. The aim is to identify their strengths, limitations, and trade-offs to guide responsible and efficient model selection for intelligent healthcare systems.

The literature considered spans the period 2000–2025, with an emphasis on 2023–2025 to capture recent developments and emerging trends in artificial intelligence and healthcare. Earlier seminal works were also retained when they provided essential conceptual or methodological insights. Eligible documents included journal or conference publications (IEEE, Elsevier, Springer, etc.) offering comparative or experimental studies on supervised algorithms. The selection focused on works addressing either general algorithmic comparisons or healthcare-related applications such as diagnosis, prediction, and triage. Only English-language papers were considered, and studies were included when they implemented at least one of the six targeted algorithms and presented measurable performance outcomes.

The literature search was conducted across multiple databases—SciSpace, IEEE Xplore, Google Scholar, PubMed, and SpringerLink—using combined keywords such as "supervised classification", "machine learning", "healthcare", "model comparison", "performance", "interpretability", and "decision support". Approximately thirty papers were initially identified, and fifteen were retained after critical reading and qualitative filtering to ensure methodological consistency and representativeness of current research trends.

Each algorithm was then analyzed through a multi-criteria evaluation grid comprising nine key dimensions inspired by established literature in AI for healthcare: overall accuracy, sensitivity (recall), robustness to noise, tolerance to missing data, interpretability, training time, prediction time, risk of overfitting, and efficiency on small datasets. Each criterion was rated on a five-point scale based on evidence from the reviewed publications. These ratings were subsequently summarized in a comparative table and visualized as a radar chart (see Section 6), facilitating a cross-sectional interpretation of algorithmic profiles and their practical suitability for healthcare contexts.

3 Fundamentals of Supervised Classification Methods

Supervised classification is a fundamental approach to machine learning, in which an algorithm learns to predict a label or class from an annotated dataset. It is based on a supervised learning process using input-output pairs, enabling the construction of a model that can be generalized to new observations [5]. In this section, we present the main classical classification algorithms, detailing their operation, advantages and limitations, in order to lay the foundations for the comparison developed in the following sections.

Naive Bayes (NB) is a probabilistic algorithm based on Bayes' theorem, combined with the assumption of conditional independence between attributes. It is particularly appreciated for its simplicity, speed of training, and robust performance in contexts such as text classification or elementary medical diagnosis [1]. However, this independence assumption is rarely verified in practice, which can adversely affect the accuracy of the model on complex, correlated data.

The Decision Tree (DT) is a tree structure in which each internal node represents a test on a characteristic, each branch a possible test value, and each leaf an output class.

Its main advantage lies in its interpretability, which makes it a widely used model in fields where transparency of decisions is essential, such as medicine [1]. However, it is sensitive to overlearning, particularly in the presence of noise or unbalanced data.

Random Forest (RF) is an ensemble method consisting of multiple decision trees, built on random subsamples of data and variables. This technique improves the robustness, accuracy and stability of predictions, at the cost of a relative loss in interpretability compared with a single tree [6]. RF is frequently used in complex medical applications for multivariate detection or prediction [2].

k-Nearest Neighbors (k-NN) is a non-parametric algorithm that assigns a class to an unlabeled instance based on the majority classes among its k nearest neighbors. It is simple to implement and efficient on small, well-separated data sets. However, it is costly in terms of prediction time and sensitive to dimensionality and noise, which can limit its effectiveness on complex real data [5].

Support Vector Machine (SVM) is an optimization method that aims to find the separating hyperplane maximizing the margin between classes. Thanks to the use of kernel functions (kernels), it is able to deal efficiently with non-linear separations. SVMs are renowned for their robustness on small datasets and their performance in a variety of contexts, including tumor detection and medical signal analysis [4, 7]. Nevertheless, their computational cost can become high on large datasets, and their interpretation remains limited, which can pose a problem in sensitive areas such as healthcare.

Although often perceived as a simple model, Logistic Regression (LR) remains an essential reference for binary classification. It offers a direct interpretation of coefficients, models the probability of class membership, and is a frequent starting point in clinical studies [1]. However, its performance is generally inferior in non-linear or complex multi-class contexts.

In short, each of these algorithms has specific characteristics that influence its effectiveness depending on the nature of the data, the constraints of the environment and the application objectives. An in-depth comparison based on rigorous criteria will enable us, in the following sections, to identify the models best suited to each context of use, particularly in healthcare.

4 Comparison Criteria for Supervised Classification Algorithms

The evaluation of supervised classification algorithms relies on a set of key criteria frequently discussed in the literature [8–12]:

- **Accuracy and related metrics** – Overall correctness of predictions; complemented by recall, precision, F1-score, and ROC-AUC for unbalanced datasets [9, 10].
- **Algorithmic complexity and execution time** – Training and inference costs determining feasibility in real-time or embedded systems [8, 11].
- **Interpretability** – Transparency of decision-making processes, crucial in regulated domains such as healthcare [4, 11].
- **Robustness to noise and missing data** – Model stability when facing imperfect or incomplete inputs [12].
- **Adaptability to small datasets** – Capability to maintain performance with limited data [5, 8].

- **Overfitting tendency** – Risk of poor generalization due to excessive adaptation to training data [10].

These criteria serve as the foundation for the comparative analysis presented in the following sections.

5 Areas of Application for Supervised Classification Algorithms

Supervised classification algorithms have established themselves as key tools in a multitude of fields where automated decision-making is essential. Their ability to learn from labeled data and generalize their predictions to new cases makes them fundamental components of modern intelligent systems. This section explores their role in a variety of application contexts, starting with generalist fields, before focusing on uses in healthcare, a sector particularly sensitive to the challenges of accuracy, interpretability and robustness.

5.1 General Applications

In the education sector, supervised models are mobilized to predict academic success, detect students at risk of dropping out, or recommend personalized pedagogical paths. Decision trees, logistic regression and SVMs are frequently used for their ability to model complex relationships while maintaining good interpretability [5].

In finance, classification is exploited for credit scoring, fraud detection and customer behavior analysis. The work of [8] shows that SVM and Random Forest models are particularly effective on massive transactional data, offering both robustness and accuracy, even in contexts of unbalanced classes. Their integration into real-time IoT systems also requires particular attention to prediction speed and memory footprint.

In digital marketing, classification is used for recommendation systems, sentiment analysis and churn prediction [10]. highlight the adaptability of models such as Naive Bayes or LightGBM to textual, noisy and evolving data, typical of digital environments.

5.2 Applications in the Healthcare Field

Healthcare is a field where supervised classification algorithms have demonstrated a direct impact on the quality of care, diagnostic efficiency and hospital management. These models intervene at several levels, from prevention to treatment, via medical decision support.

A recent review by [13] developed a Random Forest-based model capable of predicting the length of hospitalization of patients treated for cerebral aneurysms with an AUC of 0.928, illustrating the usefulness of these approaches in hospital planning and resource allocation.

A recent review by [14] identifies SVM, Random Forest and neural network algorithms as among the most widely used in assisted diagnosis systems, personalized medicine and patient follow-up. Their integration into CDSS (Clinical Decision Support Systems), however, requires a balance between robustness and explainability.

In the context of classifying non-communicable diseases (e.g. diabetes, COVID-19), [15] report accuracies approaching 100% using models such as decision trees, Random Forest and SVM. These results testify to the effectiveness of classical classifiers, while underlining the importance of pre-processing and variable selection.

For its part, [16] demonstrates that the combination of an artificial neural network (ANN) with logistic regression enables diabetes types 1 and 2 to be classified with an accuracy in excess of 96%, underlining the potential of hybrid approaches for complex diagnoses.

In intensive care units, [17] have used supervised models to improve early detection of sepsis, proposing high-throughput operational deployment strategies.

Finally, several recent works highlight the limitations of strictly supervised approaches in real clinical environments, due to data variability, noise or lack of generalizability [18]. suggest avenues towards hybrid and semi-supervised strategies to improve robustness, while [19] stress the need for ongoing model evaluation to guarantee long-term reliability in medical applications.

6 Comparative Synthesis of Supervised Classification Algorithms

To assess the performance of supervised classification algorithms in intelligent systems, particularly in healthcare, we have established a multi-criteria analysis grid rated on a scale of 1 (poor) to 5 (excellent), according to recent studies [14, 15, 20–22]. This summary is presented in the form of a comparative table, accompanied by a radar chart illustrating the performance profiles of the six models analyzed.

Table 1. Multi-criteria comparison of supervised healthcare algorithms (scale from 1 to 5).

Criteria	Naive Bayes	k-NN	Decision Tree	Random Forest	SVM	Logistic Regression
Overall accuracy	3	4	4	5	5	4
Sensitivity (Recall)	4	3	3	5	5	4
Noise robustness	2	2	3	5	4	3
Tolerance to missing data	2	2	4	4	2	3
Interpretability	3	2	5	3	2	4
Training time	5	2	1	3	2	4
Prediction time	5	1	5	4	3	5
Risk of overfitting	5	2	2	4	3	5
Efficiency on small datasets	4	5	4	3	4	4

This table highlights the relative strengths and weaknesses of each algorithm according to key criteria adapted to intelligent systems, particularly in healthcare. For example, Random Forest stands out for its overall robustness, while Naive Bayes excels in speed, and decision trees offer maximum interpretability. Conversely, models such as k-NN or SVM, while effective on certain criteria, have limitations in the presence of noise or missing data. This multi-criteria evaluation enables us to better orient our algorithmic choice according to the specific constraints of the targeted applications (real-time, explainability, imperfect data, etc.).

The numerical scores presented in Table 1 were derived from comparative analyses and empirical findings reported in recent studies [14, 15, 20, 21], and [22]. Random Forest and SVM obtained consistently high accuracies (above 95%) and strong recall on medical datasets, justifying their top scores in overall performance and sensitivity. Decision Trees were rated highest for interpretability due to their transparent rule-based structure. Naive Bayes achieved excellent training and prediction times but lower robustness and noise tolerance, mainly due to its independence assumption. k-NN performed well on small datasets yet showed reduced efficiency on larger or noisy data. Logistic Regression was considered a balanced model, combining moderate accuracy, interpretability, and stability. These quantitative values thus reflect the general consensus observed in the literature on algorithmic trade-offs between performance, robustness, and explainability in healthcare contexts.

All criteria were intentionally considered with equal importance to preserve transparency and avoid subjective bias, as the relevance of each dimension varies across healthcare applications. Future work may apply weighted multi-criteria decision analysis (MCDA) to reflect domain-specific priorities more precisely (Fig. 1).

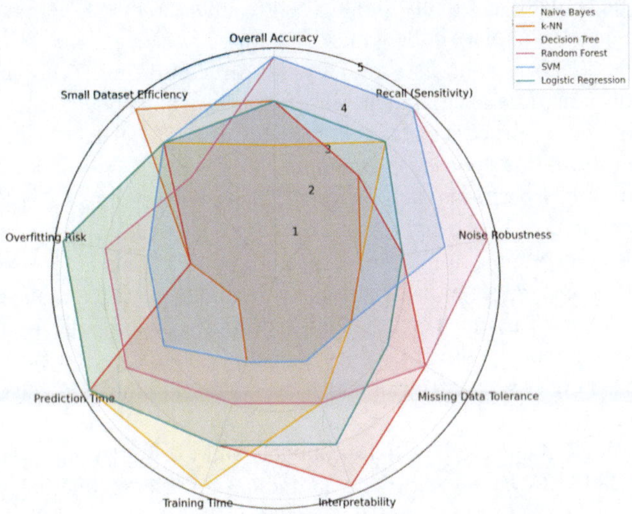

Fig. 1. Radar comparison of supervised algorithms

This graph visually illustrates the relative performance of each algorithm on ten key dimensions. Random Forest dominates overall, with excellent accuracy, robustness and sensitivity, but moderate interpretability. SVM performs very well in accuracy, but is limited in training speed and transparency. Decision Tree excels in interpretability, while Naive Bayes stands out for its speed. k-NN performs well on small datasets, but suffers in prediction time. Finally, logistic regression stands out as a good overall compromise.

6.1 Discussion of the Results

The results of the multi-criteria comparison confirm Random Forest (RF) and Support Vector Machine (SVM) as the benchmark models in medical contexts, combining excellent accuracy (often >95%), robustness against noise, and stability on varied datasets [22, 23]. RF has demonstrated its superiority notably in hospital outcome prediction tasks [13] or the classification of EEG signals in immersive environments for mental health [22]. These performances testify to its ability to handle complex, multidimensional and heterogeneous data.

However, this technical efficiency comes up against the barrier of interpretability, crucial for ethical clinical adoption, regulated and supported by medical staff. This is where simpler models such as the decision tree (DT) or logistic regression (LR) regain their relevance: they offer a direct reading of the decision rules or coefficients, which facilitates the processes of medical validation, audit or regulatory interpretability [14, 24].

The work of [15] illustrates that, properly calibrated and integrated, these classic models can achieve accuracies close to 100% in the classification of chronic diseases (including COVID-19), without recourse to complex architectures or massive resources. This demonstrates that data quality, preprocessing, and variable selection can often be more decisive than the choice of model itself.

Naive Bayes (NB), despite its simplicity and structural limitations (independence assumption, low noise tolerance), retains a useful place in embedded systems, automatic sorting solutions or pre-processing tasks in medical NLP [20]. It excels where resources are limited and speed of response is paramount.

Conversely, k-NN performs well on small, clean datasets but faces limitations in scalability, computational cost, and noise sensitivity. This makes it unsuitable for use in large-scale hospital systems, or in the presence of imperfect data [21].

A cross-sectional reading of the results also highlights the fact that the effectiveness of an algorithm depends as much on its technical profile as on its application context. For example, in the automatic coding of causes of death (ICD-10), the objective is not only precision, but also reliability of recall, since every misclassification can affect public health statistics [21].

From a strategic point of view, hybrid models and integrated interpretability techniques (e.g. SHAP, LIME) are set to play an increasing role, reconciling performance and transparency. In addition, emerging constraints linked to computational sustainability, notably in mobile healthcare systems or low-resource connected hospitals, mean that model frugality must be considered as a new selection criterion [20, 23].

From a healthcare perspective, the comparative findings highlight that the choice of classifier must align with specific clinical objectives. For early disease detection — where minimizing false negatives is vital — models with high recall and robustness such as Random Forest or SVM are most appropriate, as they can identify subtle patterns in heterogeneous data. Conversely, in the classification and long-term monitoring of chronic diseases, interpretability and model stability are prioritized, making Decision Trees or Logistic Regression more suitable for explaining decisions to clinicians. In hospital management contexts, where real-time prediction and scalability are crucial, lightweight algorithms such as Naive Bayes or optimized ensemble methods remain relevant due to their speed and efficiency. This multidimensional alignment between algorithmic profiles and healthcare needs ensures that model performance metrics are not viewed in isolation, but rather through the lens of clinical impact, ethical transparency, and operational feasibility.

In summary, there is no "perfect" model but rather algorithmic profiles suited to specific contexts. An appropriate choice depends on balancing performance, interpretability, robustness, and operational constraints according to user needs and application requirements.

7 Prospects and Recommendations

Future work should address domain-specific challenges such as managing class imbalance in rare diseases, ensuring real-time inference in critical care environments, and preserving data privacy through federated learning [13, 16, 20]. Developing explainable hybrid systems that combine classical transparency with deep learning adaptability remains a promising direction for reconciling accuracy and interpretability [22, 23]. Furthermore, cross-institutional studies are needed to validate the long-term robustness and generalizability of medical AI solutions across diverse healthcare contexts.

8 Conclusion

This study provided a comparative overview of supervised classification algorithms, highlighting their usefulness in intelligent healthcare systems. Using a multi-criteria grid, we highlighted the high performance of models such as Random Forest and SVM, particularly in critical medical tasks. However, their lack of explainability limits their adoption in sensitive clinical contexts. Conversely, simpler models such as decision trees offer appreciable transparency, although often at the expense of performance.

Although this review is based on a rigorous approach, certain limitations need to be recognized: It is not based on a proper experimental foundation (no direct implementation was carried out). The number of algorithms evaluated is limited to conventional methods. Comparative scores may vary according to datasets, hyperparameter settings and selected metrics. Despite these limitations, this methodology provides an integrated, operational and well-argued reading of recent literature, sufficient to inform algorithmic decisions in critical contexts such as the digital transformation of healthcare.

References

1. Zhou, X.: A study of machine learning applications in healthcare. Appl. Comput. Eng. **102**, 128–133 (2024)
2. Irani, H., Ghahremani, Y., Kermani, A., Metsis, V.: Time series embedding methods for classification tasks: a review. Expert. Syst. **42**(11), e70148 (2025)
3. Dhawale, C.A., Dhawale, K.A.: A review on application of reinforcement learning in healthcare. In: Dobhal, D., Sharma, S., Purohit, K., Nautiyal, L., Singh, K. (eds.) Cyber Trafficking, Threat Behavior, and Malicious Activity Monitoring for Healthcare Organizations, pp. 105–119. IGI Global, Hershey, PA (2023)
4. Riahi, V., Diouf, I., Khanna, S., Boyle, J., Hassanzadeh, H.: Digital twins for clinical and operational decision-making: scoping review. J. Med. Internet Res. **27**, e55015 (2025)
5. Kotsiantis, S.B., Zaharakis, I., Pintelas, P.: Supervised machine learning: a review of classification techniques. Emerg. Artif. Intell. Appl. Comput. Eng. **160**, 3–24 (2007)
6. Breiman, L.: Random forests. Mach. Learn. **45**(1), 5–32 (2001)
7. Cortes, C., Vapnik, V.: Support-vector networks. Mach. Learn. **20**(3), 273–297 (1995)
8. Khadse, V.M., Mahalle, P.N., Biraris, S.V.: An empirical comparison of supervised machine learning algorithms for internet of things data. In: Proceedings of the 4th International Conference on Computing Communication Control and Automation (ICCUBEA 2018), pp. 1–6. IEEE, Pune (2018)
9. Kedia, A., Narsaria, M., Goswami, S., Taparia, J.: Empirical study to evaluate the performance of classification algorithms on healthcare datasets. World J. Comput. Appl. Technol. **5**(1), 1–11 (2017)
10. Patil, M.S., Patil, H.D.: Acute pain recognition using an ensemble learning method: evaluation of performance and comparison. Int. Res. J. Multidiscip. Technovation. **7**(1), 102–114 (2025)
11. Aljaaf, A.J., Al-Jumeily, D., Hussain, A.J., Fergus, P., Al-Jumaily, M., Radi, N.: A systematic comparison and evaluation of supervised machine learning classifiers using headache dataset. In: Huang, D.S., Han, K. (eds.) Advanced Intelligent Computing Theories and Applications. ICIC 2015. Lecture Notes in Computer Science, vol. 9227, pp. 596–604. Springer, Cham (2015)
12. Krizea, M., Gialelis, J., Koubias, S.: Comparative study between fuzzy inference system, adaptive neuro-fuzzy inference system and neural network for healthcare monitoring. In: Proceedings of the 8th Mediterranean Conference on Embedded Computing (MECO 2019), pp. 1–4. IEEE, Budva (2019)
13. Zhao, J., Yi, L.: Development and validation of machine learning-based model for hospital length of stay in patients undergoing endovascular interventional embolization for intracranial aneurysms. World Neurosurg. **195**, 123636 (2025)
14. Barhate, A., Tale, A., Jikar, N., Verma, P., Kumar, P., Yesankar, P.: A comprehensive review of supervised learning algorithms in healthcare applications. In: Proceedings of the 2nd DMIHER International Conference on Artificial Intelligence in Healthcare, Education and Industry (IDICAIEI 2024), pp. 1–6. IEEE, Wardha (2024)
15. Mohammed, A.J., Abdulazeez, A.M.: Healthcare diseases classification based on machine learning algorithms: a review. Int. J. Inform. Inf. Sys. Comput. Eng. (INJIISCOM). **5**(2), 218–252 (2024)
16. Kumari, S., Ranganayaki, V.C., Sathyamoorthy, K., Selvi, K., Mohanaprakash, T.A., Tamilselvi, C.: Optuna-optimized machine learning technique for accurate diabetes prediction and classification. In: Proceedings of the 4th International Conference on Sustainable Expert Systems (ICSES 2024), pp. 1478–1485. IEEE, Kaski (2024)
17. Anjana, G., Nisha, K.L., Arun Sankar, M.S.: Improving sepsis classification performance with artificial intelligence algorithms: a comprehensive overview of healthcare applications. J. Crit. Care. **83**, 154815 (2024)

18. Gu, X., et al.: Beyond supervised learning for pervasive healthcare. IEEE Rev. Biomed. Eng. **17**, 42–62 (2024)
19. Souza, C.M.P., et al.: Machine learning model evaluation on healthcare applications: a review. Seven Editora. https://sevenpubl.com.br/editora/article/view/2824 (2023). Accessed 28 Oct 2025
20. Shivakanth, G., Marapelli, B., Shivakumar Reddy, A., Manasa, D., Konda, S.: Classification of brain tumor using machine learning techniques. In: Srivastava, S., Anand, A., Kumar, A., Saini, B., Rathore, P.S. (eds.) Bio-Inspired Optimization for Medical Data Mining, pp. 129–139. Wiley, Hoboken, NJ (2024)
21. Nordin, M.B., et al.: A comparative analysis of machine-learning algorithms for automated International Classification of Diseases (ICD-10) coding in Malaysian death records. Cureus. **17**(1), e77342 (2025)
22. Zuo, M., Chen, X., Sui, L.: Evaluation of machine learning algorithms for classification of visual stimulation-induced EEG signals in 2D and 3D VR videos. Brain Sci. **15**(1), 75 (2025)
23. Saha, S.: An overview and analysis of machine learning classification algorithms in healthcare. In: Applying Machine Learning Techniques to Bioinformatics: Few-Shot and Zero-Shot Methods, pp. 39–61. Springer, Cham (2024)

Anomaly Detection for HVAC Systems in Smart Buildings: A Machine Learning Approach

Elhabyb Khaoula(✉), Baina Amine, and Bellafkih Mostafa

RAISS Lab, National Institute of Posts and Telecommunications - INPT Rabat, Rabat, Morocco
elhabyb.khaoula@doctorant.inpt.ac.ma, {baina,bellafkih}@inpt.ac.ma
http://www.inpt.ac.ma/

Abstract. Smart buildings rely heavily on predictive maintenance to ensure optimal performance and energy efficiency. An essential aspect of this maintenance involves anomaly detection, which identifies irregularities in system behavior that could signal potential failures or inefficiencies. In this study, we focus on applying two popular anomaly detection algorithms, Isolation Forest and Local Outlier Factor (LOF), to detect anomalies in a real HVAC system dataset. We evaluate the performance of these algorithms using multiple metrics, including precision, F-score, and accuracy, to assess their effectiveness in identifying deviations from normal system operation. The results demonstrate the advantages and limitations of each algorithm, providing valuable insights into their applicability in predictive maintenance for smart building systems.

Keywords: Anomaly detection · HVAC system · Smart Buildings · Machine Learning · Unsupervised Learning · energy efficiency

1 Introduction

In the era of artificial intelligence, the integration of smart buildings has become a necessity. A smart building can be defined as any structure utilizing automated processes to manage various activities, including heating, ventilation, air conditioning, lighting, security, and other systems[1]. Among these components, the HVAC system holds a pivotal role, as it is responsible for maintaining indoor comfort but also accounts for a significant portion of energy consumption [2]. This becomes particularly critical when anomalies occur, leading to inefficiencies and wasted energy.

To address these challenges, our work focuses on one of the key operational axes of smart buildings: predictive maintenance. By employing machine learning techniques for anomaly detection, we aim to identify irregularities in HVAC systems promptly and ensure their optimal functioning. This approach not only enhances energy efficiency but also aligns with the broader goal of sustainable building management within the framework of smart building innovations

A. E. Hassanien et al. (Eds.): AISDS 2025, LNDECT 286, pp. 246–252, 2027.
https://doi.org/10.1007/978-3-032-18758-1_24

it is essential to highlight that predictive maintenance is composed of three key aspects [3]: remaining useful life (RUL) estimation[4], failure prediction, and anomaly detection. Each of these components represents a critical area of research and practical contribution to ensuring the reliability and efficiency of modern systems. In this context, this paper will specifically focus on anomaly detection. It can be defined as the process of identifying unusual, unexpected, or irregular behaviors within a system or an ensemble of systems, which deviate from normal operational patterns[5]. These anomalies may indicate early signs of potential failures, inefficiencies, or non-functioning components. Unlike outright system failures, anomalies often manifest subtly, requiring advanced techniques to detect and address them effectively before they escalate into critical problems.

This paper is structured as follows: First, the introduction presents the general concept of our study and highlights its importance. The second section provides a literature review that examines existing works in the field. The third section outlines the methodology, detailing the processes and approaches employed in our contribution and including a description of the dataset used. Next, we present the results, evaluated through performance metrics. Finally, the paper concludes with a summary of our findings and outlines future perspectives for further research

2 Literature Review

This section summarizes key studies on anomaly detection in smart buildings, focusing on methodologies, datasets, and results. To provide a clear overview, the reviewed literature is presented in a Table 1, highlighting advancements and gaps in the field.

3 Methodology

3.1 Data Collection

In this study, we utilized a dataset derived from a case study conducted at Case Western Reserve University, spanning from March 3, 2018, to June 7, 2019. The dataset comprises hourly sensor measurements, offering valuable insights into the performance and condition of an HVAC system [13]. Data were collected from two key components:

- Drive End (DE): This includes critical elements such as fans, compressors, and pumps, with sensors monitoring vibration, temperature, and power consumption.
- Fan End: Sensors in this section focus on detecting fan overheating (temperature), vibrations, and energy usage patterns.

A rigorous preprocessing step was conducted to eliminate outliers and redundancies, ensuring data reliability. The analysis, as illustrated in Fig. 1, emphasizes three key parameters, with anomaly detection specifically targeting the

Table 1. Literature Review on Anomaly Detection in Smart Buildings

Article	Algorithm	Type of Building	Dataset	Key Contributions	Results
Himeur et al.[6]	K-Nearest Neighbors	Residential and Commercial	power consumption data	abnormality detection in power consumption	Achieved 92% accuracy for detecting abnormalities with low false positive rates.
Himeur et al.[7]	Review of AI techniques	General	N/A	Summarizes current AI methods for energy anomaly detection.	highlights gaps in existing techniques.
Cicero et al.[8]	IoT-enabled Smart Buildings	1 N/A	IoT sensor dataset	Proposes a deep anomaly detection system leveraging IoT data for real-time irregularity identification.	Achieved 95% precision and recall using CNN-based autoencoders.
Staer et al.[9]	Random forest, decision tree	Office Buildings	Public energy dataset	Focuses on energy optimization and management via ML-based anomaly detection.	Improved energy efficiency by 15% and reduced false alarms in anomaly detection.
karakolis et al.[10]	End-to-end Scalable Pipeline	Industrial Buildings	Industrial sensor data	Develops a scalable real-time framework for efficient anomaly detection across large-scale buildings.	Demonstrated real-time detection with <2% latency and 90% accuracy.
Erhan et al.[11]	k-means	Mixed-use Buildings	Sensor dataset	Addresses unlabeled sensor data challenges with clustering algorithms for event anomaly detection.	Detected anomalies with an F1-score of 88% and adapted well to unlabeled data.
Haque et al.[12]	Ensemble Learning	Mixed Commercial Buildings	Control system data	Combines multiple models to enhance the accuracy and efficiency of anomaly detection in control systems.	Outperformed individual models with an accuracy of 96% and reduced energy waste by 12%.

Drive End (DE), a vital component of the system's functionality and overall performance.

The figure illustrates the similar operation of the Drive End and Fan End components. However, the key distinction lies in the failure parameter, which identifies whether the component is functioning normally or experiencing a fault. This differentiation enhances the precision of anomaly detection in the HVAC system.

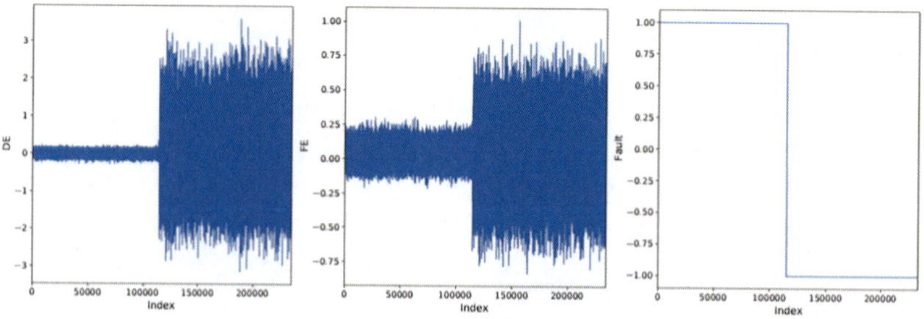

Fig. 1. Enter Caption

3.2 Model Development (Training)

To detect anomalies in the HVAC system, we chose two common anomaly detection algorithms: Isolation Forest (iForest) and Local Outlier Factor (LOF), based on the literature review. Both methods are unsupervised and effective for identifying outliers in large datasets.

1. Isolation Forest [14]: uses decision trees to isolate data points, with quickly isolated points identified as anomalies. it is presented mathematically as:

$$s(x) = 2^{-\frac{E(h(x))}{c(n)}} \tag{1}$$

with: $E(h(x)$The average path length to isolate the point x.

2. Local Outlier Factor (LOF)[15]: detects anomalies by comparing a point's local density with that of its neighbors. A point is considered an anomaly if its density is significantly lower. Presented As:

$$A = \frac{1}{|N_k(p)|} \sum_{N_k(p)} \frac{\mathrm{lrd}(o)}{\mathrm{lrd}(p)} \tag{2}$$

with: $N_k(p)$ set of k neighbors and $lrd(p)$ (local reachability density) of p

3.3 Model Evaluation (Test)

For anomaly detection, we utilize performance metrics designed to evaluate deviations from normal behavior. The key metrics include F1 score, recall, precision, and accuracy.

4 Result and Discussions

Anomaly detection relies on unsupervised algorithms to identify unusual behaviors or values in data, such as rare events, malfunctions, or irregularities in systems. In this study, a value of -1 is used to indicate a detected anomaly, particularly for algorithms like Isolation Forest or Local Outlier Factor, where each data point is classified based on its likelihood of being an anomaly. The results are presented in Table 2 for the test and the train data. Table 3 provides a detailed analysis of the anomalies detected by the selected models, offering an overview of the data points classified as abnormal in the test data. Based on the results in the table, Isolation Forest excels in recall, achieving 99.99% on the training set and 87.00% on the test set. However, its precision remains moderate (59.39% for training and 59.51% for testing), and its AUC is relatively low (65.81% for training and 65.98% for testing). On the other hand, Local Outlier Factor (LOF) shows a slightly lower precision (55.56% for training and 55.58% for testing) but outperforms Isolation Forest in AUC (70.00% for training and 70.04% for testing). LOFn's recall significantly improves on the test set (85.00%) compared to 74.00% on the training set. The F-Score for LOF is also comparable, reaching 71.43% for training and 71.45% for testing. Ultimately, while Isolation Forest excels in recall, LOF achieves a better balance with a higher AUC and improved performance in identifying anomalies on the test set. To provide more

Table 2. Performance of Algorithms for Anomaly Detection

Algorithme	La précision	Le rappel	F-score	AUC
Isolation Forest Train	59.39%	99.99%	74.52%	65.81%
Isolation Forest Test	59.51%	87.00%	74.62%	65.98%
Local Outlier Factor Train	55.56%	74.00%	71.43%	60.00%
Local Outlier Factor Test	55.58%	85.00%	71.45%	60.04%

Table 3. Comparison Between Actual and Predicted Anomalies

Algorithms	real number	prediction
Isolation Forest Train	1780	1877
Isolation Forest Test	501	594
Local Outlier Factor Train	1780	2877
Local Outlier Factor Test	501	674

details, graphical representations of the predictions will be displayed in Figs. 2 and 3.

Based on the obtained results, it is evident that unsupervised algorithms have made a key contribution to anomaly detection in HVAC systems. These algorithms offer the crucial flexibility and adaptability needed in environments where anomalies are rare and often not well defined, especially as they can operate without labeled data. Among the models tested, Local Outlier Factor (LOF) demonstrated the best performance. LOF struck a strong balance between precision, recall, and F-score, showing its effectiveness in detecting anomalies with minimal false positives. This makes LOF a highly suitable tool for applications requiring reliable detection and minimal false alarms. While Isolation Forest also performed well, particularly in recall by identifying most anomalies, its lower accuracy and tendency to produce false positives make it less reliable in certain situations where reducing unnecessary actions is critical. Thus, LOF is better

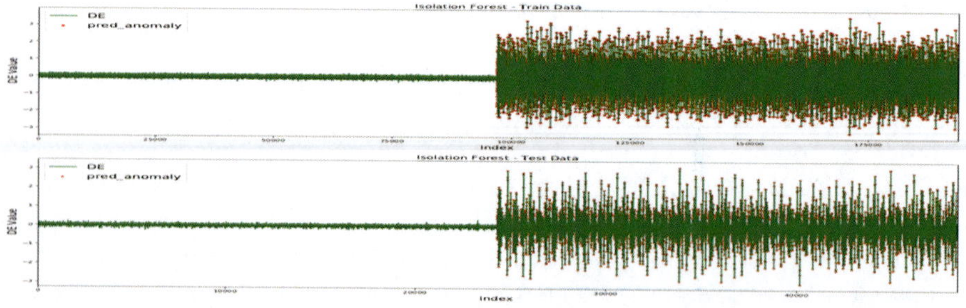

Fig. 2. Predictions of the Isolation Forest for the training and test data

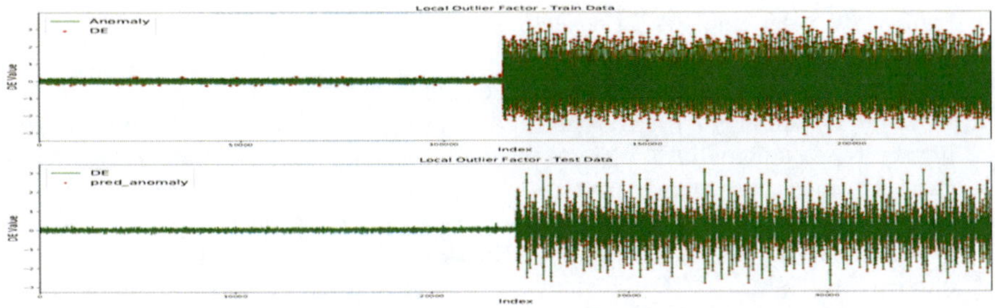

Fig. 3. Predictions of the Local Outlier Factor for the training and test data

suited for environments were minimizing false positives and achieving a more precise detection are crucial.

5 Conclusion

The findings from this study emphasize the significant role of anomaly detection algorithms, particularly Isolation Forest and Local Outlier Factor (LOF), in enhancing the performance and reliability of HVAC systems. Among the tested models, LOF demonstrated superior results in terms of balancing precision, recall, and overall detection capabilities. While Isolation Forest exhibited a high recall for detecting anomalies, its relatively low precision highlighted the potential risks of false positives in critical system monitoring. These results showcase the value of non-supervised machine learning techniques in real-world applications, especially in systems where accurate anomaly detection is essential for maintaining efficiency and avoiding costly interventions. Future work could explore hybrid approaches or fine-tuning techniques to further optimize anomaly detection in complex, dynamic environments such as HVAC systems.

References

1. Khaoula, E., Amine, B., Mostafa, B.: Machine learning and the internet of things for smart buildings: a state of the art survey. In: 2022 2nd International Conference on Innovative Research in Applied Science, Engineering and Technology (IRASET), pp. 1–10. IEEE (2022)
2. Asim, N., et al.: Sustainability of heating, ventilation and air-conditioning (HVAC) systems in buildings–an overview. Int. J. Environ. Res. Publ. Health **19**(2), 1016 (2022)
3. Nunes, P., Santos, J., Rocha, E.: Challenges in predictive maintenance–a review. CIRP J. Manuf. Sci. Technol. **40**, 53–67 (2023)
4. Khaoula, E., Amine, B., Mostafa, B.: Remanning useful life predictions using machine learning algorithms in smart buildings. In: Advances in Research and Applications, Book Chapter, Ensemble Machine Learning, pp. 197–208 (2024)
5. Samariya, D., Thakkar, A.: A comprehensive survey of anomaly detection algorithms. Ann. Data Sci. **10**(3), 829–850 (2023)

6. Himeur, Y., Alsalemi, A., Bensaali, F., Amira, A.: Smart power consumption abnormality detection in buildings using micromoments and improved K-nearest neighbors. Int. J. Intell. Syst. **36**(6), 2865–2894 (2021)
7. Himeur, Y., Ghanem, K., Alsalemi, A., Bensaali, F., Amira, A.: Artificial intelligence based anomaly detection of energy consumption in buildings: a review, current trends and new perspectives. Appl. Energy **287**, 116601 (2021)
8. Cicero, S., Guarascio, M., Guerrieri, A., Mungari, S.: A deep anomaly detection system for IoT-based smart buildings. Sensors **23**(23), 9331 (2023)
9. Sater, R.A., Hamza, A.B.: A federated learning approach to anomaly detection in smart buildings. ACM Trans. Internet Things **2**(4), 1–23 (2021)
10. Karakolis, E., Alexakis, K., Kapsalis, P., Mouzakitis, S., Psarras, J.: An end-to-end approach for scalable real time Anomaly detection in smart buildings. In: 2022 13th International Conference on Information, Intelligence, Systems and Applications (IISA), pp. 1–7. IEEE (2022)
11. Erhan, L., et al.: Smart anomaly detection in sensor systems: a multi-perspective review. Inf. Fusion **67**, 64–79 (2021)
12. Haque, N.I., Rahman, M.A., Shahriar, H.: Ensemble-based efficient anomaly detection for smart building control systems. In: 2021 IEEE 45th Annual Computers, Software, and Applications Conference (COMPSAC), pp. 504–513. IEEE (2021)
13. Data.gov: HVAC datasets (n.d.). https://catalog.data.gov/dataset?q=hvac
14. Xu, H., Pang, G., Wang, Y., Wang, Y.: Deep isolation forest for anomaly detection. IEEE Trans. Knowl. Data Eng. **35**(12), 12591–12604 (2023)
15. Xu, H., Zhang, L., Li, P., Zhu, F.: Outlier detection algorithm based on k-nearest neighbors-local outlier factor. J. Algorithms Comput. Technol. **16**, 17483026221078112 (2022)

Optimizing Intrusion Detection Using Machine Learning and Deep Learning Algorithms

Rachid Tahri(✉), Abdellah Ouammou, and Hibat Eallah Mohtadi

Faculty of Sciences and Techniques, Hassan First University, 26000 Settat, Morocco
rachid.tahritr@gmail.com, {a.ouammou,h.mohtadi}@uhp.ac.ma

Abstract. Intrusion detection is a fundamental component of network defense against increasingly sophisticated cyberattacks. Recent advances in Machine Learning (ML) and Deep Learning (DL) have improved detection accuracy, yet many approaches still face challenges of high computational cost and limited interpretability. This paper proposes a hybrid intrusion detection framework that combines ML-based feature selection with DL-based anomaly identification to enhance both efficiency and robustness. The proposed architecture integrates XGBoost with a CNNLSTM model and is evaluated on the CICIDS2017 dataset. Experimental results show that the hybrid model achieves an F1-score of 96.8% and reduces the false positive rate to 2.9%, outperforming individual ML and DL baselines. In addition, post-training quantization enables real-time deployment without significant accuracy loss.

Keywords: intrusion detection · machine learning · deep learning · hybrid framework · cybersecurity

1 Introduction

The rapid escalation of cyberattacks compels organizations to enhance the resilience of their digital infrastructures. Intrusion Detection Systems (IDSs) remain one of the most effective defenses, traditionally divided into signature-based and anomaly-based approaches. Signature systems such as *Snort* and *Suricata* compare network traffic against known attack patterns, performing well against documented threats but failing to identify zero-day or polymorphic attacks. In contrast, anomaly-based systems model normal traffic behavior and detect deviations as potential intrusions, offering generalization to unseen threats but often generating high false alarm rates and demanding significant computational resources [1].

The integration of Artificial Intelligence (AI) has reshaped IDS research. Machine Learning (ML) models such as Random Forest, Support Vector Machines (SVM), and XGBoost achieve strong classification accuracy [2–4].

A. Ouammou and H. E. Mohtadi—These authors contributed equally to this work.

A. E. Hassanien et al. (Eds.): AISDS 2025, LNDECT 286, pp. 253–262, 2027.
https://doi.org/10.1007/978-3-032-18758-1_25

However, their dependence on manual feature engineering and their limited scalability with high-dimensional data restrict generalization. Deep Learning (DL) architectures, notably Convolutional Neural Networks (CNNs) and Long Short-Term Memory (LSTM) networks, address these challenges by learning hierarchical feature representations directly from traffic data, thereby enhancing detection performance and reducing preprocessing overhead [5,6].

Despite these advances, AI-based IDSs still face several obstacles. Deep models require substantial computational power, which hinders real-time deployment, and their "black-box" nature limits interpretability [7]. Additionally, dataset imbalance, adversarial traffic manipulation, and the growing prevalence of encrypted or high-speed traffic increase the difficulty of achieving accurate and efficient detection [8,9].

To address these challenges, this paper introduces a hybrid IDS framework that combines ML-based feature optimization with DL-based classification. The proposed system integrates XGBoost-driven feature selection with a CNNLSTM classifier and employs post-training quantization for efficient, real-time inference.

The contributions of this study are threefold: (1) a systematic comparison of ML and DL methods for intrusion detection; (2) a hybrid MLDL architecture balancing accuracy, interpretability, and efficiency; and (3) empirical validation on the CICIDS2017 dataset, encompassing diverse and realistic attack scenarios [10,11].

2 Literature Review

Intrusion Detection Systems (IDSs) are generally classified as signature-based or anomaly-based. Signature approaches, used in tools like *Snort* and *Suricata*, match traffic against known attack patterns. While effective for documented threats, they fail to detect zero-day or polymorphic attacks due to reliance on frequent signature updates [12]. Anomaly-based IDSs instead learn normal traffic behavior and flag deviations as potential intrusions. Though capable of identifying novel threats, they often suffer from high false alarm rates and scalability issues [1,5]. Hybrid schemes attempt to combine both but still face limitations in throughput and adaptability for dynamic networks [7,9].

Machine Learning (ML) has automated IDS processes and improved detection accuracy. Supervised models such as Random Forest, SVM, and XGBoost perform well on labeled data, with ensembles showing strong generalization [13]. Dimensionality reduction via Recursive Feature Elimination (RFE) or Principal Component Analysis (PCA) enhances efficiency [5,12]. Unsupervised models–k-Means, Autoencoders, and One-Class SVM–detect unknown attacks without labels but yield more false positives. Reinforcement learning (e.g., DQN) supports adaptive IDSs, though its computational cost limits scalability [13,14]. Ongoing challenges include class imbalance (addressed through SMOTE or cost-sensitive learning), latency in high-throughput environments, and adversarial traffic resilience, where edge/federated learning and GAN-based augmentation show promise [2,6,7,9,15].

Deep Learning (DL) further advances IDSs by automatically extracting hierarchical features from raw data. CNNs learn spatial correlations, while LSTMs capture temporal dependencies [5,7,9,13,14]. Autoencoders reconstruct normal traffic for anomaly detection, and GANs generate synthetic samples for robust training [6,15]. CNNLSTM hybrids combine both feature types to boost accuracy. However, DL models face high computational cost, limited interpretability, and vulnerability to adversarial perturbations. Current mitigation efforts include pruning, quantization, Explainable AI (XAI), and privacy-preserving or adversarial training techniques [6,15].

Comparative analyses show that ML models are generally faster and more interpretable, while DL achieves superior accuracy on complex traffic data [12,16]. This trade-off motivates hybrid MLDL frameworks that combine ML's efficient feature reduction with DL's powerful pattern learning–aiming for scalable, accurate, and resilient IDS solutions (Table 1).

Table 1. Comparison of ML and DL approaches for IDS.

Criterion	ML	DL
Feature handling	Manual engineering (e.g., PCA, RFE)	Automatic hierarchical extraction
Accuracy	High with selected features	Superior on complex, high-dimensional data
Computational demand	Low to moderate	High (training and inference)
Interpretability	High (trees, SHAP with ensembles)	Limited; requires XAI methods
Zero-day adaptability	Restricted by labeled features	Stronger due to learned representations

The reviewed literature underscores the evolution of IDSs and the enduring balance between interpretability and performance, providing the foundation for the hybrid MLDL framework proposed in the next section.

3 Methodology

3.1 Dataset Selection

Reliable IDS evaluation requires datasets with diverse attacks, realistic traffic, and balanced classes. While NSL-KDD and UNSW-NB15 are briefly discussed for context, all experiments in this study use the full CICIDS2017 dataset, which includes modern attack types such as brute force, botnets, port scans, and DDoS [3–5]. Other datasets like UNSW-NB15 and TON-IoT are cited for comparison due to their IoT focus or newer attack profiles but suffer from class imbalance [7,9,11].

Preprocessing employed Recursive Feature Elimination (RFE) and Principal Component Analysis (PCA) for dimensionality reduction [15,17], MinMax or Z-score scaling for normalization, one-hot encoding for categorical features, and SMOTE or cost-sensitive methods to address imbalance [6,10]. The dataset was split into 70% training, 15% validation, and 15% testing sets.

To prevent data leakage, SMOTE was applied only to the training subset. Minority classes such as Heartbleed, Infiltration, and Web Attack were oversampled to match the average count of major attacks. Table 2 presents the class distribution before and after oversampling.

Table 2. Class distribution in CICIDS2017 before and after SMOTE.

Class	Original Samples	After SMOTE
Normal	227,309	227,309
DDoS	459,027	459,027
Web Attack	21,451	60,000
Infiltration	36,905	60,000
Heartbleed	5,000	60,000
Botnet	29,002	60,000
Brute Force	74,568	74,568
Port Scan	158,930	158,930

3.2 Machine Learning Approach

The ML pipeline integrates feature engineering, model training, and optimization. Classical supervised algorithms–Random Forest (RF), SVM, and XGBoost–are applied for anomaly detection, with tree-based ensembles offering strong predictive power [5,7,13]. Unsupervised approaches, including k-Means clustering, Autoencoders, and One-Class SVM, are employed to uncover patterns in unlabeled data, though their use often leads to increased false positives [5,12]. In certain configurations, hybrid strategies are adopted, such as features extracted by Autoencoders combined with XGBoost classifiers [12].

Hyperparameter tuning is conducted using Grid Search and Bayesian optimization [5]. To improve scalability and robustness, we incorporate optimization strategies such as quantization and pruning, and we consider federated deployment scenarios for distributed IDS architectures [6,9,15].

3.3 Deep Learning Approach

Deep Learning enables direct feature learning from network flows with minimal preprocessing. CNNs extract spatial correlations, while LSTMs capture temporal dependencies in sequential data. Autoencoders reconstruct normal traffic for anomaly detection, and GANs generate adversarial examples to enhance robustness. Hybrid CNNLSTM models integrate spatial and temporal learning for improved detection [5,7,9,14].

A lightweight GAN was trained to perturb selected statistical features (e.g., packet size variance, flow duration), generating synthetic attack flows mixed with real data (9:1 ratio) during CNNLSTM fine-tuning to strengthen adversarial resilience. Despite their strong accuracy, DL models remain computationally demanding. Efficiency and robustness are improved through pruning, post-training quantization, Explainable AI (XAI), and adversarial or privacy-preserving training [6,7,15].

3.4 Implementation Details

Figure 1 shows the workflow of the proposed hybrid IDS, covering data preprocessing, feature selection, model training, and evaluation.

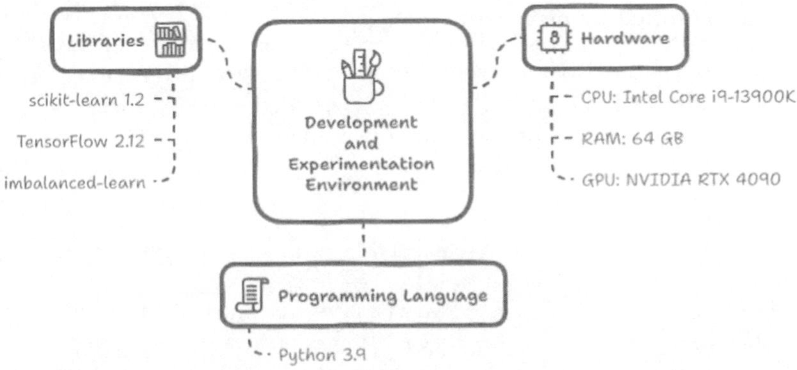

Fig. 1. Workflow of the proposed hybrid IDS and experimental setup.

The system combines efficient feature selection with deep temporal modeling. Development was conducted in Python 3.9 using `scikit-learn` 1.2 (RF, XGBoost, SVM, RFE, PCA), `TensorFlow` 2.12 (CNN, LSTM, Autoencoder), and `imbalanced-learn` for SMOTE. Experiments ran on an Intel Core i9-13900K CPU, 64 GB RAM, and an NVIDIA RTX 4090 GPU (24 GB).

Preprocessing included null-value removal, MinMax scaling, one-hot encoding, and SMOTE applied only to training data to prevent leakage. XGBoost-based RFE reduced 80 features to the 50 most relevant.

The ML baseline used XGBoost with `n_estimators=200`, `max_depth=10`, and `learning_rate=0.1`. The CNN comprised two convolutionpooling layers, followed by an LSTM with dropout. The hybrid model fed the selected features into the CNNLSTM, combining spatial extraction and temporal learning for better detection.

Post-training FP16 quantization reduced model size by about 50% and doubled inference speed while maintaining accuracy, confirming suitability for real-time IDS deployment.

3.5 Evaluation Metrics

Model performance is assessed with Accuracy, Precision, Recall, F1-Score, and AUC-ROC. PrecisionRecall curves are included to provide insight under imbalanced conditions [7]. Operational considerations include false positive/false negative rates, inference latency, and memory footprint to evaluate deployability [9,15]. Adversarial robustness is examined using attack-oriented evaluation metrics [6]. Results are reported with 95% confidence intervals, and significance testing ($p < 0.05$) is applied when comparing competing models.

4 Experimental Results and Discussion

4.1 Performance Comparison

The proposed hybrid IDS was evaluated against classical ML and DL baselines using the CICIDS2017 dataset. The data was divided into 70% for training, 15% for validation, and 15% for testing. Table 3 summarizes the performance metrics for all baseline and hybrid models.

Table 3. Performance of baseline and hybrid models on CICIDS2017.

Model	Acc (%)	Prec (%)	Rec (%)	F1 (%)	FPR (%)
Random Forest	93.5	90.1	89.2	89.6	6.5
XGBoost	95.3	92.8	91.5	92.1	5.2
SVM (RBF)	88.7	84.3	82.1	83.2	8.9
CNN	96.8	95.2	93.9	94.5	3.8
LSTM	96.5	94.8	93.5	94.1	4.1
Hybrid (XGB + CNN–LSTM)	**98.2**	**97.1**	**96.5**	**96.8**	**2.9**
Quantized Hybrid	**98.0**	**96.8**	**96.2**	**96.5**	**3.1**

As expected, deep learning architectures outperformed traditional ML classifiers in both accuracy and generalization. The CNNLSTM hybrid with XGBoost feature selection achieved the best results–98.2% accuracy, 96.8% F1-score, and the lowest 2.9% false positive rate (FPR)–demonstrating the synergy between ML-based feature reduction and DL's spatio-temporal modeling. Among ML models, XGBoost ranked highest, confirming its strength on structured data. The quantized hybrid retained strong performance (F1 = 96.5%), showing that optimization enables real-time deployment with minimal accuracy loss.

The hybrid IDS also reduced inference latency to 12.4 ms per batch, a 2.1× improvement over the baseline CNNLSTM (26.3 ms). Paired t-tests across five random seeds confirmed statistically significant gains in F1-score and AUC ($p < 0.05$) (Fig. 2).

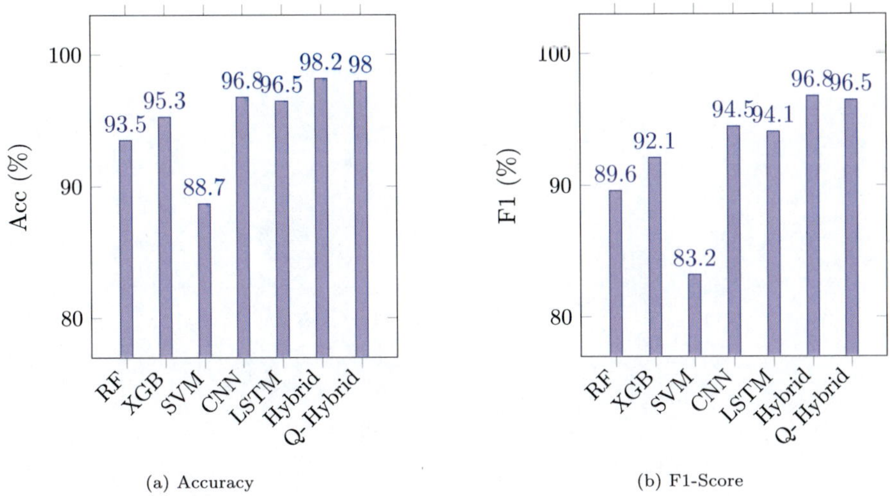

Fig. 2. Comparison of model performance on CICIDS2017.

4.2 ROC and PR Analysis

Figure 3 illustrates the ROC curves for the main models. The hybrid CNNLSTM with XGBoost-based feature selection reached an AUC close to 0.98, surpassing standalone CNN and XGBoost classifiers. This confirms that the hybrid system achieves better discrimination between benign and malicious traffic, particularly in reducing false positives.

4.3 Confusion Matrix Analysis

To better understand per-class performance, Fig. 4 presents the normalized confusion matrix of the hybrid model on CICIDS2017. The diagonal dominance indicates strong classification capability across both majority and minority attack categories. In particular, minority classes such as *Heartbleed* and *Web Attack* achieve recall values above 90%, confirming that the use of SMOTE and hybrid featuretemporal modeling effectively mitigates class imbalance.

4.4 Strengths and Limitations

ML- and DL-based IDSs surpass traditional signature systems by detecting both known and unknown threats [9]. DL architectures such as CNNs and LSTMs effectively model high-dimensional traffic and capture complex patterns without manual feature engineering. Hybrid models further enhance performance by combining spatial and temporal learning [15].

However, several challenges remain. Deep models require substantial computational resources, limiting their use in lightweight environments compared to Random Forest or XGBoost [5]. Their limited interpretability also hinders

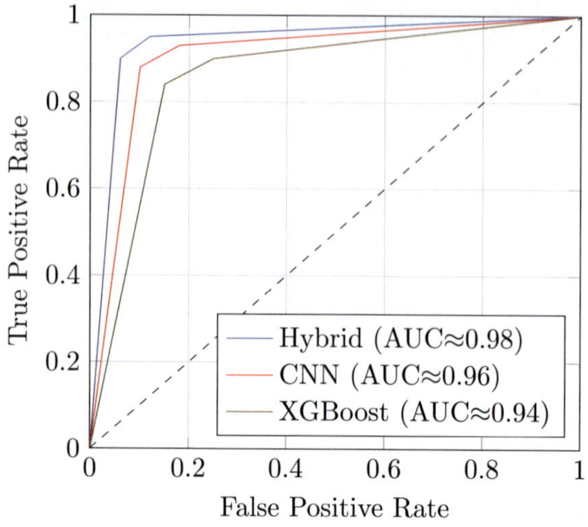

Fig. 3. ROC curves for selected models on CICIDS2017.

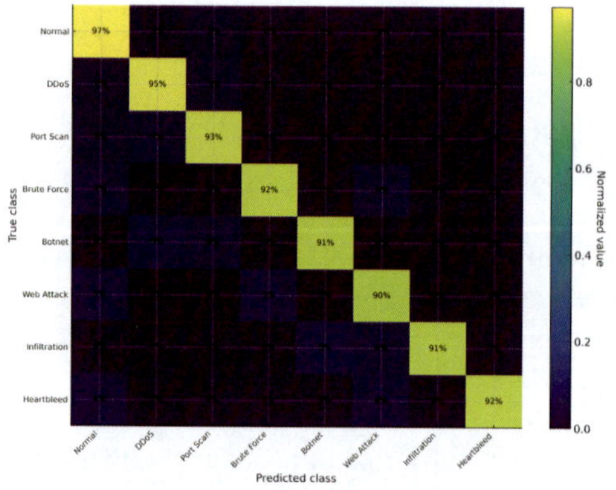

Fig. 4. Normalized confusion matrix for the hybrid IDS on CICIDS2017.

analyst insight, motivating the use of Explainable AI (XAI) [7]. Moreover, IDS datasets are typically imbalanced–normal traffic greatly exceeding attacks–causing biased training. Techniques like SMOTE and cost-sensitive learning reduce this bias but cannot eliminate it entirely [12].

4.5 Optimized Hybrid Framework

The integration of ML and DL allows for a balanced approach. In our framework, feature selection through RFE and PCA ensures compact and informative inputs, while XGBoost provides efficient classification. CNNLSTM layers capture spatialtemporal dependencies, improving the detection of subtle attack patterns. To enhance resilience, GAN-based adversarial training is used to simulate sophisticated evasions, while quantization and pruning make the model efficient enough for real-time application. Future extensions will focus on federated learning for decentralized deployment and Explainable AI (XAI) integration to enhance transparency.

5 Conclusion and Future Work

The proposed hybrid intrusion detection framework combines Machine Learning and Deep Learning to improve accuracy and robustness. Using XGBoost-based feature selection with a CNNLSTM model, it unites efficient feature reduction and spatio-temporal pattern learning. Tests on CICIDS2017 achieved 98.2% accuracy, 96.8% F1-score, and a 2.9% false positive rate. Post-training quantization halved the model size and doubled inference speed, confirming real-time applicability.

Beyond accuracy, the system enhances efficiency, interpretability, and resilience to adversarial traffic. Its feature optimization and lightweight design make it suitable for large-scale, time-critical environments, though dataset bias, resource use, and limited explainability remain challenges.

Future work will explore federated learning for distributed, privacy-preserving training and integrate Explainable AI to boost transparency and trust. Real-world deployment on live traffic and emerging attacks will test scalability and adaptability, advancing hybrid MLDL IDS toward scalable, interpretable, and resilient cyber defense.

References

1. Sommer, R., Paxson, V.: Outside the closed world: on using machine learning for network intrusion detection. In: IEEE Symposium on Security and Privacy, pp. 305–316 (2010)
2. Ouammou, A., Zaaloul, A., Hanini, M., Bentahar, A.: Modeling decision making to control the allocation of virtual machines in a cloud computing system with reserve machines. IAENG Int. J. Comput. Sci. **48**(2), 285–293 (2021)
3. Tahri, R., Jarrar, A., Lasbahani, A., Balouki, Y.: A comparative study of machine learning algorithms on the UNSW-NB 15 dataset. In: ITM Web of Conferences, vol. 48, p. 03002. EDP Sciences (2022)
4. Lasbahani, A., Tahri, R., Jarrar, A., Balouki, Y.: A new centralized detectionbased process for evaluating anomalies and analyzing the first causes using machine learning and web semantic. iJOE **19**(03), 113 (2022)
5. Rajkumar, K.: Semi-supervised deep-ELM for DDoS attack detection and mitigation using the optimallink model in IoT networks. Comput. Sec. (2025)
6. Nassa, V.K., Vasuki, P.: Iot botnet detection from software-defined network using American zebra optimization algorithm with SSRNN-ELM. Int. J. Comput. Intell. Syst. (2025)
7. Tokgoz, E.: Artificial bee colony optimization techniques' utilization for intrusion detection systems' analysis. In: IEEE 4th International Conference on AI in Cybersecurity (2025)

8. Ezzidani, A., Ouammou, A., Hanini, M., Ben Tahar, A.: A SMDP approach to evaluate the performance of a vehicular cloud computing system with prioritize requests. Math. Model. Eng. Prob. **8**(6) (2021)
9. Albogami, N.N.: Intelligent deep federated learning model for enhancing security in internet of things-enabled edge computing environment. Sci. Rep. (2025)
10. Tahri, R., Balouki, Y., Jarrar, A., Lasbahani, A.: Intrusion detection system using machine learning algorithms. In: ITM Web of Conferences, vol. 46, p. 02003. EDP Sciences (2022)
11. Ouammou, A., Hanini, M., Tahar, A.B., El Kafhali, S.: Analysis of a m/m/k system with exponential setup times and reserves servers. In: Proceedings of the 4th International Conference on Big Data and Internet of Things, pp. 1–5 (2019)
12. Hatamleh, H., Alnaser, A.M.A., Saloum, S.S., Sharadqeh, A.: Software-defined networking-internet of things security with novel image-based authentication and artificial intelligence-powered two-stage intrusion detection. Technologies **13**(2), 55 (2025)
13. Zhang, T., Tang, X., Wang, J., Liu, J.: Network security management in heterogeneous networks. Electronics **14**(3), 568 (2025)
14. Asutkar, S.M., Goje, S.S., Asutkar, G.M.: Study of various AI-based security models for threat analysis in communication and IoT networks. In: AIP Conference Proceedings (2025)
15. Edozie, E., Shuaibu, A.N., Sadiq, B.O.: Artificial intelligence advances in anomaly detection for telecom networks. Artif. Intell. Rev. (2025)
16. Mohtadi, H.E., Ouammou, A., Hanini, M., Haqiq, A.: Resilient vehicular fog computing networks: an analytical approach to system reliability under breakdown and vacation interruptions. Clust. Comput. **28**(2), 108 (2025)
17. Ouammou, A., Tahar, A.B., Hanini, M., El Kafhali, S.: Modeling and analysis of quality of service and energy consumption in cloud environment. Int. J. Comput. Inf. Syst. Ind. Manag. Appl. **10**, 9–9 (2018)

From Malware to Intelligence: Attributing APT Malware to Nation-State Actors

Younes Ben Ammou[1(✉)], Ghizlane Orhanou[1], and Lahoucine Ballihi[2]

[1] LabMiA-SI, Faculty of Science of Rabat, Mohammed V University in Rabat, Rabat, Morocco
younes.benammou.yba@gmail.com

[2] LRIT, Faculty of Science of Rabat, Mohammed V University in Rabat, Rabat, Morocco

Abstract. Recently, state-sponsored cyber-attacks have surged, with Advanced Persistent Threats (APTs) frequently bypassing conventional security defenses and operating undetected. These groups employ sophisticated malware and advanced techniques, making both detection and attribution extremely difficult. Identifying the state behind an APT is critical, as it provides actionable intelligence that can inform strategic decisions and real-world responses. To address this challenge, we propose a novel approach that leverages static malware features to attribute APT campaigns to their likely state sponsors. By analyzing these features, we uncover distinctive patterns, enabling inference of their national affiliation. Our method achieves a classification accuracy of 95.9%, surpassing traditional techniques. This approach offers a precise understanding of APT activity and equips organizations and policymakers with insights to implement targeted defenses and proactive measures against state-backed cyber threats.

Keywords: Cyber Attack · Cyber Threat Actor · Advanced Persistent Threats · Cyber Threat Intelligence · Malware Attribution

1 Introduction

In today's digital era, the sharp increase in cyberattacks has created new levels of complexity and urgency for cybersecurity. Among the most concerning threats are state-backed intrusions carried out by Advanced Persistent Threats (APTs). These groups are known for their persistence and high degree of sophistication, constantly adapting their tactics, techniques, and procedures to evade detection. By exploiting advanced malware and tools, they manage to bypass security defenses and remain hidden for extended periods. The widespread rise of APT campaigns has significantly raised the risks for both organizations and nations, putting critical infrastructure, intellectual property, and even national security at stake.

Attributing such attacks is especially difficult because of the secrecy and technical sophistication involved. Yet, accurate attribution is vital, as it helps uncover the sponsoring states behind these operations, thereby supporting the design of effective counterstrategies and policies. Traditional attribution methods, which depend heavily on manual investigation, are often slow and prone to human error. To overcome these

A. E. Hassanien et al. (Eds.): AISDS 2025, LNDECT 286, pp. 263–272, 2027.
https://doi.org/10.1007/978-3-032-18758-1_26

limitations, machine learning has become a valuable alternative, offering scalable and systematic analysis of massive datasets to identify unique behavioral patterns that link specific malwares to their sponsoring states.

1.1 Motivations, Objectives, and Contributions

Traditional attribution approaches primarily rely on dynamic analysis, behavioral profiling, and sophisticated forensic techniques, which demand substantial computational resources and are vulnerable to evasion mechanisms such as anti-sandboxing and polymorphism. An alternative approach is to leverage system API calls, which can be extracted through static analysis yet still capture key aspects of malware behavior, objectives, and coding style. These API patterns may serve as reliable indicators of the malware's origin or sponsoring entity. By exploiting API-based signals, it is possible to achieve rapid, scalable, and cost-efficient early attribution, reducing the need for extensive behavioral monitoring.

The key goals of this study are:

- To investigate the predictive capability of system API calls as standalone features for early malware-to-state attribution;
- To assess multiple machine learning models for malware-to-state attribution.

1.2 Outline

This paper is structured as follows. Section 2 introduces the technical background on malware and state-sponsored attacks and reviews relevant approaches in the literature. Section 3 details the methodology, including dataset, feature extraction and selection, and the machine learning models employed, along with the experimental setup, evaluation metrics, and discussion of the results. Section 4 provides a critical discussion of the findings, addressing the implications and limitations. Finally, Section 5 concludes the paper by summarizing the key contributions and results, and outlines potential directions for future research.

2 Technical Background and Related Work

This section provides the necessary technical context on APTs, their operational methods, and the importance of malware attribution in defending against them. It also presents previously proposed approaches in the literature.

2.1 Technical Background

Advanced Persistent Threats (APTs) are long-term, well-resourced cyber operations often linked to state actors. They employ stealth, persistence, and evolving tactics, techniques, and procedures (TTPs) to compromise high-value targets over extended periods. Because of their sophistication, attributing such attacks to specific states - especially via malware artifacts - is particularly challenging [1, 10].

Malware refers to software specifically designed to cause harm, such as espionage, data exfiltration, or sabotage. In state-sponsored campaigns, malware often serves as the central artifact for attribution. Careful analysis of malware code, behavior, and command-and-control (C2) infrastructure can reveal unique traits that link it to sponsoring states. Recent research emphasizes the importance of both static and dynamic analysis with robust models to strengthen attribution accuracy [2].

State-sponsored cyberattacks are operations where a nation-state directly or indirectly funds, directs, or supports cyber activities. Unlike cybercrime motivated primarily by profit, state-sponsored campaigns pursue geopolitical, strategic, or military goals, such as intelligence gathering, disruption of adversary operations, or destabilization of critical infrastructure. Recent studies confirm that these attacks are increasingly global in scope and rely on highly customized malware [3].

Cyber Attribution is the process of identifying responsibility for cyberattacks, which may involve individuals, APT groups, or sponsoring states [4]. Attribution combines technical evidence (e.g., malware code features, IoCs, network traces), behavioral data (TTPs), and contextual intelligence [11].

Malware Attribution to State focuses on linking malware samples directly to sponsoring states. This process leverages:

- **Code, binary and behavior analysis**: identifying unique sequences of API calls or byte-level patterns [5, 12].
- **Infrastructure analysis**: tracking C2 servers and domains back to regions or entities linked to states.
- **Victimology and geopolitical context**: correlating targeted sectors or countries with national interests [1].
- **Explainable models**: enabling analysts to justify attribution claims with interpretable evidence [5].

2.2 Related Work

Several studies have explored various methodologies to address the problem of malware and software authorship attribution. Sangwoo Lee and Jung Wook Cho [6] proposed a malware authorship attribution model based solely on runtime module analysis. Their method achieved an accuracy of 87.1% across six malware author groups, with 89% precision, 87% recall, and an F1-score of 87%. Using automated static analysis tools, they extracted the list of runtime modules and dynamic link libraries loaded during the execution of malware files on Windows systems. In contrast, [7] presented an approach based on Cyber Threat Intelligence (CTI) reports, which involves extracting features from three modalities: attribute types, natural language text, and topological relationships. These features are fused to form a multidimensional feature vector for each CTI report, which is then processed using multilevel heterogeneous graph attention networks. Similarly, [8] collected dynamic information from a large malware dataset associated with seven Advanced Persistent Threat (APT) groups. They applied the Term Frequency–Inverse Document Frequency (TF-IDF) technique to select relevant features, addressed class imbalance using the Synthetic Minority Oversampling Technique (SMOTE), and performed attribution using a Random Forest classifier. Meanwhile, [9] extracted Program Dependence Graph (PDG) features from multiple programming languages, including

C++, Java, and C#. They identified the most significant PDG features with TF-IDF, mitigated class imbalance using SMOTE, and trained a deep learning model for source code authorship attribution.

3 Methodology

Figure 1 presents the overall framework of the proposed methodology, with the subsequent sections detailing its key components and the essential steps of its implementation.

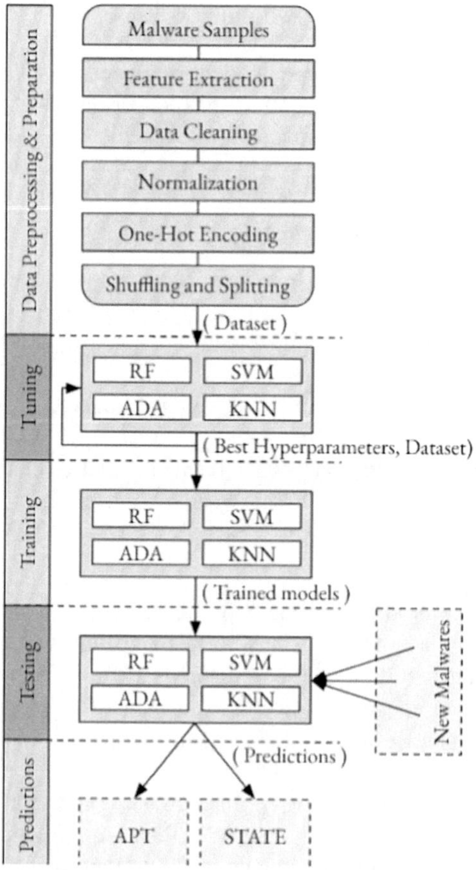

Fig. 1. Overall proposed methodology

3.1 Data Preprocessing and Preparation

The dataset for this study was constructed from malware hashes linked to known states. Corresponding binaries were obtained from reputable repositories and sandboxes, including VirusTotal and GitHub. Only Portable Executable (PE) files with extractable characteristics were retained. A custom multi-threaded engine, built using the Pefile library, was employed to extract static features efficiently from each binary, enabling large-scale analysis within a short time frame. All extracted attributes were organized into a structured database to facilitate querying and subsequent analysis. The final dataset comprises over 2,800 malware samples attributed to five sponsoring states, with distributions varying across countries (Table 1).

Table 1. Dataset distribution

Country	Samples
China	1 293
Russia	556
North Korea	268
Pakistan	345
United States of America	395

3.2 Model Selection

The machine learning models were selected based on their suitability for multiclass classification and their demonstrated effectiveness in cybersecurity applications. Four models were chosen: Random Forest (RF), AdaBoost (AB), Support Vector Machine (SVM), and K-Nearest Neighbors (KNN). Together, these models represent a combination of ensemble methods, distance-based algorithms, and margin-based classifiers. The selection criteria emphasized scalability to large datasets, interpretability, and robustness against class imbalance.

3.3 Model Optimization and Training

The model Optimization process was structured to optimize performance while efficiently exploring the hyperparameter space. Initially, RandomizedSearchCV (RCV) was applied to conduct a broad, randomized search across hyperparameter values, enabling diverse configurations to be assessed at a reduced computational cost. Based on these results, GridSearchCV (GCV) was then employed to perform an exhaustive search within the most promising ranges, systematically evaluating all parameter combinations to refine them to their optimal values. This two-stage process balanced exploration with precision, ensuring effective fine-tuning of the models.

The following hyperparameters were tuned:

- **Random Forest (RF)**: number of estimators (n_estimators), maximum depth (max_depth), and feature selection (max_features);
- **AdaBoost (AB):** learning rate (learning_rate) and number of estimators;
- **Support Vector Machine (SVM):** kernel type, regularization parameter (C), and polynomial degree;

- **K-Nearest Neighbors (KNN):** number of neighbors (n_neighbors) and distance metrics (p).

For hyperparameter optimization, training, and evaluation (Fig. 2), the dataset was split into 80% training and 20% testing, with randomized splitting to preserve uniform sample distribution. The training portion was further subjected to 5-fold cross-validation, ensuring robust evaluation and reliable performance estimates for both model selection and final predictions.

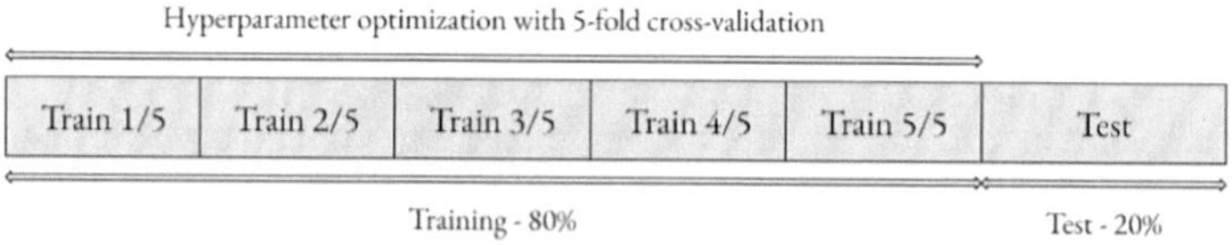

Fig. 2. Overview of the dataset split

3.4 Evaluation Metrics

To comprehensively evaluate the performance of machine learning models, the following metrics were used:

- **Accuracy,** the proportion of correctly classified samples to total samples. Mathematically, it is defined in Eq. (1). A higher accuracy value indicates that the model performs well across all classes; however, it may be misleading if the dataset is imbalanced.

$$Accuracy = \frac{TP + TN}{TP + TN + FP + FN} \tag{1}$$

Where: TP, TN, FP, and FN represent true positives, true negatives, false positives, and false negatives, respectively.

- **Precision,** measures the proportion of true positives out of all predicted positives. It is expressed in Eq. (2), it indicates how reliable the model's positive predictions are. A higher precision means that the model produces fewer false positives.

$$Precision = \frac{TP}{TP + FP} \tag{2}$$

- **Recall,** calculates the proportion of true positives out of all actual positives. It is given by Eq. (3). High recall means the model successfully captures most of the relevant (positive) instances, but it may come at the cost of more false positives.

$$Recall = \frac{TP}{TP + FN} \tag{3}$$

- **F1 Score,** is the harmonic mean of precision and recall. It is computed using Eq. (4). Useful when the dataset is imbalanced, as it combines both precision and recall into a single measure of performance.

$$F1 = 2 \times \frac{Precision \times Recall}{Recall + Recall} \tag{4}$$

- **Training Time:** Training time measures the computational time required to train the model. It is measured in seconds.
- **Inference Time:** Inference time assesses the time taken by the model to make predictions on new data. It is also measured in seconds per prediction.

3.5 Results Analysis

The performance of the selected machine learning models was analyzed, Tables 2 and 3 present the results of hyperparameter optimization for state identification using RCV and GCV, respectively. Furthermore, Table 4 summarizes the performance of the four models in the testing phase after hyperparameter optimization.

Table 2. Hyperparameter optimization results for APT group attribution (RandomizedSearchCV)

Model	Best Parameters	BestAccuracy
RF	n estimators: 250, min samples split: 2, min samples leaf: 1, max features: 'sqrt', max depth: 30, bootstrap: False	0.9473
Adaboost	n estimators: 300, learning rate: 1.0, algorithm: 'SAMME'	0.7138
SVM	kernel: 'rbf', gamma: 'scale', degree: 5, C: 100.0	0.9447
KNN	weights: 'distance', p: 2, n neighbors: 3, algorithm: 'brute'	0.9277

Table 3. Hyperparameter optimization results for APT group attribution (GridSearchCV)

Model	Best Parameters	BestAccuracy
RF	bootstrap: False, min samples leaf: 1, max depth: 40, max features: 'sqrt', min samples split: 2, n estimators: 235	0.9490
Adaboost	algorithm: 'SAMME', learning rate: 1.1, n estimators: 330	0.7382
SVM	C: 120, degree: 4, gamma: 'scale', kernel: 'rbf'	0.9451
KNN	algorithm: 'brute', n neighbors: 2, p: 2, weights: 'distance'	0.9218

Table 4. Final state attribution performance metrics for Random Forest, AdaBoost, support vector machine, and K-nearest neighbors

Model	Fit time(s)	Pred time(s)	Accuracy	Precision	Recall	F1-Score
RF	3.3727	0.0239	95.80%	95.70%	95.80%	95.70%
Adaboost	29.7075	0.9981	80.40%	79.90%	80.40%	79.60%
SVM	2.7873	1.6333	95.90%	95.90%	95.90%	95.90%
KNN	0.0006	2.5402	95.20%	95.20%	95.20%	95.20%

Support Vector Machine emerged as the most reliable model, with consistently high-performance during optimization. With a radial basis function (RBF) kernel and optimized parameters: {C: 120, degree: 4, gamma: 'scale', kernel: 'rbf'}, it achieved accuracies of 93.25% for state attribution during optimization. During the evaluation, it maintained strong performance with accuracy of 95.9% (state attribution). These results underscore the capability of SVM to generalize well, especially in high-dimensional feature spaces such as those derived from API system calls.

Random Forest also demonstrated consistently high performance during hyperparameter optimization with an accuracy of 93.12%, with parameters: {n_estimators: 235, max_depth: 40, bootstrap: False}. During the evaluation phase, Random Forest achieved 95.8% accuracy. This demonstrates Random Forest's ability to handle complex relationships between features effectively, likely due to its ensemble nature and robustness to overfitting.

K-Nearest Neighbors exhibited stable performance, achieving accuracy of 91.29% for state attribution with parameters: {n_neighbors: 2, p: 2, weights: 'distance'}. During the evaluation, its performance increased and 95.2%.

AdaBoost showed notable improvements when optimized. The optimal parameters were: {n_estimators: 330, learning_rate: 1.1, algorithm: 'SAMME'}. During the evaluation phase, AdaBoost achieved 80.4% for state attribution. Its performance lagged behind other models, likely due to its tendency to overfit smaller datasets or imbalanced class distributions.

4 Discussion

The analysis of API system calls highlights their strong potential as reliable features for malware attribution to nation-state actors. Leveraging machine learning to examine these behavioral traces can significantly improve the accuracy, scalability, and automation of attribution processes. Several aspects underscore their importance:

Distinct Behavioral Patterns: API system calls reflect the precise operations a piece of malware is designed to perform. The ordering, repetition, and combination of these calls often create distinctive behavioral patterns that act as "signatures" for specific threat groups. For instance, calls associated with file access, registry edits, or particular network activities may recur in samples tied to the same actor, supporting attribution efforts.

Tracing Tool Usage Back to Actors: Many adversaries rely on custom-built frameworks or widely used penetration-testing tools. These leave consistent traces in the system calls invoked during execution. Malware generated with platforms like Cobalt Strike or Metasploit, for example, often exhibits recognizable system call footprints, which analysts can use to connect artifacts back to their originators.
Resilience Against Code Obfuscation: While attackers frequently disguise or obfuscate their code, the underlying need to interface with the operating system cannot be fully hidden. Because API calls remain necessary for fundamental operations, they provide a resilient feature set for attribution even when adversaries deploy sophisticated evasion strategies.
Operational Intent Revelation: Examining the types of system calls used can also shed light on the broader goals of the operation. Heavy reliance on HTTP communication may signal data theft, repeated encryption function calls suggest ransomware, while persistent registry changes may reveal espionage-driven objectives.

Taken together, these findings illustrate that API system calls not only support accurate actor attribution but also provide defenders with valuable context about the purpose and intent of malicious campaigns. Their integration into cyber threat intelligence pipelines enhances both the effectiveness of attribution and the overall ability of organizations to anticipate, understand, and counter advanced threats.

4.1 Limitations

While this study demonstrates the potential of machine learning for malware-to-state attribution, several limitations should be noted:

Class Imbalance: Attribution datasets often contain uneven representation of attacker groups, with some APTs or nation-state actors appearing far more frequently than others. This imbalance can bias models toward the dominant classes, and while techniques such as re-sampling or class-weight adjustments help, they do not fully resolve the issue.
Open World Assumption: Most attribution approaches operate under the assumption that all relevant threat actors are present in the training data. In practice, however, new or previously unknown groups frequently emerge. This open-world reality poses a major challenge, as models may misclassify or fail to attribute when confronted with novel actors.

5 Conclusion and Perspectives

This study explored the use of machine learning models to enhance cyber threat attribution by analyzing API system calls from malware samples. The experiment results demonstrated the effectiveness of classification models, such as Random Forest and Support Vector Machine, for malware-to-state attribution. The key findings, contributions, and recommendations are summarized as follows:

Support Vector Machines, demonstrated strong and stable performance, leveraging kernel methods to effectively capture non-linear relationships and distinguish between states, achieving the highest accuracy 95.9% for state attribution.

Future research should aim to integrate dynamic analysis to overcome challenges posed by obfuscation and polymorphism in malware samples, as well as to explore additional attribution models. In conclusion, this study demonstrates the potential of system API calls combined with machine learning for cyber threat attribution. By bridging the gap between manual investigation and automated systems, these approaches enhance analysts' ability to respond more effectively, thereby strengthening organizational defense against Advanced Persistent Threats in an increasingly hostile cyber environment.

References

1. Rani, N., Shukla, S.K.: AURA: a multi-agent intelligence framework for knowledge-enhanced cyber threat attribution. arXiv preprint arXiv:2506.10175 (2025)
2. Sun, Y., Fang, H., Wang, Z., Wang, X., Chen, H.: RoMA: robust malware attribution via byte-level adversarial training with global perturbations and adversarial consistency regularization. arXiv preprint arXiv:2502.07492 (2025)
3. Bae, C., Jeong, S., Kang, D., Kim, D., Kim, H.K.: Threat behavior textual search by attention graph isomorphism. arXiv preprint arXiv:2404.10944 (2024)
4. Quintero-Bonilla, S., Rey, A.M.: A new proposal on the advanced persistent threat: a survey. Appl. Sci. **10**(11), 3874 (2020). https://doi.org/10.3390/app10113874
5. Alshammari, R., Zincir-Heywood, A.N.: Bon-APT: detection, attribution, and explainability of APT malware using temporal segmentation of API calls. Comput. Secur. **140**, 103862 (2024). https://doi.org/10.1016/j.cose.2024.103862
6. Lee, S., Cho, J.W.: Malware authorship attribution model using runtime modules based on automated analysis. JOIV: Int. J. Inform. Visualization. **6**(1–2), 214–220 (2022). https://doi.org/10.30630/joiv.6.1-2.941
7. Xiao, N., Lang, B., Wang, T., Chen, Y.: APT-MMF: an advanced persistent threat actor attribution method based on multimodal and multilevel feature fusion. Comput. Secur. **144**, 103960 (2024). https://doi.org/10.1016/j.cose.2024.103960
8. Li, S., Zhang, Q., Wu, X., Han, W., Tian, Z.: Attribution classification method of APT malware in IoT using machine learning techniques. Secur. Commun. Netw. **2021**, 9396141 (2021). https://doi.org/10.1155/2021/9396141
9. Ullah, F., Wang, J., Jabbar, S., Al-Turjman, F., Alazab, M.: Source code authorship attribution using hybrid approach of program dependence graph and deep learning model. IEEE Access. **7**, 141987–141999 (2019). https://doi.org/10.1109/access.2019.2943639
10. Prasad, N., Diro, A., Warren, M., Fernando, M.: A survey of cyber threat attribution: challenges, techniques, and future directions, 104606. Comput. Secur. **157** (2025). https://doi.org/10.1016/j.cose.2025.104606
11. Basnet, A.S., Ghanem, M.C., Dunsin, D., Kheddar, H., Sowinski-Mydlarz, W.: Advanced persistent threats (APT) attribution using deep reinforcement learning. Digit. Threat. Res. Pract. **6**(3), 14 (2025). https://doi.org/10.1145/3736654
12. Li, J., Liu, J., Zhang, R.: Advanced persistent threat group correlation analysis via attack behavior patterns and rough sets. Electronics. **13**(6), 1106 (2024). https://doi.org/10.3390/electronics13061106

Transformer-Based Detection Method for Temporal Dynamics in Malicious Encrypted Traffic

Min Zhang(✉)

School of Network & Communication Engineering Chengdu Technological University, Chengdu, Sichuan 611730, China
zhangmhf@cdtu.edu.cn

Abstract. Artificial intelligence and intelligent algorithms drive advancements in cybersecurity defense, yet the widespread adoption of encryption protocols renders malicious traffic stealthy, rendering traditional methods ineffective. Existing deep learning models suffer from limitations in temporal awareness and capturing long-term dependencies. To address the challenge of identifying malicious traffic in encrypted scenarios, this study proposes a novel detection model for temporal dynamic changes in malicious encrypted traffic. This model integrates dynamic heterogeneous graphs with Transformer architecture. It constructs heterogeneous graphs by segmenting traffic through temporal snapshots, aggregates spatial features via a heterogeneous graph attention network, and leverages multi-head self-attention mechanisms to capture temporal dependencies. Results demonstrate that the model converges in approximately 25 training epochs, achieving a final loss as low as 0.03. In long-time-series scenarios, its F1 score improved from 0.955 to 0.979, achieving over 95% detection capability against unknown attacks and an alarm accuracy rate of 97.5%, significantly outperforming comparison models. This research overcomes technical bottlenecks in encrypted traffic detection, providing precise technical support for real-time network monitoring. It holds significant practical implications for enhancing network defense systems against complex encrypted threats.

Keywords: Malicious encrypted traffic · Transformer · Dynamic heterogeneous graph · Sequential detection · Graph attention network

1 Introduction

The rapid iteration of artificial intelligence and intelligent algorithms has driven the intelligent upgrade of cybersecurity defenses while intensifying the complexity of offensive-defensive interactions. In recent years, the use of Transport Layer Security (TLS) encryption protocol has become increasingly common, and its encrypted channels provide certain convenience for the hidden behavior of malicious traffic. Some cyber-attack activities, such as distributed denial of service (DDoS) attacks and botnet communication, can utilize the characteristics of "invisible content" to bypass traditional defenses.

A. E. Hassanien et al. (Eds.): AISDS 2025, LNDECT 286, pp. 273–281, 2027.
https://doi.org/10.1007/978-3-032-18758-1_27

Traditional attack detection methods are mainly based on rule or signature matching, and have a strong dependence on predefined feature databases. They are not only helpless in the face of unknown new attack behaviors, but also completely ineffective once in an encrypted scenario. This is a major weakness in current network security [1]. In recent years, the emergence and development of deep learning technology have provided prospects for addressing this challenge. Deep learning techniques mainly involve extracting statistical features of traffic flow through convolutional neural networks, capturing basic temporal relationships through recurrent neural networks, and simulating relationships between objects through graph neural networks (GNNs) [2, 3]. However, relevant practical applications have shown that current deep learning techniques still have significant shortcomings. For example, it often lacks flexibility in temporal structure and still relies on inherent data training. GNN is accustomed to capturing static slices of changing traffic, lacking consideration for heterogeneous entity relationships and unable to capture the trend changes of traffic on the timeline. Therefore, the dynamic trajectory of malicious behavior is difficult to effectively capture, which leads to a significant lack of detection accuracy and generalization ability of traditional deep learning techniques [4]. To improve the dynamic capture performance of network attack behavior and the review performance of encrypted traffic, this study proposes a Transformer based time dynamic detection method for malicious encrypted traffic. This method models dynamic heterogeneous graphs and learns tenses based on self-attention. This study overcomes the technical challenge of identifying malicious traffic in encrypted environments, provides technical support for real-time network monitoring, and improves the effectiveness of defense systems. This study innovatively constructed a temporal heterogeneous graph to capture spatiotemporal features, and introduced Transformer to simulate long-term dependency relationships using its multi head self-attention mechanism. The method proposed in this study effectively improves the detection accuracy and generalization ability of network attack behavior, and avoids the decryption step.

2 Research Foundation

Malicious encrypted traffic is characterized by hidden content and unclear behavior. This includes attacks like ransomware communications. Therefore, the main challenge is to extract effective identification features without decryption [5]. GNN fundamentals encompass definitions of graphs, heterogeneous graphs, and meta-paths. Their core logic involves iteratively updating node embeddings by aggregating neighboring node information to characterize entity relationships [6]. Transformer models, centered on multi-head self-attention mechanisms, capture features from multiple representational subspaces by concurrently computing sequence element correlations, thereby overcoming traditional temporal modeling limitations [7].

3 Malicious Encrypted Traffic Temporal Dynamics Detection Method Based on Dynamic Heterogeneous Graphs and Transformers

3.1 Traffic Modeling Based on Dynamic Heterogeneous Graphs

Malicious encrypted traffic fundamentally constitutes a data flow involving dynamic interactions among multiple entities. The core challenge lies in the difficulty of simultaneously capturing the heterogeneous relationships between entities and their dynamic evolution over time. Consequently, research must establish a modeling framework tailored to these characteristics, with dynamic heterogeneous graphs emerging as a key technical approach. By segmenting traffic sequences into temporal snapshots and integrating multi-type node information via Heterogeneous Graph Attention Networks (HAN), both spatial topology and temporal dynamic features can be captured simultaneously. The specific modeling architecture is illustrated in Fig. 1.

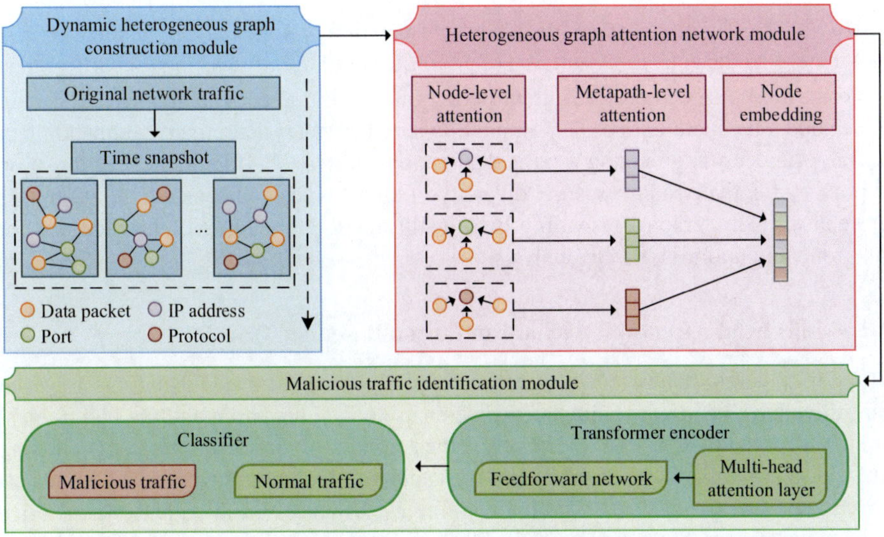

Fig. 1. Traffic modeling architecture based on dynamic heterogeneous graphs

As illustrated in Fig. 1, this study constructs a comprehensive dynamic heterogeneous graph traffic modeling framework designed to simultaneously capture both the spatial topology and temporal evolution features within encrypted traffic. First, raw network traffic is partitioned into a series of ordered temporal snapshots, each corresponding to a static heterogeneous graph. Entities such as packets, IP addresses, ports, and protocols serve as nodes in the graph, with their connectivity represented by edges of different types. Subsequently, these temporal snapshot graphs are fed into the HAN module. This module employs a hierarchical attention mechanism to aggregate node information. At the node-level attention stage, node embeddings based on specific meta-paths are obtained by learning the importance of different neighboring nodes [8]. The node-level attention score is calculated as shown in Eq. (1).

$$s_{i,j}^{\phi} = \mathrm{FFN}\left(W_a h_i + W_b h_j\right) \tag{1}$$

In Eq. (1), $s_{i,j}^{\phi}$ represents the attention score between node i and node j under meta-path ϕ; FFN denotes the feedforward neural network; W_a and W_b are trainable weight matrices; h_i and h_j are the initial feature embeddings of nodes i and j, respectively; ϕ is the predefined meta-path. Then, through meta-path-level attention, the embeddings from different semantic paths are weighted and fused to generate a vector that comprehensively represents the node features at the current time slice. The calculation formula for node embedding aggregation based on the meta-path is shown in Eq. (2).

$$\boldsymbol{h}_i^{\phi} = \sigma\left(\sum_{j \in \mathcal{N}_i^{\phi}} \alpha_{i,j}^{\phi} \boldsymbol{h}_j\right) \tag{2}$$

In Eq. (2), $\boldsymbol{h}_i^{\phi}$ denotes the embedded representation of node i under the meta-path ϕ; $\sigma(\cdot)$ represents the activation function (e.g., *ReLU*) used to introduce nonlinear features; $\alpha_{i,j}^{\phi}$ is the normalized node-level attention weight; $\mathcal{N}_i^{\phi}$ is the set of neighboring nodes for node i under the meta-path ϕ. The summation term generates the semantic embedding of the target node by aggregating weighted neighbor features. This hierarchical aggregation approach effectively addresses the diversity of entity types and semantic relationships in heterogeneous graphs, providing high-quality node embedding representations for subsequent temporal dynamic analysis.

3.2 Multi-head Attention Mechanism Fusion Based on Transformers

Through dynamic heterogeneous graph modeling, node feature embeddings for each time slice have been generated using time-snapshot segmentation and the hierarchical attention mechanism of HAN. However, the temporal dynamic dependencies among these discrete embeddings remain poorly captured. Traditional recurrent neural networks and their variants suffer from low training efficiency and struggle to characterize long-range temporal associations. To address this, the study introduces the Transformer model. Node embeddings from each time snapshot are treated as sequential inputs. After incorporating relative position encoding, the multi-head self-attention mechanism integrates temporal information. The specific implementation of this technical approach is illustrated in Fig. 2.

As depicted in Fig. 2, the core of the Transformer temporal modeling mechanism is the multi-head self-attention mechanism. The research first transforms the temporal sequence data of traffic into a format processable by the model by treating the node embeddings of each temporal snapshot as a "word" in the sequence and superimposing relative position encoding [9]. This sequence then enters the Transformer encoder module, where the multi-head self-attention layer concurrently calculates correlations among all elements in the sequence. This captures dynamic patterns of traffic across different time slices from multiple distinct representational subspaces [10]. For instance, one attention head may focus on patterns of periodic traffic fluctuations, while another concentrates on deviations between anomalous behavior and normal patterns. The layer's output then undergoes residual connections and layer normalization, effectively addressing vanishing gradients and stabilizing training. A nonlinear transformation is then applied through a feedforward network layer, further enhancing the model's expressive power. The stacked processing of the multilayered Transformer encoders allows the model to generate a robust representation vector. This vector integrates the spatial and temporal characteristics of each time slice. It then serves as the final discriminant basis for detecting malicious traffic before the data is passed to the downstream classifier.

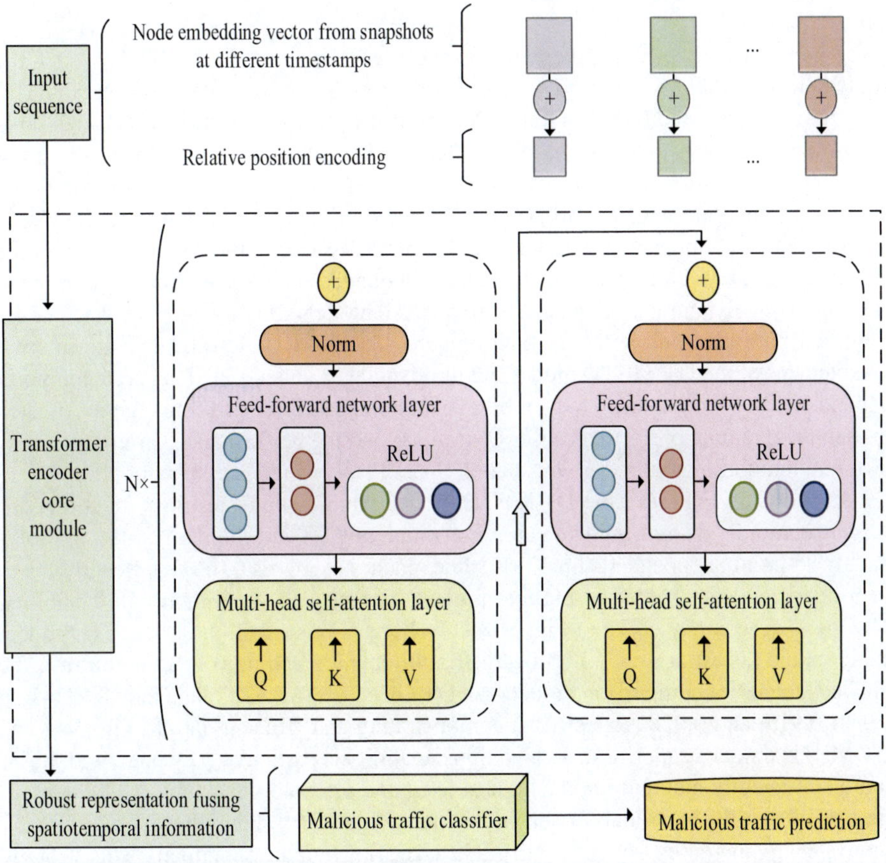

Fig. 2. Schematic of the Transformer-based temporal modeling mechanism

The overall operational flow of the research model is as follows: First, the original encrypted traffic is segmented into time snapshots using a fixed time window. For each snapshot, a heterogeneous graph is constructed containing nodes (IP, port, packet, etc.) and multi-class associated edges. Heterogeneous graph attention networks are then used to hierarchically aggregate node information and generate embeddings. Subsequently, these embeddings are combined with relative position encoding before feeding into a Transformer. The Transformer simultaneously captures temporal correlations through multi-head self-attention, followed by residual connections, layer normalization, and a feedforward network. This process outputs robust vectors integrating spatio-temporal features for classifiers to identify malicious traffic. The proposed model is named the Dynamic Heterogeneous Graph and Transformer-Based Malicious Encrypted Traffic Temporal Dynamic Detection Model (DT-METD).

4 Performance Validation and Application Effectiveness of the DT-METD Model

To comprehensively evaluate the performance advantages of the DT-METD model, the study selected three representative deep learning models as comparative benchmarks: the Dynamic Directed Traffic Graph Convolutional Network (DDT-GCN), the Long Short-Term Memory with Attention (LSTM-ATT), and the Convolutional Temporal Network (Convolutional Temporal Convolutional Network, Conv-TCN). (Convolutional Temporal Convolutional Network, Conv-TCN). Experiments utilized the CIC-IDS 2017 dataset, which contains encrypted traffic samples of various attack types. During data preprocessing, raw traffic was segmented into sequences using fixed time windows and structured into dynamic heterogeneous graphs. All models were trained and tested under a unified hardware and software environment. The study first validated the internal mechanisms of the DT-METD model and its advantages in handling long-term temporal dependencies. By comparing the temporal capture capabilities and convergence of each model, the dynamic performance during training and the performance variations across different temporal lengths were evaluated. Specific test results are shown in Fig. 3.

As shown in Fig. 3(a), the DT-METD model exhibits the fastest loss function convergence, rapidly decreasing from an initial value of 0.38 and stabilizing after approximately 25 training epochs. Its final loss value reaches as low as 0.03, significantly lower than other models, indicating highly efficient training and superior fitting capability. The training accuracy curve in Fig. 3(b) corroborates this trend: the DT-METD model achieves the fastest accuracy improvement, ultimately reaching 0.982. In contrast, the final accuracies of comparison models—DDT-GCN, LSTM-ATT, and Conv-TCN—are 0.941, 0.915, and 0.928 respectively, demonstrating a significant gap. In Fig. 3(c), the DT-METD model's superior generalization capability is particularly evident, with its F1 score consistently maintaining the highest level and ultimately reaching 0.979, demonstrating its stability and reliability on unseen data. Figure 3(d) shows that DT-METD's F1 score steadily increases from 0.955 to 0.979 as the time window lengthens. This strongly demonstrates that its core Transformer module effectively captures long-range temporal dependencies. In contrast, other models exhibit slight F1 score declines beyond 15 s, highlighting DT-METD's unique advantage in detecting temporal dynamic changes. To validate DT-METD's practical value in cybersecurity defense, real-time DDoS attack detection experiments were conducted using the CIC-DDoS 2019 dataset. This dataset contains diverse encrypted DDoS attack traffic, enabling realistic simulation of malicious scenarios. DT-METD was compared against two traditional detection methods, with results summarized in Table 1.

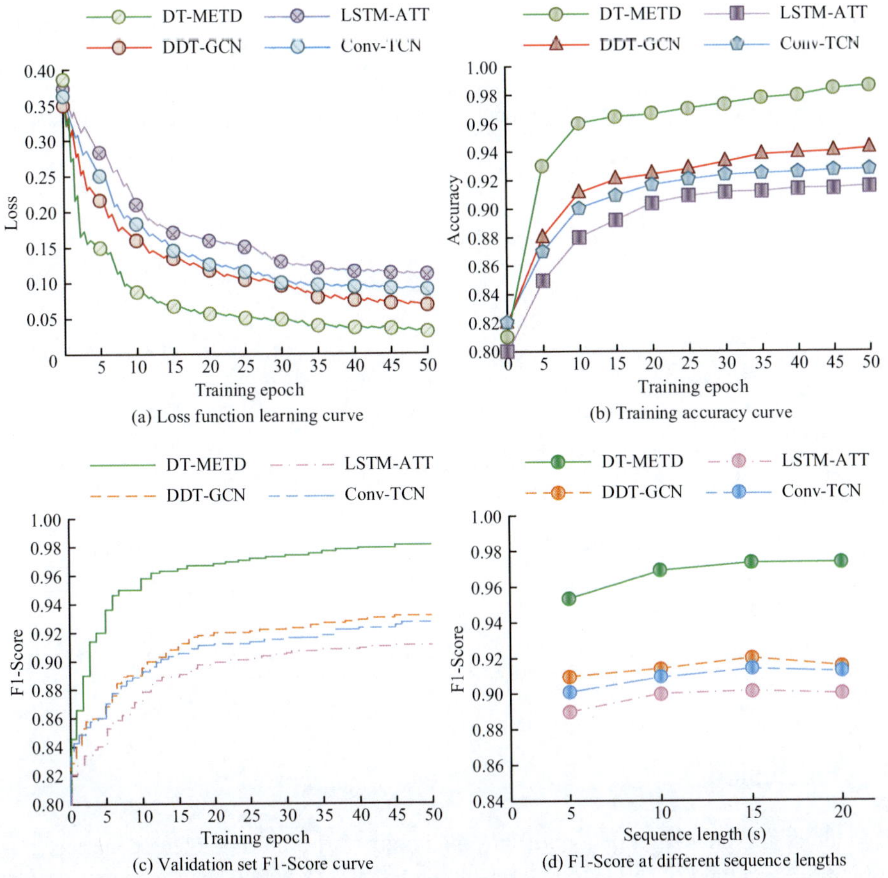

Fig. 3. Comparison of temporal capture capability and convergence among different models

As shown in Table 1, the DT-METD model demonstrates significant advantages in practical DDoS attack detection applications. Although it slightly lags behind traditional signature- and rule-based methods in average detection latency and packet processing rate, its overall performance far surpasses the comparison models. The DT-METD model has a false positive rate of only 1.8%, which is much lower than the signature (12.4%) and rule-based approaches (8.5%). Its false negative rate is also low at 2.1%, which contrasts with traditional methods (55.6% and 32.7% respectively). The DT-METD model achieves an alert accuracy of 97.5% and remains effective against over 95% of unknown attacks. This data shows the model has both high detection accuracy for encrypted traffic and strong generalization capabilities against complex threats. The DT-METD model sacrifices some processing speed. However, it gains detection accuracy and exceptional adaptability to unknown threats. This makes it a more useful defense solution against malicious encrypted traffic.

Table 1. Comparison of DT-METD model and traditional detection methods in DDoS attack detection

Detection Metric	DT-METD Model	Signature-Based Detection	Rule-Based IDS
Average Detection Latency(ms)	15.2	8.9	12.5
False Positive Rate(%)	1.8	12.4	8.5
False Negative Rate(%)	2.1	55.6	32.7
Alert Accuracy(%)	97.5	45.1	68.3
Ability to Handle Unknown Attacks(%)	95+	0	<5
Average Resource Consumption(CPU%)	45.6	15.8	25.4
Packet Processing Rate(pkts/s)	820	2500	1850
Scalability	Excellent	Poor	Medium
Effectiveness on Encrypted Traffic	Excellent	Poor	Poor
Model Deployment Complexity	High	Low	Low

5 Conclusion

To address the challenge of identifying malicious traffic in encrypted scenarios, this study proposes the DT-METD model, which integrates dynamic heterogeneous graphs with Transformers. The model constructs dynamic heterogeneous graphs by segmenting traffic through time snapshots, aggregates spatial features using HAN, and leverages Transformers to capture temporal dependencies. During performance validation, the DT-METD model showed better overall performance than all reference models. It achieved an accuracy of 0.982 and an F1 score of 0.979. As the time window became longer, its F1 score showed a steady upward trend. DDoS simulation tests show the model has a false positive rate of only 1.8% and a false negative rate of 2.1%. The alert accuracy is 97.5%. The model can defend against over 95% of unknown attacks. In summary, this study validates the effectiveness of combining dynamic heterogeneous graphs with Transformers, substantially enhancing the accuracy and generalization of encrypted traffic detection. However, limitations remain, such as a detection latency of 15.2 ms, 45.6% CPU consumption, and relatively high deployment complexity. Future work will focus on optimizing model light weighting to reduce resource consumption and detection latency, thereby improving real-time deployment adaptability.

References

1. Fu, C., Li, Q., Xu, K.: Flow interaction graph analysis: unknown encrypted malicious traffic detection. IEEE/ACM Trans. Netw. **32**(4), 2972–2987 (2024)
2. Yang, Y. et al.: A network traffic classification method based on dual-mode feature extraction and hybrid neural networks. IEEE Trans. Netw. Serv. Manag. **20**(4), 4073–4084 (2023)
3. Achterberg, J., Akarca, D., Strouse, D.J., Duncan, J., Astle, D.E.: Spatially embedded recurrent neural networks reveal widespread links between structural and functional neuroscience findings. Nat. Mach. Intell. **5**(12), 1369–1381 (2023)
4. Gao, X., Jia, X., Li, Y., Xiong, H.: Dynamic scenario representation learning for motion forecasting with heterogeneous graph convolutional recurrent networks. IEEE Rob. Autom. Lett. **8**(5), 2946–2953 (2023)
5. Yuan, T., Chung, H.M., Fu, X.: PP-MART: efficient privacy-preserving multi-agent reinforcement learning for cooperative intelligence in communications. IEEE Netw. **38**(5), 196–203 (2023)
6. Ferriol-Galmés, M., et al.: RouteNet-Fermi: network modeling with graph neural networks. IEEE/ACM Trans. Netw. **31**(6), 3080–3095 (2023)
7. Liu, Y., et al.: A survey of visual transformers. IEEE Trans. Neural Netw. Learn. Syst. **35**(6), 7478–7498 (2023)
8. Chen, H., Hong, P., Han, W., Majumder, N., Poria, S.: Dialogue relation extraction with document-level heterogeneous graph attention networks. Cogn. Comput. **15**(2), 793–802 (2023)
9. Sharma, N., Sharma, K.P., Mangla, M., Rani, R.: Breast cancer classification using snapshot ensemble deep learning model and t-distributed stochastic neighbor embedding. Multimed. Tools Appl. **82**(3), 4011–4029 (2023)
10. Vasanthi, P., Mohan, L.: Multi-head-Self-Attention based YOLOv5X-transformer for multi-scale object detection. Multimed. Tools Appl. **83**(12), 36491–36517 (2024)

Advanced AI Models and Emerging Applications

Multi-agent Q-Learning with Tensorized Q-Function for Adaptive Genetic Algorithms in High-Dimensional Classification

Basma El Mourtji[1(✉)], Tayeb Ouaderhman[2], and Hasna Chamlal[2]

[1] Department of Mathematics and Informatics, Mathematics, Data Science, and Emerging Technologies Laboratory, Faculty of Sciences and Techniques, Hassan First University of Settat, Morocco
basma.elmourtji@uhp.ac.ma

[2] Department of Mathematics and Informatics, LIS Laboratory, Faculty of Sciences Ain Chock, Hassan II University, Casablanca, Morocco
{tayeb.ouaderhman,hasna.chamlal}@univh2c.ma

Abstract. Evolutionary algorithms require careful tuning of control parameters such as mutation and crossover rates to efficiently explore complex search spaces. This paper proposes a novel multi-agent reinforcement learning framework that dynamically adapts these parameters based on population-level feedback. Each agent observes a discretized state vector composed of key statistics–namely, fitness and diversity–that capture essential aspects of the evolutionary dynamics. The corresponding Q-values are stored in a tensorized Q-function, where each dimension reflects a discretized component of the state. This structured representation enables Q-learning agents to better estimate the utility of actions under varying population conditions, effectively balancing exploration and exploitation throughout the genetic search. Experimental evaluation on high-dimensional biological datasets shows that this adaptive strategy improves the stability of feature selection and achieves competitive classification accuracy compared to static parameter settings.

Keywords: Feature Selection · High-Dimensional Classification · Reinforcement Learning · Genetic Algorithm · Multi-Agent Q-Learning

1 Introduction

Feature selection is a critical step in high-dimensional classification, especially in bioinformatics and genomics, where the number of features often exceeds the number of samples. Its goal is to identify a compact and informative subset that improves predictive accuracy while reducing model complexity and overfitting (Chamlal et al., 2021; 2023). Traditional approaches–filter, wrapper, and embedded methodshave been extensively studied (Ang et al., 2015), but suffer from limitations such as ignoring feature interactions, high computational

T. Ouaderhman and H. Chamlal—Contributing authors.

A. E. Hassanien et al. (Eds.): AISDS 2025, LNDECT 286, pp. 285–294, 2027.
https://doi.org/10.1007/978-3-032-18758-1_28

cost, or model dependence. Heuristic approaches, particularly Genetic Algorithms (GAs), provide flexible alternatives by iteratively improving candidate solutions via selection, crossover, and mutation. However, GA performance is highly sensitive to control parameters, notably crossover (p_c) and mutation (p_m) probabilities, which govern the exploration-exploitation balance.

Several enhanced GA-based methods have been proposed to improve convergence and feature relevance. Memetic strategies like MBEGA (Zhu et al., 2007) use Markov blanket-based filtering, while IG-SGA (Salem et al., 2017) integrates Information Gain with genetic search. Other hybrids include ADGA (Akhavan and Hasheminejad, 2023), and MFRAG (Xie et al., 2022), combining domain-specific operators or ensemble classifiers. Yet, these approaches largely rely on fixed or manually tuned parameters, limiting adaptability in dynamic optimization environments.

Reinforcement Learning (RL) offers a natural framework for adaptive parameter control. Q-learning (Hu et al., 2023), a model-free RL algorithm, iteratively updates action values based on rewards and has been extended to Multi-Agent RL (MARL) (Hu et al., 2024) for decentralized decision-making. MAQL frameworks have been successfully applied in communications (Shams et al., 2014), and traffic optimization (Zhang et al., 2015). However, their effectiveness depends critically on the design of the *state representation*. Traditional approaches often combine population statistics such as fitness and diversity into a single scalar (Chen et al., 2020), which introduces hyperparameter sensitivity and may obscure state transitions as GAs converge. To address this, we propose a vectorized state representation in which fitness and diversity are modeled as separate, discretized dimensions. The resulting state-action values are stored in a tensorized Q-table, enabling agents to autonomously adjust crossover and mutation rates, capturing nuanced population dynamics without manual weighting.

Building on this representation, we introduce **TQGA** (Tensorized Q-learning for Genetic Algorithm adaptation), a multi-agent framework in which each agent tunes a specific GA parameter based on observed population statistics. This approach facilitates adaptive and informed parameter control, improving both convergence and feature selection performance across multiple high-dimensional datasets.

2 Problem Formulation

Feature selection aims to identify a compact and informative subset $\mathcal{S} \subseteq \mathcal{X} = \{x_1, x_2, \ldots, x_n\}$ that maximizes classification performance while minimizing the number of selected features. This can be formulated as:

$$\max_{\mathcal{S} \subseteq \mathcal{X}} f(\mathcal{S}) \quad \text{subject to} \quad |\mathcal{S}| \ll n, \tag{1}$$

where $f(\mathcal{S})$ quantifies the predictive relevance of subset $\mathcal{S}$ using a classifier. Genetic Algorithms (GAs) are widely used for this task due to their population-based exploration. Each candidate solution is represented by a binary vector $c \in \{0, 1\}^n$ indicating selected features. The GA evolves the population using crossover and mutation operators, parameterized by probabilities p_c and p_m. These parameters are crucial to balancing exploitation and exploration. Improper values may degrade performance: overly aggressive p_m can introduce noise, while a low p_c may slow convergence.

Q-learning has been proposed to adapt p_c and p_m during the search. However, a major limitation lies in the simplistic representation of the environment's state. Prior approaches, such as (Chen et al., 2020), often compress multiple descriptors (e.g., fitness and diversity) into a single scalar via weighted aggregation, introducing hyperparameter sensitivity and potentially masking important dynamics. For example, during convergence, fitness may increase while diversity decreases, yet their scalar combination can remain unchanged, obscuring meaningful state transitions. More generally, conventional Q-learning struggles to encode the non-linear dependencies between interdependent state dimensions, which can lead to inaccurate stateaction value estimates and suboptimal policy updates. This issue is compounded by the implicit assumption of a stationary environment, whereas in practice–especially in evolutionary search–the state distribution evolves continuously over time (Ma et al., 2024).

3 Proposed Approach

In this section, we introduce a GA-Q-learning hybrid framework, where the genetic algorithm acts as the environment and the reinforcement learning agent adaptively tunes key parameters p_c and p_m. At each time step t, the GA is in a state $s_t \in S$, observed by the agent, which then selects an action $a_t \in A$ corresponding to parameter adjustments according to its policy. The GA applies this action through its evolutionary operators, producing a new state s_{t+1} and a reward r_t reflecting the resulting population fitness: $r_t = f(s_t, a_t, s_{t+1})$, where f denotes the fitness evaluation function. The agent updates its Q-values from the (s_t, a_t, r_t, s_{t+1}) tuple, reinforcing actions that yield positive rewards and penalizing those with negative outcomes. Through repeated interaction, the agent learns to optimize p_c and p_m based on accumulated experience, enhancing the GA's adaptability and performance over time.

State Representation. We model the GA population state at time t as a vector $s_t = [f(P_t), d(P_t)]$, where $f(P_t)$ and $d(P_t)$ denote normalized measures of fitness and diversity, respectively. The fitness of an individual c is given by:

$$f(c) = \begin{cases} \frac{\sum_{i=1}^{n} w_i \cdot c_i}{\sum_{i=1}^{n} c_i} & \text{if } \sum_{i=1}^{n} c_i > 0, \\ 0 & \text{otherwise,} \end{cases} \tag{2}$$

with w_i representing feature relevance scores (Chamlal et al., 2023, El Mourtji et al., 2025). Population fitness $f(P)$ is computed as the normalized interquartile range of individual fitness values:

$$f(P) = \frac{\mathrm{IQR}(\{f(c) \mid c \in P\})}{\max f(c) - \min f(c) + \epsilon}. \tag{3}$$

Diversity $d(P)$ is the average normalized Hamming distance (D_H) between all pairs in P:

$$d(P) = \begin{cases} \frac{1}{n} \cdot \frac{D_H(P)}{m} & \text{if } N \geq 2, \\ 0 & \text{otherwise.} \end{cases} \tag{4}$$

To discretize the continuous $[0, 1]$ range of $f(P)$ and $d(P)$, we divide each dimension into N_I intervals of width $\Delta I = 1/N_I$, producing discrete states $I(1), \ldots, I(N_I)$. For example, $N_I = 5$ yields intervals of length 0.2, so a population with $[f(P) = 0.75, d(P) = 0.32]$ maps to state $[4, 2]$.

Unlike scalar combinations of fitness and diversity (Chen et al., 2020), which require weighted coefficients and may mask changes during convergence, the vector-based representation avoids hyperparameter tuning and captures nonlinear interactions more effectively, leading to a more accurate and adaptive state description.

Action Set Representation. In the proposed multi-agent framework, one agent controls the crossover probability (p_c) and another controls the mutation probability (p_m). Each agent selects from a discrete set of predefined actions. For the crossover agent, the action set is

$$A_c = \{a_{c1}, a_{c2}, a_{c3}\}, \quad \text{with } a_{c1} = 0.7,\ a_{c2} = 0.8,\ a_{c3} = 0.9,$$

and for the mutation agent:

$$A_m = \{a_{m1}, a_{m2}, a_{m3}\}, \quad \text{with } a_{m1} = 0.1,\ a_{m2} = 0.2,\ a_{m3} = 0.3.$$

When an agent selects an action, the corresponding parameter is directly assigned. For example, choosing a_{c3} sets $p_c = 0.9$, and selecting a_{m1} sets $p_m = 0.1$. This indexed action representation keeps the policy discrete and interpretable while allowing precise control over GA parameters.

Reward Design. The reward signals guide the agents based on improvements in classification accuracy and population diversity. For the crossover agent, the reward combines the relative gain in the best individual's fitness and the feature reduction rate:

$$r_c = \eta_1 \frac{\max f(c_t) - \max f(c_{t-1})}{\max f(c_{t-1})} + (1 - \eta_1)\left(1 - \frac{\sum_j c_t^{*,j}}{n}\right), \tag{5}$$

where $\eta_1 \in [0, 1]$ balances fitness and reduction.
Similarly, the mutation agent's reward is based on average fitness improvement and diversity:

$$r_m = \eta_2 \frac{\sum_i f(c_t^i) - \sum_i f(c_{t-1}^i)}{\sum_i f(c_{t-1}^i)} + (1 - \eta_2)\left(1 - \frac{\sum_i \sum_j c_t^{i,j}}{N \times n}\right), \tag{6}$$

with $\eta_2 \in [0, 1]$.

Positive rewards are assigned when classification accuracy improves across generations, promoting effective parameter adjustment.

Action Selection Strategy. Agents select actions based on their estimated values $Q(s, a)$, balancing exploration and exploitation. Formally, the probability of choosing an action a at state s_t can be expressed generically as:

$$P(a \mid s_t) = \frac{F(Q(s_t, a))}{\sum_{a' \in A} F(Q(s_t, a'))}, \tag{7}$$

where $F(\cdot)$ is a monotonically increasing function mapping estimated action values to selection probabilities, and A is the set of possible actions.

This formulation highlights that actions with higher estimated values are more likely to be selected, while still allowing exploration of less-promising actions, ensuring adaptive learning in a dynamic environment.

Q-Table Representation. To capture the joint effect of fitness and diversity on parameter tuning, the Q-values are stored in a multi-dimensional table (tensor) indexed by discretized fitness and diversity states and possible actions:$Q \in \mathbb{R}^{N_{I_f} \times N_{I_d} \times N_a}$. This structure enables the agents to distinguish between different population conditions and adjust parameters accordingly, balancing exploitation of high-fitness states and maintenance of genetic diversity.

Learning Update. At each generation t, agents:

1. Observe the current state s_t (fitness and diversity),
2. Select actions a_t to adjust p_c and p_m,
3. Apply GA with chosen parameters and observe new state s_{t+1} and rewards r_t,
4. Update the Q-values via the standard Q-learning rule:

$$Q(s_t, a_t) \leftarrow Q(s_t, a_t) + \alpha\big[r_t + \gamma \max_{a} Q(s_{t+1}, a) - Q(s_t, a_t)\big],$$

where α and γ are learning rate and discount factor, respectively.

4 Experiments and Results

4.1 Experimental Protocol

We evaluated TQGA on a set of widely used benchmark datasets (see Table 1) to assess its generalizability. Experiments were conducted with three classifiers–SVM, Naïve Bayes (NB), and Random Forest (RF)–and performance was measured using classification accuracy. While most parameters were fixed (Table 2), key GA parameters, namely crossover and mutation probabilities, were adaptively tuned by the Q-learning agents.

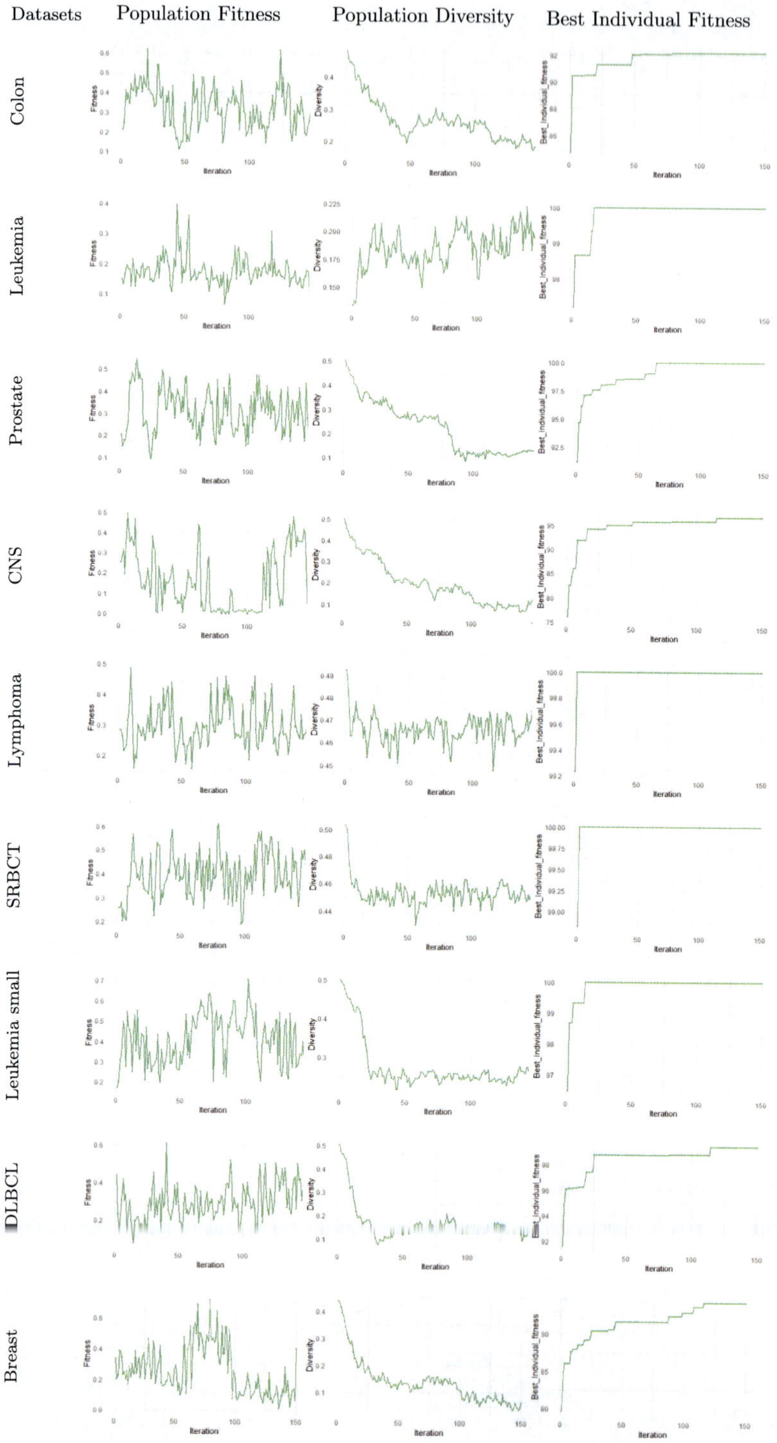

Fig. 1. Evolution of population fitness, diversity, and best fitness over iterations for each dataset using the TQGA method with the SVM classifier.

Table 1. Summary of dataset characteristics.

Dataset	No. of features	No. of instances	Classes
Colon	2000	62	Binary (2)
Leukemia	7128	72	Binary (2)
Prostate	12600	102	Binary (2)
CNS[1]	7133	60	Binary (2)
Lymphoma	4026	62	Multiclass (3)
SRBCT[2]	2308	83	Multiclass (4)
Leukemia small	3571	72	Binary (2)
DLBCL[3]	7070	77	Binary (2)
Breast	4348	78	Binary (2)

Table 2. Main fixed parameters used in the experiments.

Parameter	Value
Population size	30
States per dimension	5
Discount factor γ	0.9
Elitism rate	0.5
Balances fitness and reduction	0.5

4.2 Convergence Behavior and Learning Dynamics

To illustrate the internal dynamics of TQGA, we track three indicators across generations using SVM: average population fitness, genetic diversity, and best individual fitness (see Fig. 1). Initially, the population exhibits high diversity and moderate fitness, promoting exploration. As the algorithm progresses, fitness rises while diversity declines, reflecting effective exploitation. Notably, the Leukemia dataset shows slower convergence with sustained early diversity, likely due to a higher number of informative features. These trends are consistent across multiple runs.

4.3 Ablation Study

To evaluate the impact of key components in our approach, we conduct an ablation study comparing five variants derived incrementally from a baseline genetic algorithm (GA) with fixed crossover and mutation probabilities:

- Classic GA: fixed crossover $p_c = 0.7$ and mutation $p_m = 0.1$;
- Heuristic GA: adaptive probabilities based on population diversity relative to initial diversity;
- GAQL: Q-learning agents tuning probabilities with a static learning rate and greedy policy;
- GAQL+Exploration: GAQL extended by a dynamic exploration mechanism linked to population diversity;
- TQGA: full model integrating all adaptive components and tensor-structured Q-learning.

All variants share identical experimental setups. Table 4 reports mean classification accuracy (Acc) with standard deviation (SD) and feature reduction rate (Rr) over 10 independent runs on nine datasets.

To assess statistical significance of accuracy differences across methods on the nine datasets, we used the Friedman test with average ranks shown in Table 3. Lower rank indicates better performance.

The Friedman test statistic ($\chi^2_F = 27.58$, $p \approx 1.5 \times 10^{-5}$) rejects the null

Table 3. Comparison of feature selection methods: (a) Average ranks over nine datasets; (b) Holm-Bonferroni post-hoc test comparing TQGA against other methods.

(a) Average Ranks

Method	Avg. Rank
TQGA	**1.28**
GAQL + Exp	2.22
Heuristic GA	3.17
GAQL	3.44
Classic GA	4.89

(b) Holm-Bonferroni Test

Method	z	p	Adj. α
Classic GA	4.85	< 0.0001	0.0125
GAQL	2.91	0.0037	0.0167
Heuristic GA	2.53	0.0113	0.0250
GAQL + Exp	1.27	0.2051	0.0500

Table 4. Comparison of feature selection methods: Classic GA, Heuristic GA, GA with Q-learning (GAQL), GAQL with dynamic exploration, and TQGA. Metrics are accuracy (Acc ± SD) and reduction rate (Rr).

Classifier	Method	Metric	Colon	Leuk	Pros	CNS	Lym	SRBCT	LeukemiaS	DLBCL	Breast
SVM	Classic GA	Acc	88.22±5.22	98.22±0.74	98.11±2.18	92.42±4.22	98.71±1.25	98.99±1.15	97.72±2.75	92.12±2.46	90.43±4.02
		Rr	0.9857	0.9965	0.9852	0.9911	0.9822	0.9614	0.9835	0.9932	0.9876
	Heuristic GA	Acc	91.28±4.78	99.22±0.91	99.43±0.44	95.72±3.18	98.91±0.95	99.21±0.55	100±0.00	99.78±0.23	93.13±2.05
		Rr	0.9793	0.9983	0.9885	0.9910	0.9806	0.9731	0.9887	0.9943	0.9867
	GAQL	Acc	95.58±2.98	100±0.00	99.42±0.32	97.44±2.22	99.90±0.10	98.98±1.05	99.12±0.49	99.46±0.30	92.44±0.99
		Rr	0.9735	0.9956	0.9827	0.9922	0.9860	0.9722	0.9845	0.9937	0.9899
	GAQL+Exp	Acc	95.85±2.40	100±0.00	99.79±0.21	97.23±2.86	100±0.00	99.81±0.15	99.14±0.51	100±0.00	92.73±1.02
		Rr	0.9802	0.9959	0.9868	0.9932	0.9866	0.9731	0.9851	0.9946	0.9887
	TQGA	Acc	98.87±2.35	100±0.00	100±0.00	99.40±0.59	100±0.00	100±0.00	100±0.00	100±0.00	94.33±0.97
		Rr	0.9877	0.9997	0.9888	0.9935	0.9930	0.9692	0.9971	0.9952	0.9886
NB	Classic GA	Acc	90.85±2.03	99.14±0.54	92.23±1.55	88.12±1.40	98.22±1.60	99.04±1.05	94.72±2.95	97.65±1.79	89.81±3.09
		Rr	0.9699	0.9957	0.9944	0.9919	0.9864	0.9790	0.9829	0.9922	0.9850
	Heuristic GA	Acc	92.24±1.29	99.64±0.46	95.23±0.72	91.88±1.54	99.18±1.78	100±0.00	95.66±2.60	100±0.00	89.20±2.69
		Rr	0.9687	0.9988	0.9952	0.9915	0.9873	0.9792	0.9817	0.9906	0.9898
	GAQL	Acc	92.17±2.48	99.27±0.17	95.68±0.98	92.01±0.91	100±0.00	99.22±0.49	97.14±1.99	98.75±1.07	88.75±2.00
		Rr	0.9644	0.9966	0.9951	0.9922	0.9843	0.9822	0.9899	0.9919	0.9933
	GAQL+Exp	Acc	92.85±2.40	100±0.00	94.68±1.12	92.48±1.74	100±0.00	100±0.00	97.04±2.09	99.37±1.99	90.44±2.22
		Rr	0.9711	0.9942	0.9956	0.9921	0.9860	0.9935	0.9901	0.9929	0.9939
	TQGA	Acc	94.27±1.10	100±0.00	97.58±1.24	93.98±1.54	100±0.00	100±0.00	100±0.00	100±0.00	90.94±2.91
		Rr	0.9735	0.9995	0.9956	0.9932	0.9860	0.9935	0.9919	0.9932	0.9926
RF	Classic GA	Acc	90.85±3.22	98.81±0.64	94.89±2.66	87.88±2.15	98.21±0.79	99.75±0.12	96.88±1.95	97.71±2.50	88.21±2.71
		Rr	0.9720	0.9944	0.9944	0.9909	0.9920	0.9804	0.9813	0.9877	0.9839
	Heuristic GA	Acc	91.57±2.88	99.26±0.96	97.00±2.17	89.74±2.44	99.91±0.77	100±0.00	97.62±2.15	98.75±1.25	87.91±3.29
		Rr	0.9690	0.9922	0.9961	0.9928	0.9930	0.9922	0.9829	0.9909	0.9903
	GAQL	Acc	93.17±2.43	100±0.00	96.16±1.12	89.98±2.50	100±0.00	99.45±0.42	98.72±1.71	97.52±1.81	91.04±2.42
		Rr	0.9756	0.9912	0.9941	0.9911	0.9913	0.9770	0.9843	0.9932	0.9878
	GAQL+Exp	Acc	94.85±2.80	100±0.00	95.26±0.99	90.88±2.64	100±0.00	100±0.00	99.72±0.15	97.41±1.88	94.20±1.79
		Rr	0.9881	0.9941	0.9956	0.9918	0.9926	0.9802	0.9836	0.9953	0.9881
	TQGA	Acc	97.18±2.83	100±0.00	98.14±0.69	92.89±1.88	100±0.00	100±0.00	99.92±0.07	99.14±0.29	95.10±1.87
		Rr	0.9886	0.9952	0.9968	0.9926	0.9930	0.9911	0.9855	0.9962	0.9877

hypothesis of equal performance. Post-hoc analysis confirms TQGA significantly outperforms Classic GA, GAQL, and Heuristic GA, while differences with GAQL+Exploration are not statistically significant, indicating benefits from dynamic exploration.

5 Conclusion

We proposed a reinforcement learningbased feature selection framework that integrates a genetic algorithm with multi-agent Q-learning to adapt crossover and mutation probabilities. A tensor-based state representation encodes population fitness and diversity, enabling decentralized and context-aware operator control.

Experiments on nine diverse transcriptomic datasets demonstrated consistently high classification accuracy with substantial feature reduction. An ablation study confirmed the role of each component, and Friedman ranking with HolmBonferroni validation established the robustness and significance of the gains.

Although effective on the tested datasets, the method may incur high computational costs on very large-scale data. Future work will address scalability, while exploring multi-objective formulations and more expressive Q-function approximators to further enhance decision quality.

References

Chamlal, H., Ouaderhman, T., Mourtji, E.L., B.: Multicriteria approaches based on a new discrimination criterions for feature. Selection (2021). https://doi.org/10.1016/j.knosys.2023.110420

Chamlal, H., Ouaderhman, T., Mourtji, B.E.: Feature selection in high dimensional data: a specific preordonnances-based memetic algorithm. Knowl. Based Syst. **266**, 110420 (2023). https://doi.org/10.1016/j.knosys.2023.110420

Ang, J.C., Mirzal, A., Haron, H., Hamed, H.N.A.: Supervised, unsupervised, and semi-supervised feature selection: a review on gene selection. IEEE/ACM Trans. Comput. Biol. Bioinf. **13**(5), 971–989 (2015)

Zhu, Z., Ong, Y.-S., Dash, M.: Markov blanket-embedded genetic algorithm for gene selection. Pattern Recogn. **40**(11), 3236–3248 (2007)

Salem, H., Attiya, G., El-Fishawy, N.: Classification of human cancer diseases by gene expression profiles. Appl. Soft Comput. **50**, 124–134 (2017)

Akhavan, M., Hasheminejad, S.M.H.: A two-phase gene selection method using anomaly detection and genetic algorithm for microarray data. Knowl. Based Syst. **262**, 110249 (2023)

Xie, W., Fang, Y., Yu, K., Min, X., Li, W.: MFRAG: multi-fitness rankaggreg genetic algorithm for biomarker selection from microarray data. Chemom. Intell. Lab. Syst. **226**, 104573 (2022)

Hu, K., et al.: Overview of underwater 3D reconstruction technology based on optical images. J. Mar. Sci. Eng. **11**(5), 949 (2023)

Hu, K., et al.: A review of research on reinforcement learning algorithms for multi-agents. Neurocomputing 128068 (2024)

Shams, F., Bacci, G., Luise, M.: Energy-efficient power control for multiple-relay cooperative networks using q-learning. IEEE Trans. Wireless Commun. **14**(3), 1567–1580 (2014)

Zhang, X., Li, H., Peng, J., Liu, W.: A cooperative q-learning path planning algorithm for origin-destination pairs in urban road networks. Math. Probl. Eng. **2015**(1), 146070 (2015)

Chen, R., Yang, B., Li, S., Wang, S.: A self-learning genetic algorithm based on reinforcement learning for flexible job-shop scheduling problem. Comput. Ind. Eng. **149**, 106778 (2020)

Ma, Z., Chen, J., Guo, H., Ma, Y., Gong, Y.-J.: Auto-configuring exploration-exploitation tradeoff in evolutionary computation via deep reinforcement learning (2024). https://doi.org/10.48550/arXiv.2404.08239

El Mourtji, B., Ouaderhman, T., Chamlal, H.: A new filter based-instance selection for high dimensional data classification. Eng. Appl. Artif. Intell. **158**, 111082 (2025)

AI Methods for the Heston–American Free-Boundary PDE: A Comparative Study and ID/OOD Robustness Evaluation

Zakaria Elbayed(✉) and Abdelmjid Qadi Elidrissi

Laboratory for Analysis and Modeling of Systems and Decision Support, Center for Doctoral Studies in Science and Technology and Medical Sciences, Hassan First University, National School of Applied Sciences Berrchid, Berrechid, Morocco
{z.elbayed.doc,abdelmjid.qadi-elidrissi}@uhp.ac.ma

Abstract. We study data–driven surrogates for pricing American put options under the Heston model. Our approach enforces *hard* no–arbitrage through a capped reconstruction of the time value (TV), $U = \Phi + (1 - \Phi)\sigma(z)$, which guarantees $U \in [\Phi, 1]$ by construction, and promotes *PDE–consistent* shape constraints (nondecreasing in time to maturity, nonincreasing in log–moneyness) via finite–difference penalties inspired by the variational inequality (VI) for American options. We benchmark three learning paradigms on clean synthetic datasets: point–wise multilayer perceptrons (MLP), point–wise Transformers, and a volumetric 3D Fourier Neural Operator (FNO). On out–of–distribution (OOD) scenarios, FNO–3D achieves the lowest error on grids (RMSE $\approx$ 0.212), while point– wise models remain competitive and extremely fast per query (MLP: ID RMSE $\approx$ 0.0636, OOD RMSE $\approx$ 0.311; Transformer: ID RMSE $\approx$ 0.0706, OOD RMSE $\approx$ 0.320), with zero arbitrage in all cases. Beyond empirical results, we justify the surrogate design through hard bounds and an informal consistency argument combining monotone, stable, and consistent VI labels with universal approximation. We release datasets, splits, configurations, and checkpoints to facilitate reproducibility.

Keywords: American options · Heston model · Variational inequality · No–arbitrage · Fourier Neural Operator · Transformer · Deep learning for PDEs

1 Introduction

American option pricing under stochastic volatility remains a central question in computational finance and scientific machine learning. Under the Heston dynamics [5], the American price solves a free–boundary variational inequality (VI) in (ℓ, v, τ), with $\ell = \log(S/K)$ and time–to–maturity τ. The early–exercise boundary (EEB) induces limited regularity (only piecewise smoothness), which makes robust approximation challenging across parameters and market regimes [6]. Classical numerical methods (finite differences, PSOR) deliver reliable oracles but become costly on large parameter grids or when entire price *surfaces* are required; neural surrogates promise millisecond–level responses yet may violate financial structure (arbitrage bounds, monotonicities) and degrade under distribution shift.

A. E. Hassanien et al. (Eds.): AISDS 2025, LNDECT 286, pp. 295–304, 2027.
https://doi.org/10.1007/978-3-032-18758-1_29

Problem and Approach. We learn the *time value* (TV) of a unit–strike American put and reconstruct the price as,

$$U(\ell, v, \tau) = \Phi(\ell) + (1 - \Phi(\ell))\,\sigma(z), \qquad \Phi(\ell) = \max(1 - e^{\ell}, 0),$$

so that $U \in [\Phi, 1]$ holds *by construction* (hard no–arbitrage). This reparameterization improves conditioning near short maturities and in the EEB neighborhood. To inject VI structure without heavy residual minimization, we add light finite– difference (FD) shape penalties on mini–batches enforcing $\partial_\tau U \geq 0$ and $\partial_\ell U \leq 0$. Training labels are produced by a monotone, stable, and consistent FD scheme for the Heston VI [2, 6], used as a high–quality oracle.

Models and Evaluation. We compare three paradigms: (a) a pointwise MLP, (b) a pointwise Transformer (Fourier feature tokenization + self–attention), and (c) a volumetric 3D Fourier Neural Operator (FNO) on 48^3 grids [9]. Pointwise models target ultra–low–latency quotes at arbitrary (X,P); FNO outputs full (ℓ,v,τ) volumes in one pass, valuable for risk aggregates and calibration. We adopt a unified protocol with a clean in–distribution (ID) split and two out– of–distribution (OOD) regimes (domain shift in (ℓ,v,τ) and parameter shift in $(\kappa,\theta,\sigma,\rho,v_0,r,q)$), and we report accuracy on TV (RMSE/MAE) together with finance sanity checks (arbitrage counts).

Contributions.

- **Finance–aware surrogates with hard guarantees.** A capped TV reconstruction enforces $U \in [\Phi,1]$ exactly; lightweight FD penalties encode VI monotonicities, stabilizing learning around the EEB while avoiding heavy PDE residuals.
- **A unified, cross–paradigm benchmark.** We evaluate MLP, Transformer, and 3D FNO under a consistent ID/OOD protocol, clarifying accuracy– latency trade–offs (pointwise speed vs. grid OOD accuracy).
- **Strong and interpretable results.** On OOD grids, 3D FNO attains the lowest error (RMSE ≈ 0.212); pointwise models remain competitive (MLP: OOD ≈ 0.311, Transformer: OOD ≈ 0.320) with small ID errors (MLP ≈ 0.0636, Transformer ≈ 0.0706), and zero arbitrage across all runs.
- **Consistency perspective.** Given monotone/stable/consistent VI labels and universal approximation, empirical minimizers with hard capping and FD monotonicities converge (in a viscosity sense) to a monotone limit compatible with the VI solution [2, 6] (informal statement).
- **Reproducibility.** We release cleaned datasets, fixed splits, training configurations, and checkpoints to facilitate verification and reuse.

2 Related Work

American Options as Variational Inequalities. Pricing American options via obstacle problems / variational inequalities (VIs) is classical [6, 13]. For Heston [5], monotone–stability–consistency arguments guarantee convergence of finite–difference schemes to the viscosity solution [2], and remain the reference oracle when accuracy is paramount.

Deep Learning for American Option Pricing (2020–2025). Several works use supervised networks to accelerate American pricing from trusted labels (e.g. FD/PSOR), reporting near–instantaneous quotes once trained; see, e.g., feedforward surrogates for American options and empirical speed/accuracy trade-offs [1]. Residual-based routes (PINNs/DGM) have also been applied to option pricing under Black–Scholes/Heston, but training can be delicate near the free boundary and at short maturities; recent papers discuss stability and sampling strategies [3, 8]. Our approach follows the *constrained supervised* line: learn from a stable FD oracle, enforce finance structure hard, and add light VI shape constraints.

Arbitrage–Free Learning and Shape Constraints. In data-driven volatility modelling, arbitrage-free design has been enforced either by architecture/parameterization (e.g. IVS factor models or generators) or by losses with explicit constraints; see *arbitrage-free* IVS prediction/generation [10–12] and recent work on detecting static arbitrage with NNs [11]. In neural pricing, hard bounds and monotonicities (with respect to strike/maturity) help preserve financial sanity and improve extrapolation; our capped time-value and FD monotonicity penalties fall into this stability-by-design family.

Operator Learning for PDEs. Neural operators learn maps between function spaces and have emerged as strong PDE surrogates. The Fourier Neural Operator (FNO) introduced resolution-invariant global convolution in Fourier space [9], while the Neural Operator framework formalized approximation and discretization invariance [7]. PINO augments FNO with physics residuals across resolutions [8]. Benchmarks such as PDEBench [14] stress-test generalization and out-of-distribution (OOD) robustness, a perspective we adopt via our ID/OOD protocols.

Transformers for PDE Surrogates. Beyond sequence modeling, Transformers have been adapted to PDE fields via spectral token mixers (AFNO) that replace quadratic self-attention by Fourier-domain mixing [4], and via scalable Transformer backbones for regular grids. These results motivate our inclusion of a pointwise Transformer (feature tokenization + self-attention) alongside an operator learner (3D FNO) and a tabular MLP.

Positioning. Compared to PINNs/PDE-residual training, we rely on a monotone/stable/consistent FD oracle and enforce finance structure with *hard* noarbitrage and *light* VI-consistent shape penalties. Relative to prior American pricing surrogates, we (i) evaluate three paradigms (MLP, Transformer, 3D FNO), (ii) under a unified ID/OOD protocol (domain and parameter shift), and (iii) report both accuracy and arbitrage metrics, clarifying when pointwise vs. operator learners are preferable.

3 Mathematical Approach

Under the risk–neutral measure, S follows Heston [5]:
$\frac{dS_t}{S_t} = (r-q)dt + \sqrt{V_t}dW_t^S,\quad dV_t = \kappa(\theta - V_t)dt + \sigma\sqrt{V_t}dW_t^V,\quad d\langle W^S, W^V\rangle_t = \rho dt$, with $V_t \geq 0$. We use normalized variables $\ell = \log(S/K)$, variance v, and time to maturity $\tau = T - t$. The unit–strike American put payoff is $\Phi(l) = \max(1 - e^l, 0)$; the normalized price U satisfies $U \in [\Phi, 1]$, with time value $\mathrm{TV} = U - \Phi \in [0, 1 - \Phi]$.

Variational Inequality (VI). The American price is the viscosity solution of the obstacle problem

$$\mathrm{minn}\partial_\tau U - \mathrm{L}^{\mathrm{H}} U + rU, U - \Phi^{\mathrm{o}} = 0, U(\cdot, \cdot, 0) = \Phi, U \in [\Phi, 1], \tag{1}$$

with Heston operator (in (ℓ, v))

$$L^H f = \frac{1}{2} \upsilon\, \partial_{ll} f + \rho\sigma\sqrt{\upsilon}\, \partial_{l\upsilon} f + \frac{1}{2}\sigma^2 \upsilon\, \partial_{\upsilon\upsilon} f + \left(r - q - \frac{1}{2}\upsilon\right) \partial_l f + \kappa(\theta - \upsilon)\partial_\upsilon f,$$

and standard boundary conditions [2, 6, 13].

Hard no–arbitrage via capping. We approximate the time–value by a network output z_θ and reconstruct

$$\mathrm{TV}^\theta = (1 - \Phi)\sigma(z_\theta), \quad U_\theta = \Phi + TV_\theta, \tag{2}$$

where σ is the logistic sigmoid.

Theorem 1. (Hard no–arbitrage). For any $\mathrm{z}_\theta \in \mathrm{R}$, the reconstruction (2) guarantees $\mathrm{U}_\theta \in [\Phi, 1]$.

Proof. Since $\sigma \in (0,1)$, we have $0 \le (1 - \Phi)\sigma(z_\theta) \le 1 - \Phi$. Adding Φ yields $U_\theta \in [\Phi, 1]$.

Shape Constraints via FD Penalties. For American puts, one expects $\partial_\tau U \ge 0$ and $\partial_\ell U \le 0$. We promote these with finite–difference (FD) penalties on subsamples, with small steps $\varepsilon_\tau, \varepsilon_\ell > 0$:

$$\mathcal{P}_\tau(\theta) = \mathbb{E}\Big[(\partial_\tau U_\theta)^-\Big] \approx \mathbb{E}\Big[\Big(\tfrac{U_\theta(\ell,v,\tau+\varepsilon_\tau) - U_\theta(\ell,v,\tau-\varepsilon_\tau)}{2\varepsilon_\tau}\Big)^-\Big]$$

$$\mathcal{P}_\ell(\theta) = \mathbb{E}\Big[(\partial_\ell U_\theta)^+\Big] \approx \mathbb{E}\Big[\Big(\tfrac{U_\theta(\ell+\varepsilon_\ell,v,\tau) - U_\theta(\ell-\varepsilon_\ell,v,\tau)}{2\varepsilon_\ell}\Big)^+\Big],$$

with $x^- = \max(-x, 0)$ and $x^+ = \max(x, 0)$.

Proposition 1 (Informal consistency) Assume (i) labels from a monotone, stable, and consistent FD scheme for (1) so that $\mathrm{U}_\mathrm{h} \to \mathrm{U}^\star$ uniformly as mesh $\mathrm{h} \to 0$ [2]; (ii) a universal model class on compact sets; and (iii) empirical risk combining prediction and FD penalties with weights $(\lambda_\tau, \lambda_\ell) \to \infty$. Then empirical minimizers $\mathrm{U}_{\theta\mathrm{n}}$ converge in L^2 (on compacts) to a monotone limit U^∞ consistent with the viscosity solution; if the VI solution is unique, $\mathrm{U}^\infty = \mathrm{U}^\star$.

4 Methodology

4.1 Task and Inputs

We learn $\mathrm{TV}(\ell, v, \tau; \kappa, \theta, \sigma, \rho, v_0, r, q) \in [0, 1 - \Phi(\ell)]$ and reconstruct $U = \Phi + \mathrm{TV}$ via (2). Pointwise inputs: $X = (\ell, v, \tau)$; parameters: $P = (\kappa, \theta, \sigma, \rho, v_0, r, q)$.

4.2 Synthetic Datasets and Splits

We generate three clean banks:

- **ID points**: 200,000 labeled tuples (X,P) with TV and exercise indicator $\mathbb{1}[\text{TV} > 0]$.
- **OOD–domain points**: 60,000 tuples with enlarged (ℓ,v,τ) ranges (e.g., $\ell \in [-3,3]$, higher v, longer τ).
- **OOD–parameters points**: 60,000 tuples outside the training hyper–box in $(\kappa,\theta,\sigma,\rho,v_0,r,q)$.

For volumetric learning we construct 48^3 grids with 11 input channels and a TV target: 32 ID grids and 24 OOD grids (12 domain + 12 parameters). Labels come from a stable monotone FD solver for the Heston VI [**?**].

4.3 Architectures

MLP (pointwise, capped). A fully–connected network on $[\ell,v,\tau,P] \in \mathrm{R}^{10}$ (e.g., widths 512–512–256) outputs a logit z used in (2).

Transformer (pointwise, capped). Each scalar is embedded with Fourier features and mapped to tokens of width d; a [CLS] token plus L self–attention blocks, then a linear head producing z and the capped reconstruction (2).

FNO–3D (volumetric). A 3D Fourier Neural Operator [**?**] takes the 11– channel tensor on a 48^3 lattice and outputs a TV volume; the price volume is $U = \Phi + \text{TV}$ channelwise.

4.4 Training Objective and Regularization

On pointwise data we minimize

$$\mathcal{L}(\theta) = \mathbb{E}\big[\ell_{\text{pred}}(\text{TV}_\theta, \text{TV}^\star)\big] + \lambda_\tau\, \mathcal{P}_\tau(\theta) + \lambda_\ell\, \mathcal{P}_\ell(\theta) + \lambda_{\text{exc}}\, \mathbb{E}\big[\mathbb{1}_{\{\text{exercise}\}} \quad (U_\theta - \Phi)^2\big], \tag{3}$$

where ℓ_{pred} is Huber or L2 on TV (or BCE on ratio $r = \text{TV}/(1-\Phi)$). For FNO–3D we use an L^2 volumetric loss; optional FD penalties may apply on slices.

Table 1. Main results on TV (lower is better). Arbitrage shows counts ($U < \Phi$ / $U > 1$). FNO–3D is evaluated on OOD grids

Model	OOD RMSE	OOD MAE	ID RMSE	Arbitrage	Notes
MLP (cap+FD)	0.311	0.1266	0.0636	0 / 0	Pointwise
Transformer (cap+FD)	0.320	0.1411	0.0706	0 / 0	Pointwise
FNO–3D (48^3)	**0.212**	**0.097**	—	0 / 0	Grid–based

4.5 Training and Compute

Features are standardized using ID statistics. We use AdamW with cosine schedule, EMA, and early stopping. We oversample the payoff kink ($|\ell| \leq 0.1$) and short maturities ($\tau \leq 0.25$). All results use a fixed random seed. Hardware: *¡GPU/CPU, VRAM¿*. Parameter counts (approx.): MLP *¡M¿*, Transformer *¡M¿*, FNO–3D *¡M¿*.

Oracle and Numerical Quality. Labels are produced by a monotone, stable, and consistent finite–difference scheme for the Heston VI [**?**,2] with standard boundary conditions: $U(\ell \to +\infty, \cdot, \tau) \to 0$, $U(\ell \to -\infty, \cdot, \tau) \to 1$, reflecting/absorbing at $v = 0$. This ensures convergence to the viscosity solution and yields clean synthetic ground truth for supervised learning.

5 Results

Protocol and Metrics. We evaluate models on the time value (TV) target and reconstruct prices $U = \Phi + \text{TV}$; this enforces $U \in [\Phi, 1]$ by construction, so we always report *arbitrage counts* ($U < \Phi$, $U > 1$) alongside error metrics. For pointwise surrogates (MLP and Transformer) we use an ID holdout split and a combined OOD split (domain+parameters). For FNO–3D we evaluate on OOD grids of size 48^3. Errors are root mean–square (RMSE) and mean absolute (MAE) on TV; figures visualize U for financial interpretability.

High-Level Reading of Table 1. On OOD scenarios, the volumetric FNO–3D achieves the lowest RMSE/MAE, consistent with its nonlocal operator bias and spectral parameterization on grids. Pointwise models remain competitive in accuracy and *excel in latency*: they deliver instant per-query quotes, which is attractive for trading/risk micro services. On the ID split, both pointwise surrogates reach small errors ($\approx$ 0.06–0.07 RMSE), highlighting that the capped TV target is well-conditioned in-distribution. Across all models and splits, arbitrage counts are zero, corroborating the hard bound $U \in [\Phi, 1]$.

Qualitative Comparison of Price Surfaces. Figure 1 contrasts the learned $U(\ell, \tau)$ at median variance. MLP and Transformer produce smooth, financially consistent surfaces with the expected increase in τ and decrease in ℓ. FNO reproduces global structure on slices and sharpens curved isocontours near the early-exercise boundary (EEB), which helps OOD generalization on grids.

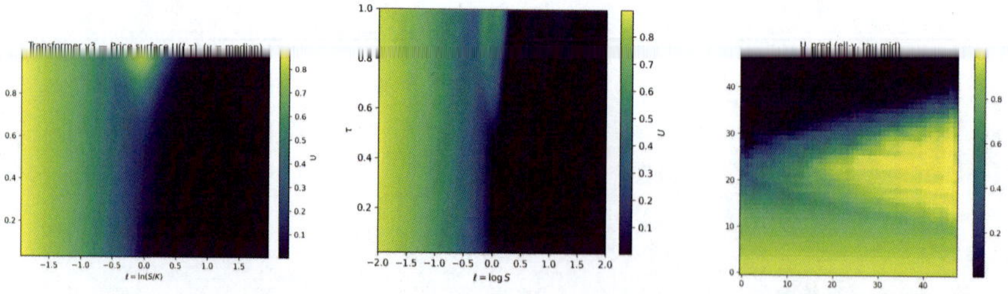

Fig. 1. Comparison of price surfaces $U(\ell, \tau)$ at median v across paradigms. **Left**: Transformer, **Middle**: MLP, **Right**: FNO-3D

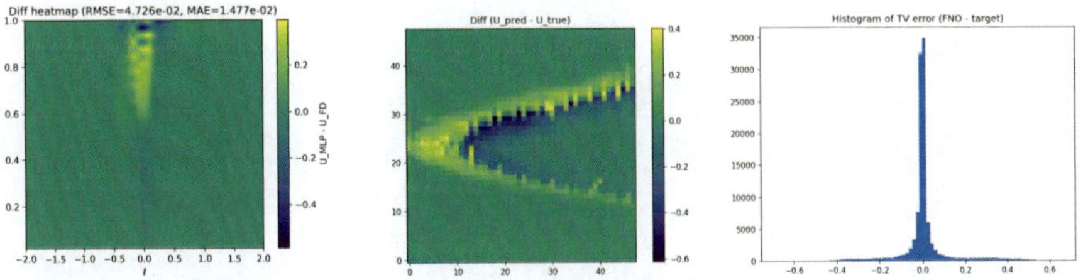

Fig. 2. Error structure on a representative slice. **Left**: MLP vs. FD oracle (diff heatmap). **Middle**: FNO vs. FD on the same slice. **Right**: TV error histogram (FNO)

Error Geometry and EEB Localization. Figure 2 focuses on error maps and distribution. For MLP, the error heatmap (vs. the FD oracle) concentrates near the EEB where the obstacle induces non-smoothness; this is a classical difficulty for American options. FNO exhibits low, spatially coherent errors on the same slice, and its TV error histogram shows a tight bulk with thin tails—consistent with Table 1.

One-Dimensional Slices: Monotonicities and Conditioning. Figures 3 and 4 inspects U along ℓ (for several τ) and along τ (for several ℓ). Both MLP and Transformer display $\partial_\tau U \geq 0$ and $\partial_\ell U \leq 0$ across tested ranges, in line with VI theory and our finite-difference (FD) penalties. The capped reconstruction stabilizes the small-τ regime and prevents overshoots above 1.

No-Arbitrage and Shape Constraints: Sanity Checks. All evaluated configurations yielded zero arbitrage violations (counts in Table 1), due to the hard capping $U \in [\Phi, 1]$. Monotonicity penalties further reduce spurious non-monotone behavior; visually, slices and heatmaps do not exhibit systematic violations. This combination—hard bounds + soft FD penalties—keeps the surrogates faithful to the VI structure while avoiding heavy residual-based training.

Practical Trade-Offs. Pointwise surrogates (MLP, Transformer) are best when ultra-low-latency scalar quotes are needed (e.g., pricing a stream of (X,P)). FNO provides full (ℓ, v, τ) volumes in one pass, useful for risk aggregates and calibration, and achieves the strongest OOD accuracy on grids. Hence, model choice depends on the downstream workload: quoting vs. surface generation.

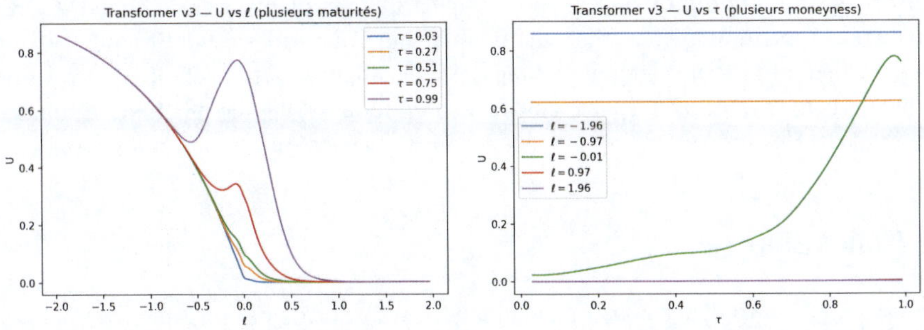

Fig. 3. 1D cuts confirming shape constraints. **Left**: U vs. ℓ for several τ (Transformer). **Right**: U vs. τ for several ℓ (Transformer). Similar trends observed for MLP (not shown)

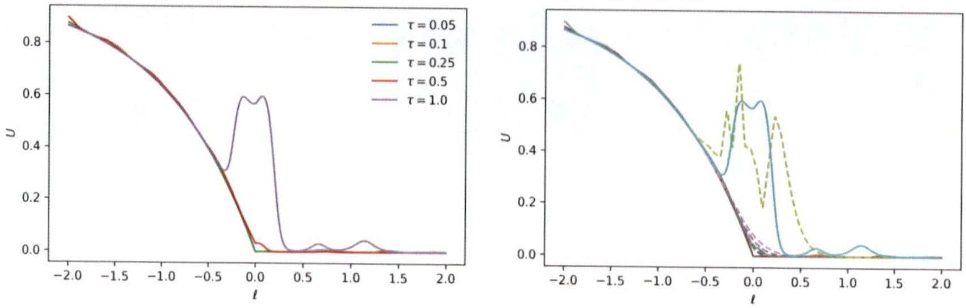

MLP — U vs. ℓ MLP vs. FD oracle on $U(\ell)$

Fig. 4. 1D cuts of U confirm expected monotonicities and the stabilizing effect of the capped TV reconstruction (a) MLP — U vs. ℓ and (b) MLP vs. FD oracle on $U(\ell)$

Limitations of This Evaluation. All datasets are synthetic and derived from a convergent FD oracle; this introduces a solver-dependent bias. Our OOD regimes are substantial but do not cover all tail configurations (very high variance, very long maturities). FNO was trained on 48^3 grids, a tractability compromise; higher resolutions may further improve accuracy at higher compute cost.

6 Discussion

Where Each Paradigm Excels. FNO–3D yields the lowest OOD error on volumetric grids, consistent with its nonlocal spectral bias; pointwise MLP/Transformer remain competitive in–distribution and are operationally attractive for many *scalar* quotes (low latency/memory). Under domain/parameter shift, FNO better reconstructs full surfaces, while pointwise models degrade gracefully near the ID manifold. A pragmatic deployment is hybrid: pointwise engine for quoting; periodic FNO refresh for risk, calibration, and diagnostics.

Error Geometry and Constraints. Errors concentrate near the early–exercise boundary (EEB), where the VI solution is only piecewise smooth. Capping the time value confines $U \in [\Phi, 1]$ and stabilizes short maturities; finite–difference penalties on $\partial_\tau U \geq 0$ and $\partial_\ell U \leq 0$ suppress residual violations without heavy PDE residual minimization.

Sensitivity and Limitations. The most impactful ingredients were: capped TV, FD penalties, oversampling near the payoff kink/short maturities, and EMA/early stopping; architecture variants altered rankings less than these structural choices. Limitations include oracle (FDM) bias, sparse coverage at extreme OOD, the 48^3 FNO resolution, and scope to Heston American puts (extensions to barriers/jumps are future work).

7 Conclusion

Summary. We introduced constrained, finance-aware surrogates for American put pricing under Heston, combining a *hard* no-arbitrage cap $U \in [\Phi, 1]$ with *soft* finite-difference penalties that encode VI-consistent shape constraints. Within a unified

ID/OOD benchmark, a volumetric 3D FNO achieved the strongest OOD accuracy on grids (RMSE ≈ 0.212), while pointwise MLP and Transformer offered competitive errors (ID RMSE ≈ 0.06–0.07, OOD ≈ 0.31–0.32) and excellent per-query efficiency. All models satisfied arbitrage bounds by construction.

Key takeaways. (i) Modeling the *time value* and capping it in $[0, 1 - \Phi]$ is a robust way to hard-enforce financial bounds and stabilize learning near the EEB. (ii) Simple FD penalties are sufficient to steer monotonicities without resorting to heavy PDE residuals. (iii) Operator-learning (FNO) is particularly effective when full surfaces are needed and the evaluation domain shifts, while pointwise surrogates remain preferable for ultra-low-latency quoting.

Outlook. Three directions are especially promising: (1) *Market integration*—joint training/calibration on market panels with microstructure-aware noise models and embedded constraint layers (e.g., isotonic or lattice components for certified monotonicity). (2) *Model breadth*—extension to other stochastic volatility and jump-diffusion settings, exotic payoffs (barriers, baskets), and path-dependent features via lifted state. (3) *Resolution and UQ*—higher-resolution FNOs, multiresolution schemes, and uncertainty quantification (e.g., conformal risk bounds) to provide calibrated error bars alongside quotes.

Reproducibility. We release cleaned ID/OOD datasets and grids, fixed splits, training configurations, and checkpoints to facilitate verification and reuse. This provides a transparent baseline for future constrained surrogates and operator learners under American-style free-boundary dynamics.

References

1. Anderson, M. and Golden, R.M.: Accelerated American option pricing with deep neural networks. SSRN Working Paper (2023)
2. Barles, G., Souganidis, P.E.: Convergence of approximation schemes for fully nonlinear second order equations. Asymptot. Anal. **4**, 271–283 (1991)
3. Gatta, V., Orlando, G., Mainardi, F.: Meshless methods for American option pricing through physics-informed neural networks. Eng. Anal. Bound. Elem. **151**, 68–82 (2023). https://doi.org/10.1016/j.enganabound.2023.02.040
4. Guibas, J., et al.: Adaptive Fourier neural operators: efficient token mixing for transformers. arXiv preprint (2021)
5. Heston, S.L.: A closed-form solution for options with stochastic volatility with applications to bond and currency options. Rev. Financ. Stud. **6**(2), 327–343 (1993)
6. Jaillet, P., Lamberton, D., Lapeyre, B.: Variational inequalities and the pricing of American options. Acta Appl. Math. **21**, 263–289 (1990)
7. Kovachki, N.B., Li, Z., Azizzadenesheli, K., Bhattacharya, K., Stuart, A.M., Anandkumar, A.: Neural operator: learning maps between function spaces with applications to PDEs. J. Mach. Learn. Res. **24**(251), 1–97 (2023)
8. Li, Z., Kovachki, N., Bhattacharya, K., Stuart, A.M., Anandkumar, A.: Physics informed neural operator for learning partial differential equations. Commun. ACM (Research Highlights). **1**(3), 1–27 (2024). https://doi.org/10.1145/3648506
9. Li, Z., et al.: Fourier neural operator for parametric partial differential equations. In: International Conference on Learning Representations (ICLR) (2021)

10. Nard, I.D., Marazzina, D., Scotti, S.: Deep smoothing of the implied volatility surface. In: Advances in Neural Information Processing Systems (NeurIPS) Workshops (2020)
11. Neufeld, A., Sester, L.: Neural networks can detect model-free static arbitrage strategies. Appl. Math. Optim. **90**(2), 41 (2024)
12. Ning, B., Wang, J., Chen, W.: Arbitrage-free synthesis of the implied volatility surface via variational autoencoders. SIAM J. Financ. Math.
13. Pham, H.: Continuous-Time Stochastic Control and Optimization with Financial Applications, Stochastic Modelling and Applied Probability, vol. 61. Springer (2009). https://doi.org/10.1007/978-3-540-89500-8
14. Takamoto, M., et al.: PDEBench: An extensive benchmark for scientific machine learning. In: NeurIPS 2022 Datasets and Benchmarks Track (2022)

Optimal Base Station Placement for UAV Relative Localization: A Greedy-Fuzzy Logic Approach

Asma Boulkaid[1,2(✉)], Aurelio Bermúdez[1], Rafael Casado[1], Luis Orozco-Barbosa[1], and Abdelkrim Haqiq[2]

[1] Albacete Research Institute of Informatics, University of Castilla-La Mancha, Campus Universitario, s/n, 02071 Albacete, Spain
asma.boulkaid@alu.uclm.es,
{aurelio.bermudez,rafael.casado,luis.orozco}@uclm.es

[2] Computer Networks, Mobility and Modelling Laboratory (IR2M), Hassan First University, Faculty of Sciences and Techniques, Settat, Morocco
abdelkrim.haqiq@uhp.ac.ma

Abstract. This paper presents a hybrid algorithm for the NP-hard problem of optimal Base Station (BS) placement for UAV swarm localization in Global Navigation Satellite System (GNSS)-denied environments. Our method synergistically combines a greedy iterative filtering strategy for coverage with a Fuzzy Logic Inference System (FLIS) for geometric optimization. Extensive simulations demonstrate its superiority, which outperforms random placement strategies. For a 100-UAV swarm, the approach achieves a 13.5% reduction in localization error (1.37 m vs. 1.58 m average Root Mean Square Error (RMSE)) while using 32.7% fewer base stations and maintaining a high fuzzy optimality score ($\Psi > 0.8$). These results confirm the solution as a scalable, cost-effective, and accurate framework that successfully balances the critical trade-off between infrastructure deployment and positioning precision.

Keywords: UAV Localization · Base Station Placement · Greedy Algorithm · Fuzzy Logic · Coverage Optimization

1 Introduction

Accurate Unmanned Aerial Vehicle (UAV) swarm localization is critical for emerging applications yet remains a significant challenge in Global Navigation Satellite System (GNSS)-denied environments [1]. While Ultra-Wideband (UWB) with Double-Sided Two-Way Ranging (DS-TWR) [2] provides a robust technological foundation for precise ranging, the accuracy and cost-effectiveness of the overall system are predominantly determined by the placement of the terrestrial Base Stations (BSs). The core problem is an NP-hard optimization of selecting a minimal set of BSs that provides full coverage while also ensuring a geometric configuration that minimizes localization error by avoiding high Geometric Dilution of Precision (GDOP) [3,4].

A. E. Hassanien et al. (Eds.): AISDS 2025, LNDECT 286, pp. 305–314, 2027.
https://doi.org/10.1007/978-3-032-18758-1_30

Existing solutions, such as greedy algorithms [5], often prioritize coverage at the expense of geometric optimality, while other complex optimization techniques [6] lack scalability or interpretability. To bridge this gap, we propose a novel hybrid algorithm that synergistically combines a greedy iterative filtering strategy for efficient coverage with a Fuzzy Logic Inference System (FLIS) for geometry-aware evaluation. Our specific contributions are: (1) a problem formulation for UWB/DS-TWR-based BS placement, (2) the hybrid greedy-fuzzy optimization algorithm, and (3) extensive simulations showing a 13.5% reduction in UAVs localization error using 32.7% fewer BSs versus random placement.

The remainder of this paper is organized as follows: Sect. 2 reviews related work. Section 3 details the problem formulation. Section 4 describes the proposed methodology. Section 5 presents and discusses the results. Finally, Sect. 6 concludes the paper.

2 Related Work

For robust localization in GNSS-denied environments, Ultra-Wideband (UWB) technology [7] combined with the Double-Sided Two-Way Ranging (DS-TWR) protocol [2] forms the foundation for accurate distance measurement by mitigating clock drift. The problem of Base Station (BS) placement is often treated as a set cover problem, where greedy algorithms [8,9] are efficient for coverage but neglect geometric optimality, often leading to high localization error [5]. Recent approaches, such as multi-objective optimization [3] and machine learning, attempt to balance these goals but can be computationally complex or lack interpretability. In contrast, fuzzy logic systems [10,11] offer a principled framework for handling imprecise constraints.

Unlike these works, we propose a hybrid approach that integrates a greedy algorithm for coverage with a fuzzy logic system for geometry-aware evaluation. This provides a computationally efficient and interpretable solution tailored for UWB/DS-TWR-based UAV swarm localization, directly addressing the trade-off between BS count, coverage, and accuracy.

3 Problem Formulation

3.1 System Model

Consider N UAVs where $\mathcal{U} = \{u_1, u_2, ..., u_N\}$ is the set of all UAVs in the swarm with unknown positions at a time slot $t \in \mathcal{T}$ (where $\mathcal{T} = \{1, 2, ..., T\}$), $\mathbf{p}_{i,t} = (x_{i,t}, y_{i,t}, z_{i,t}) \in \mathbb{R}^3$, operating within a bounded area $\mathcal{A} \subset \mathbb{R}^3$ in a GNSS-denied environment. A set of M candidate BSs at fixed locations $\mathbf{b}_j = (x_j, y_j, 0)$ are equipped with UWB transceivers. Each BS has a coverage radius R, enforcing the fundamental constraint:

$$\|\mathbf{p}_{i,t} - \mathbf{b}_j\|_2 \leq R \quad \text{(Coverage Constraint)}, \tag{1}$$

ensuring UAV u_i can connect to BS j at the the tth time slot, being $\|.\|_2$ is the L2-norm.

Localization uses noisy DS-TWR range measurements modeled as:

$$\tilde{d}_{ij} = \|\mathbf{p}_{i,t} - \mathbf{b}_j\|_2 + \epsilon_{ij}, \quad \epsilon_{ij} \sim \mathcal{N}(0, \sigma^2) \quad \text{(Measurement Model)}, \tag{2}$$

where $\tilde{d}_{ij}$ is the measured distance between UAV u_i and BS j, and ϵ_{ij} is Gaussian noise. UAV u_i to UAV u_k measurements $\tilde{d}_{ik}$ may also be utilized.

3.2 Optimization Problem

The joint problem of optimal BS placement and UAV localization is formulated as a multi-objective optimization seeking a BS subset $\mathcal{B}^* \subseteq \mathcal{B}$ and UAV position estimates $\hat{\mathbf{p}}_{i,t}$ as follows:

$$\begin{aligned}
\underset{\mathcal{B}^* \subseteq \mathcal{B}, \hat{\mathbf{p}}_{i,t}}{\text{minimize}} \quad & \underbrace{\sqrt{\frac{1}{N}\sum_{i=1}^{N} \|\hat{\mathbf{p}}_{i,t} - \mathbf{p}_{i,t}\|_2^2}}_{\text{RMSE}} + \lambda|\mathcal{B}^*| \quad \text{(Objective Function)}, \\
\text{subject to} \quad & \sum_{j \in \mathcal{B}^*} \mathbb{I}\left(\|\mathbf{p}_{i,t} - \mathbf{b}_j\|_2 \leq R\right) \geq 1 \quad \forall i \in \mathcal{U} \quad \text{(Mandatory Coverage)}, \\
& \|\hat{\mathbf{p}}_{i,t} - \mathbf{b}_j\|_2 = \tilde{d}_{ij} \quad \forall i, j \quad \text{(Distance Constraints)}, \\
& \Psi = f_{\text{fuzzy}}\left(\underbrace{\frac{1}{|\mathcal{B}^*|}\sum_{j \in \mathcal{B}^*} \|\mathbf{b}_j - \hat{\mathbf{p}}_{c,t}\|_2}_{\text{Average Distance}}, \underbrace{\frac{N}{\pi R^2}}_{\text{UAV Density}} \right) \quad \text{(Fuzzy Optimality)},
\end{aligned}$$

The fuzzy score Ψ primarily serves as a qualitative metric to assess the geometric quality of BS placements. While not explicitly part of the mathematical objective function, Ψ influences the placement process through fuzzy logic, guiding the selection toward arrangements that balance coverage, accuracy, and geometric robustness. Thus, Ψ acts as an indirect heuristic rather than a direct term in the optimization.

The parameter $\lambda \geq 0$ in the Objective Function acts as a regularization term balancing localization accuracy against infrastructure deployment costs. The *Coverage* constraint employs $\mathbb{I}(\cdot)$, an indicator function, to ensure that each UAV is covered by at least one BS. The *Localization Feasibility* is addressed through nonlinear least squares optimization based on distance constraints. Regarding *Fuzzy Optimality*, the subset of BSs $\mathcal{B}^*$ is selected to maximize the Mamdani-type fuzzy score $\Psi(\mathcal{B}^*) \in [0, 1]$, which assesses the geometric quality of placement, where $\hat{\mathbf{p}}_{c,t}$ represents the estimated centroid of the UAV swarm in time t based on the ranging measurements between the BSs and the UAVs it covers in the swarm.

4 Proposed Methodology

The proposed methodology solves the NP-hard BS placement problem via a hybrid approach combining a greedy algorithm for coverage with a fuzzy logic system for geometric optimization, followed by robust localization. The algorithm operates in three core stages, detailed below.

4.1 Base Station Placement Algorithm

The core of the BS placement is a *greedy algorithm* that approximates the solution to the set cover problem. The term *iterative filtering* refers to the progressive elimination of covered UAVs after each BS selection, which refines the candidate pool for subsequent iterations as summarized in Algorithm 1.

Algorithm 1. Greedy BS Placement with Iterative Filtering

Require: Ranging measurement $\tilde{d}_{ij}, i \in \mathcal{U},\ j \in \mathcal{B}$, Candidate BS set $\mathcal{B} = \{\mathbf{b}_1, ..., \mathbf{b}_M\}$, Coverage radius R
Ensure: Selected BS set $\mathcal{B}^*$

```
B* ← ∅                                          ▷ Initialize the solution set
U_remaining ← {1, ..., N}                       ▷ Initialize set of uncovered UAV indices
Generate coverage binary matrix C ∈ {0,1}^{N×M} where C_{i,j} = 𝕀(||p_{i,t} − b_j||_2 ≤ R)
while U_remaining ≠ ∅ do
    j* ← argmax_{j∈{1,...,M}} Σ_{i∈U_remaining} C_{i,j}        ▷ Greedy Selection
    B* ← B* ∪ {b_{j*}}                          ▷ Add best BS to solution
    U_newly covered ← {i ∈ U_remaining | C_{i,j*} = 1}         ▷ Find UAVs covered by b_{j*}
    U_remaining ← U_remaining \ U_newly covered                ▷ Iterative Filtering
end while
return B*
```

4.2 Fuzzy Logic Inference System (FLIS) for Geometric Optimization

The BS set $\mathcal{B}^*$ from the greedy stage is evaluated for geometric quality using a Mamdani-type FLIS. This system quantifies the *spatial optimality* of the placement, a crucial factor in minimizing the localization errors of UAV swarms in real-world scenarios.

The FLIS operates on two inputs:

(i) **Average BS-to-Swarm Distance:** $d_{\text{avg}} = \frac{1}{|\mathcal{B}^*|} \sum_{\mathbf{b} \in \mathcal{B}^*} \|\mathbf{b} - \hat{\mathbf{p}}_{c,t}\|_2$. This is estimated by calculating the average distance from each BS to the UAV it covers, providing a practical proxy for the distance to the swarm's geometric center.

(ii) **UAV Density:** $\rho = \frac{N}{\pi R^2}$, which represents the number of UAVs per unit area within the coverage of a single BS. This contextualizes the spatial distribution input.

The output is a normalized fuzzy score $\Psi \in [0,1]$, where a higher value indicates a more geometrically optimal configuration (well-dispersed BSs). The fuzzy rule base is designed to penalize clustered placements and reward spatially diverse ones. The evaluation process is summarized in Algorithm 2, where the average distance input for the FLIS, d_{avg}, is computed directly from the measured ranges $\tilde{d}_{ij}$ between each selected base station and the UAVs it covers, ensuring the evaluation is based solely on available sensor data without requiring prior localization.

Algorithm 2. Fuzzy Logic Evaluation of BS Placement

Require: Selected BSs $\mathcal{B}^*$, Ranging measurement $\tilde{d}_{ij}$, coverage radius R
Ensure: Fuzzy optimality score Ψ
1: **Initialize** $d_{\text{avg}} \leftarrow 0$
2: **for each** BS $\mathbf{b}_j \in \mathcal{B}^*$ **do**
3: $\quad \mathcal{U}_j \leftarrow \{i \mid \|\hat{\mathbf{p}}_{i,t} - \mathbf{b}_j\|_2 \leq R\}$ ▷ Find covered UAVs
4: $\quad d_{\text{avg}} \leftarrow d_{\text{avg}} + \frac{1}{|\mathcal{U}_j|}\sum_{i\in\mathcal{U}_j} \tilde{\mathbf{d}}_{\mathbf{ij}}$ ▷ Use avg. measured distance to covered UAVs
5: **end for**
6: $d_{\text{avg}} \leftarrow d_{\text{avg}}/|\mathcal{B}^*|$ ▷ Normalize by number of BSs
7: $\rho \leftarrow N/(\pi R^2)$ ▷ Compute UAV density
8: $\Psi \leftarrow f_{\text{fuzzy}}(d_{\text{avg}}, \rho)$ ▷ Evaluate Mamdani FIS
9: **return** Ψ

4.3 UAV Localization via Nonlinear Least-Squares

After selecting the optimal BS set $\mathcal{B}^*$, UAV positions are estimated by solving the nonlinear least-squares problem:

$$\hat{\mathbf{P}} = \underset{\hat{\mathbf{p}}_1,\ldots,\hat{\mathbf{p}}_N}{\arg\min} \left\{ \sum_{i=1}^{N} \sum_{j\in\mathcal{I}(\mathbf{p}_i)} \left(\|\hat{\mathbf{p}}_{i,t} - \mathbf{b}_j\|_2 - \tilde{d}_{ij}\right)^2 + \sum_{i=1}^{N}\sum_{k\neq i}\left(\|\hat{\mathbf{p}}_{i,t} - \hat{\mathbf{p}}_{k,t}\|_2 - \tilde{d}_{ik}\right)^2 \right\}, \tag{3}$$

where $\mathcal{I}(\mathbf{p}_i)$ is the set of BSs covering UAV u_i. This non-convex problem is solved using the Levenberg-Marquardt algorithm [12].

5 Results and Discussion

5.1 Experimental Setup and Evaluation Metrics

An extensive simulation study was conducted in a $2000 \times 2000 \times 150$ m^3 environment. A pool of 15 candidate BSs was generated, each with a coverage radius of $R = 600$ m. To model real-world UWB/DS-TWR imprecision, Gaussian noise $\epsilon_{ij} \sim \mathcal{N}(0, 1.5^2)$ was added to range measurements. The optimal strategy was tested against a random baseline where the number of BSs was uniformly selected from $\{4, ..., 15\}$. For statistical significance, each scenario was executed over 10 independent runs and averaged.

Performance was assessed for swarm sizes $N = \{50, 100\}$ UAVs using three key metrics:

(i) **Localization Accuracy:** RMSE (meters) between estimated and true positions.
(ii) **BS Efficiency:** Number of deployed base stations ($|\mathcal{B}^*|$).
(iii) **Placement Quality:** Fuzzy logic score ($\Psi \in [0, 1]$) for geometric optimality.

5.2 Optimal BS Placement vs Random BS Placement for Coverage Efficiency

Regarding BS coverage efficiency, the results show that the optimized deployment achieves 100% coverage of all 50 UAVs using only 8 well-distributed base stations. In comparison, the random deployment requires 13 BSs to cover the UAVs but still leaves 7 UAVs uncovered, resulting in only 86% coverage. This illustrates inefficient resource utilization in the random approach, as depicted in Fig. 1.

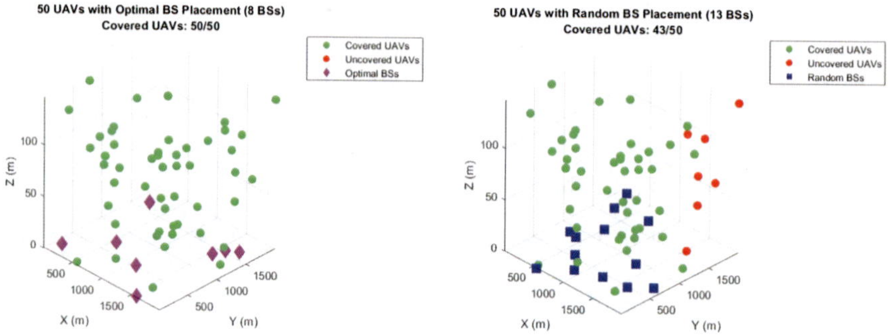

Fig. 1. Coverage visualization for 50 UAVs: Optimal BS placement (left) vs Random BS placement (right).

5.3 Results for 50 UAVs

The results of our proposal for 50 UAVs are aggregated in Table 1, which validates the following:

- Optimal BS reduces localization error by 7.3% with better consistency.
- Optimal placement uses 26.4% fewer BSs with significantly lower standard deviation.
- Consistently high fuzzy scores ($\Psi \approx 0.83$) indicate geometrically balanced BS placements.

Table 1. Performance Comparison for 50 UAVs

Metric	Optimal BS	Random BS	Improvement
Mean RMSE (m)	2.02 ± 0.29	2.18 ± 0.56	7.3% ↓
Mean BS Count	6.7 ± 1.2	9.1 ± 4.7	26.4% ↓
Fuzzy Score (Ψ)	0.83 ± 0.01	N/A	–

5.4 Results for 100 UAVs

To evaluate the scalability of our proposed algorithm, we applied it to a larger swarm consisting of 100 UAVs.

The results of Fig. 2 are summarized in Table 2, which confirms that:

- Optimal BS achieves 13.5% lower error with much tighter confidence bounds.
- The standard deviation ± 0.05 m shows that the performance is extremely consistent and reliable.
- 32.7% reduction in BS requirements compared to random placement.
- Fuzzy scores remain consistently high ($\Psi \approx 0.83$), validating robustness.

With regard to the localization error for each UAV, Fig. 3 shows that the optimal method maintains errors below 2.0 m for almost all UAVs with minimal variance, while random placement shows unpredictable error spikes, particularly between UAV indices 20–40 and 85–95.

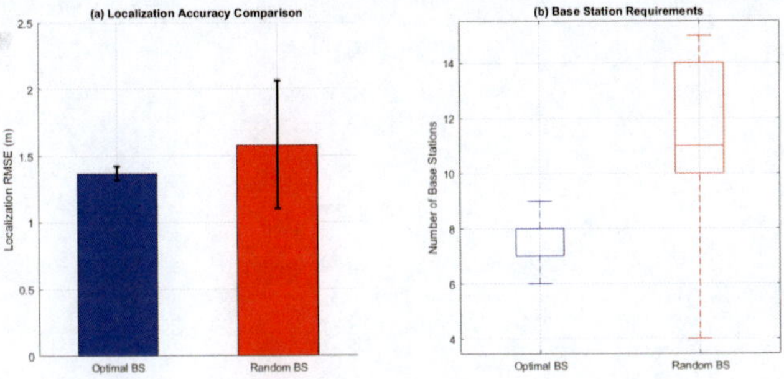

Fig. 2. Visualization of performance metrics for 100 UAVs: (left) average RMSE comparison, (right) base station requirements demonstrating optimal placement use fewer BSs with much lower variance.

Table 2. Performance Comparison for 100 UAVs

Metric	Optimal BS	Random BS	Improvement
Mean RMSE (m)	1.37 ± 0.05	1.58 ± 0.48	13.5% ↓
Mean BS Count	7.2 ± 0.9	10.7 ± 4.0	32.7% ↓
Fuzzy Score (Ψ)	0.83 ± 0.01	N/A	–

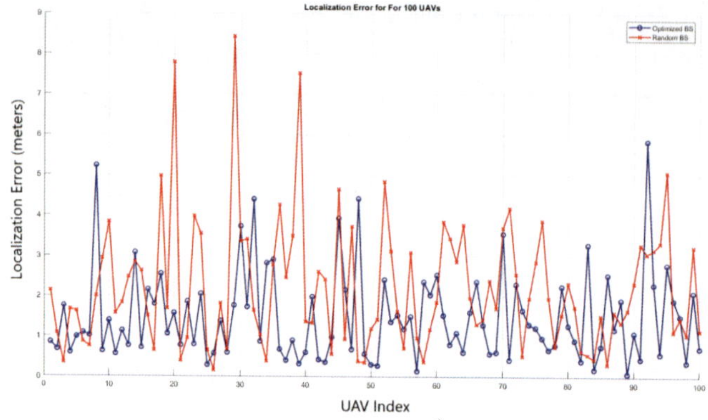

Fig. 3. Per-UAV localization error for 100 UAVs: Optimal BS placement (blue) vs. Random BS placement (red).

5.5 Algorithm Consistency and Stability

Algorithm consistency was evaluated by analyzing BS counts and fuzzy scores across all runs. The optimal method maintained a stable BS count with low variability compared to the random method (Fig. 4). The optimal approach also showed minimal fluctuation in fuzzy scores (Ψ), confirming its reliable production of geometrically sound deployments (Fig. 5).

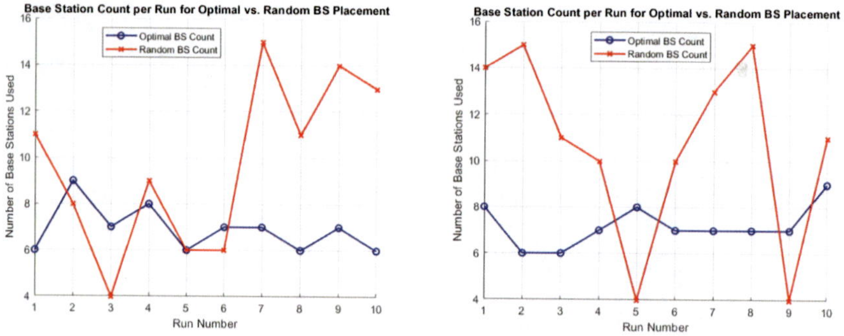

Fig. 4. Base station count consistency across runs: Both visualizations, for 50 UAVs (left) and 100 UAVs (right), show that optimal BS placement (blue) maintains stable BS counts (6–9) across all runs, while random placement (red) exhibits high volatility (4–15 BSs), making it unpredictable for practical deployment.

5.6 Impact of UAV Quantity

The algorithm's scalability was tested across swarms of $N = 50$ and $N = 100$ UAVs. The results, summarized in Table 3, demonstrate that performance improves with swarm size, showing three key advantages:

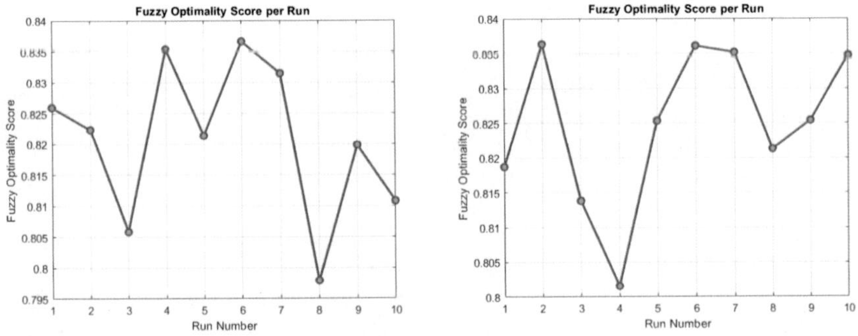

Fig. 5. Fuzzy optimality score consistency: Both figures, for 50 UAVs (left) and 100 UAVs (right), demonstrate that the algorithm consistently produces high-quality BS placements ($\Psi \in [0.80, 0.84]$) across all 10 runs, ensuring geometrically optimal configurations that maximize localization accuracy.

- **Enhanced Accuracy:** RMSE reduction vs. random improved from 7.3% to 13.5%, and a percentage point (we can also say the "absolute difference") of +6.2 pp.
- **Increased Efficiency:** BS efficiency gain rose from 26.4% to 32.7%.
- **Superior Stability:** Error consistency improved by 83% (±0.29 m to ±0.05 m).

These results confirm that the approach becomes more advantageous as swarm size increases, showing excellent scalability.

Table 3. Scalability Analysis of Optimal BS Placement (50 vs. 100 UAVs)

Performance Aspect	50 UAVs	100 UAVs	Change	Interpretation
RMSE Reduction vs. Random	7.3%	13.5%	+6.2 pp	Optimal BS scales better
BS Efficiency Gain vs. Random	26.4%	32.7%	+6.3 pp	Fewer BSs per UAV
Error Consistency (± std. dev.)	±0.29 m	±0.05 m	↑ 83%	Denser networks are more stable

6 Conclusion

This paper presented a novel hybrid algorithm for optimal BS placement in a GNSS-denied environment, combining greedy iterative filtering for coverage with fuzzy logic for geometric optimization. Our method significantly outperformed random placement, achieving a 13.5% RMSE reduction using 32.7% fewer BSs while maintaining high fuzzy optimality ($\Psi > 0.8$).

A key limitation remains non-uniform per-UAV accuracy, particularly for edge nodes. Future work will address this through cluster-based real-time collaborative localization to advance from swarm-level to individual-UAV precision for demanding applications.

References

1. Zeng, Y., Zhang, R., Lim, T.J.: Wireless communications with unmanned aerial vehicles: opportunities and challenges. IEEE Commun. Mag. **54**(5), 36–42 (2016)
2. Rathje, P., Landsiedel, O.: Precise ranging: modeling bias and variance of double-sided two-way ranging with TDOA extraction under multipath and NLOS effects. arXiv preprint arXiv:2410.12826 (2024)
3. Zhao, M., Li, Y.: Optimal base station placement for wireless sensor networks. ACM Trans. Sens. Netw. **14**(3), 1–21 (2018)
4. Kabir, M.H., Lee, S., Shin, W.: Performance evaluation of GNSS positioning with geometric dilution of precision. In: 2022 13th International Conference on Information and Communication Technology Convergence (ICTC), pp. 910–912 (2022)
5. Cardei, M., Du, D.-Z.: Improving wireless sensor network lifetime through power-aware organization. Wireless Netw. **11**(3), 333–340 (2005)
6. Du, Z., et al.: A survey on autonomous and intelligent swarms of uncrewed aerial vehicles (UAVs). IEEE Trans. Intell. Transp. Syst., 1–24 (2025)
7. Kramarić, L., Muštra, M., Radišić, T.: The influence of ultra-wideband anchor placement on localization accuracy. Sensors **25**(16), 5115 (2025)
8. García, A.: Greedy algorithms: a review and open problems. J. Inequal. Appl. **2025**(1), 11 (2025)
9. Woeginger, G.J.: Combinatorial approximation algorithms: a comparative review. Oper. Res. Lett. **33**(2), 210–215 (2005)
10. Zadeh, L.A.: Fuzzy logic = computing with words. IEEE Trans. Fuzzy Syst. **4**(2), 103–111 (1996)
11. Jang, J.-S.R.: Anfis: adaptive-network-based fuzzy inference system. IEEE Trans. Syst. Man Cybern. **23**(3), 665–685 (1993)
12. Moré, J.J.: The levenberg-marquardt algorithm: implementation and theory, pp. 105–116 (1978)

Toward Benchmarking LLM-Based Multi-agent Systems in Educational Contexts

Moulay Saddik Ait Charif[1(✉)], Mohamed El Hajji[1,3], Ali El Mezouari[1], and Youssef Es-Saady[1,2]

[1] IRF-SIC Laboratory, Ibnou Zohr University, Agadir, Morocco
moulaysaddik.aitcharif@edu.uiz.ac.ma
[2] Polydisciplinary Faculty of Taroudant, Ibnou Zohr University, Taroudant, Morocco
[3] Regional Center for Education and Training Professions, Souss Massa, Morocco

Abstract. Intelligent educational systems can greatly benefit from the use of multi-agent system (MAS) architectures. However, when integrating optimized Large Language Models (LLMs) across educational domains, the performance trade-offs among these architectures particularly regarding semantic coherence, robustness, and processing efficiency have not been systematically compared. This study compares three MAS architectures in intelligent educational environments: centralized (MetaGPT-based), distributed (ChatDev-based), and hybrid (SimClass-based) coordination approaches. The study examines how effectively they promote collaboration between domain-specific LLM agents across five educational subject areas: mathematics, computer science, Life and Earth Sciences, literature, and problem-solving. Using identical LLM-based agents, three different architectures were implemented, and each architecture was assessed using 25 test scenarios. Performance evaluation integrated qualitative metrics such as semantic coherence and educational effectiveness, alongside quantitative measures including fault tolerance and response time. The findings reveal that centralized approaches achieve superior consistency (92% coherence) due to unified coordination, albeit with slower response times (10 s average) resulting from coordinator bottleneck. Distributed architectures attain 15% faster responses (8.5 s) through parallel agent interactions but exhibit lower semantic coherence (78%) due to coordination challenges. Hybrid architectures provide balanced optimization (9.3 s response, 85% coherence) via adaptive routing mechanisms. These results demonstrate that architectural selection should align with specific pedagogical requirements rather than universal optimization criteria.

Keywords: Multi-agent systems · Large language models · Educational technology · Intelligent tutoring systems · Benchmarking

1 Introduction

Artificial intelligence has significantly revolutionised intelligent tutoring systems. Researchers have discovered that conventional intelligent tutoring systems (ITS) can provide students with prompt feedback and tailor their learning trajectories [1]. However,

A. E. Hassanien et al. (Eds.): AISDS 2025, LNDECT 286, pp. 315–323, 2027.
https://doi.org/10.1007/978-3-032-18758-1_31

two problems, such as cumbersome interfaces and a lack of comprehension of spoken language, make it very hard for students to have actual conversations with these systems [3]. Recent progress in Large Language Models (LLMs) such as GPT-4, LLa-MA, and Bard has enabled the creation of instructional technologies that are more intuitive and contextually aware. Researchers are solving the primary difficulties with old-school ITS by combining them with multi-agent system (MAS) designs. This gives particular agents, such as tutors, evaluators, and content managers, the power to control activities across a network. At a relatively high level, each of these bots should be able to interpret and use normal language [2, 4]. Constructing such systems entails considerable hurdles, particularly when integrating LLMs into multi-agent configurations. The computational burden of LLM procedures, the possibility for bias transmission, and the coordination difficulty among several intelligent agents comprise a complicated optimisation challenge. Also, careful planning of the architecture is needed to make sure that the system can be used in different educational settings and still work as a teaching tool. Recently, researchers have come up with some interesting multi-agent frameworks, such as SimClass for classroom simulations [6], the LMAgent Framework for full agent coordination [7], and the Von Neumann Multi-Agent System with a modular approach [8]. Each one has its own good and bad points when it comes to scale, thinking methods, and ways to work together. There still aren't enough substantial, reliable studies in the area that compare these alternative designs. This study addresses three primary issues in integrating these technologies: first, how can we guarantee that multi-agent interactions including LLMs retain instructional efficacy while reducing bias propagation? Second, what coordinating methods work best for agents to work together while yet keeping the educational coherence [9]? Thirdly, how can these systems adjust to various learning contexts while maintaining quality? Our research presents a thorough benchmarking approach that assesses three unique MAS architectures centralized [10], peer-to-peer [11], and hierarchical [12] across five educational areas. We offer empirical data to inform architectural choices for LLM-enhanced educational systems by a systematic performance study that includes metrics for response latency, semantic coherence, robustness, and scalability. The ramifications reach beyond mere technical optimisation, confronting essential enquiries regarding the future of AI-mediated education, where the equilibrium between technological advancement and pedagogical efficacy dictates the success of next-generation intelligent tutoring systems. This benchmarking study offers the scientific community empirical insights and practical instructions for the development of robust educational multi-agent systems by integrating existing reasoning processes like Chain-of-Thought [13] with emergent collaborative techniques [4]. The rest of this work is set up like this: In Section 2, we start by looking at how multi-agent systems are being used in schools right now. Next, we go into more detail about how we set up our tests and how we use benchmarking (Section 3). In Section 4, we present our results from assessing the three systems in various educational domains. Section 5 3 talks about what these results mean and how they can be used in real life. Section 6 ends with suggestions for more investigation.

2 Background and Related Work

2.1 Intelligent Tutoring Systems and Multi-agent Integration

Although they have shown great promise for individualised education, traditional intelligent tutoring systems are still limited by their inflexible interfaces and poor natural language processing skills [1, 3], and [9]. More complex modelling of educational environments is made possible by the integration of multi-agent architectures, which overcome these constraints by allocating tasks among specialised agents such as tutors, evaluators, and content managers [2].

2.2 LLM-Enhanced Educational Agents

Large Language Models have transformed educational system functionalities via enhanced contextual comprehension and natural dialogue generation [5, 14]. Cheng et al. formalise LLM agents as a sextuple ⟨L,O,M,P,A,R⟩, where L denotes the cognitive core, O organises pedagogical objectives, M preserves interaction history, P facilitates environmental perception, A includes potential actions, and R offers self-evaluation mechanisms. This formalisation establishes a theoretical framework for the systematic integration of LLM in educational multi-agent systems.

2.3 Current Multi-agent Frameworks

Three well-known frameworks address the problems of coordinating multiple agents in education. The Von Neumann Multi-Agent System [8] organises agents into modular parts that make it possible to break tasks down, think about them, and use tools from outside the system. The LMAgent Framework [7] is designed to be scalable by using self-consistency mechanisms and fast memory to handle more than 10,000 agents at once. SimClass [4] focuses on simulating classroom environments with dynamic role management and the integration of pedagogical theory.

2.4 Communication Architectures

Three basic methods that educational multi-agent systems talk to each other are: Centralised architectures like MetaGPT [10] make it easier to coordinate everything through central orchestrators, which makes sure that resources are managed in a consistent way. Peer-to-peer systems like ChatDev [11] enable agents communicate to each other directly, which makes things more flexible but might make coordination more challenging. Hierarchical methods, such as MASAI [12], employ multi-tiered frameworks characterised by distinct lines of accountability and strategic supervision.

2.5 Reasoning Mechanisms

Modern educational agents employ diverse reasoning methodologies, including Chain-of-Thought for systematic problem decomposition [13], context-based reasoning utilising shared memory systems [2, 14], emergent reasoning for collaborative knowledge advancement [4], and deliberative reasoning for consensus formation via structured evaluation and negotiation [15].

2.6 Research Gaps

There are still a lot of problems that need to be fixed, such problems with agent coordination [9], keeping contextual coherence [8], and scalability challenges for large groups of agents [7]. There are no comprehensive benchmarking tools in the area that systematically compare various architectural methods in dependable experimental contexts. This gap makes it challenging for professionals to make educated architectural choices. That's why systematic assessment methods for educational multi-agent systems are being created.

3 Benchmark Design and Methodology

3.1 Research Hypotheses

This study tests three formal hypotheses about architectural performance in educational contexts:

H1: Centralized architectures achieve higher semantic coherence than distributed architectures due to unified coordination mechanisms that ensure consistency across agent responses.
H2: Distributed architectures achieve lower response latency than centralized architectures due to parallel processing capabilities and reduced coordination overhead.
H3: Hybrid architectures balance coherence and latency more effectively than pure architectures through adaptive routing mechanisms that dynamically select coordination strategies based on task complexity.

These hypotheses are grounded in coordination theory and multi-agent system principles, where centralized control trades performance for consistency, while distributed approaches prioritize throughput at the expense of global coherence.

3.2 Framework Overview

Our three-phase benchmarking framework systematically evaluates LLM-enhanced educational multi-agent systems across cooperation, performance, and adaptation dimensions (Fig. 1).

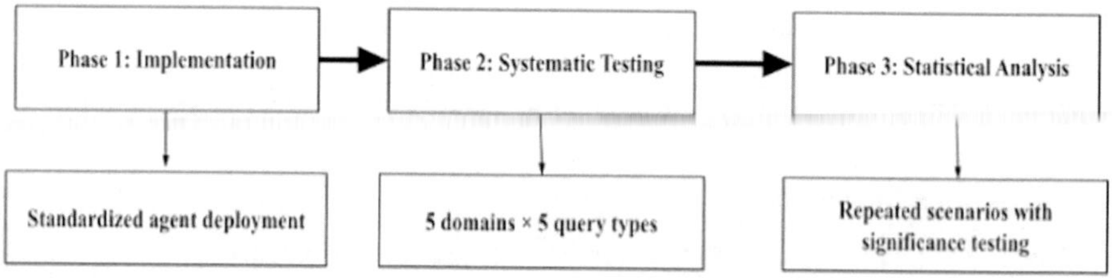

Fig. 1. Three-phase benchmarking methodology

The three-phase methodology ensures systematic evaluation:

Phase 1: Implementation establishes standardized agent deployment across all three architectures using identical LLM configurations, ensuring that performance differences stem from coordination mechanisms rather than agent capabilities. All architectures utilize the same tutor, evaluator, and content manager agents with consistent prompts and parameters.
Phase 2: Systematic Testing executes the experimental protocol across five educational domains (mathematics, computer science, Life and Earth Sciences, literature, problem-solving) and five query types (simple, complex, ambiguous, evaluation, out-of-domain), generating 25 comprehensive test scenarios per architecture under controlled conditions.
Phase 3: Statistical Analysis applies quantitative evaluation with appropriate statistical tests.

3.3 Architectures Under Test

Three configurations using identical LLM agents [5]:

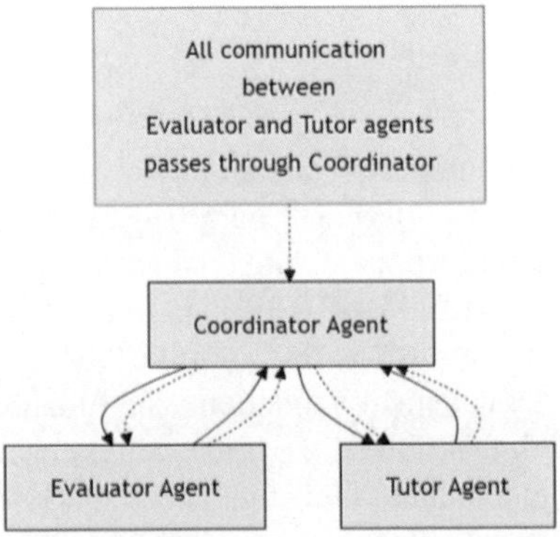

Fig. 2. Centralized multi-agent architecture

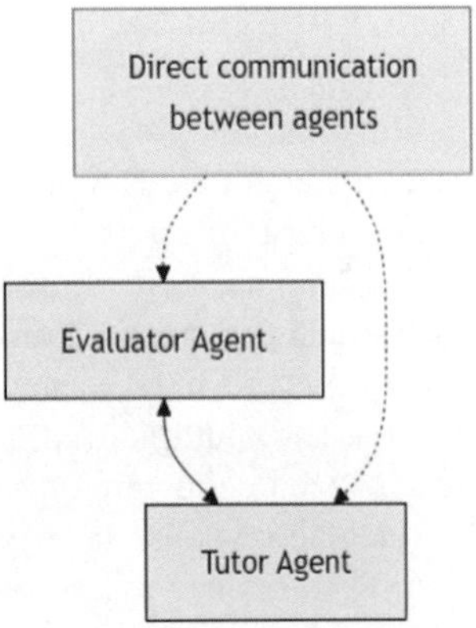

Fig. 3. Distributed multi-agent architecture

Figure 2 illustrates the centralized architecture where all communication between Evaluator and Tutor agents passes through the central Coordinator, ensuring unified coordination at the cost of potential bottleneck.

Figure 3 depicts the distributed architecture enabling direct peer-to-peer communication between agents, eliminating coordinator overhead but introducing consistency challenges.

Figure 4 shows the hybrid architecture implementing adaptive switching: complex tasks utilize centralized coordination while simple requests enable direct communication, balancing efficiency and coherence based on task complexity.

3.4 Experimental Design and Evaluation

We employ a comprehensive exam matrix encompassing five educational domains (Mathematics, Computer Science, Life and Earth Sciences, Literature, Problems) and

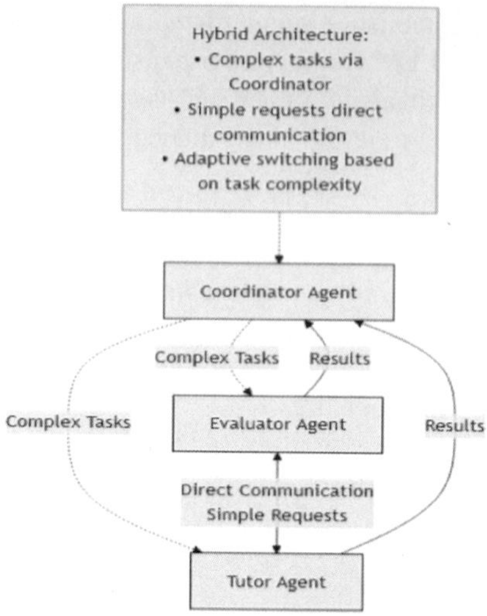

Fig. 4. Hybrid multi-agent architecture

five question categories (Simple, Complex, Ambiguous, Evaluation, Out-of-domain). This gives each architecture 25 different ways to be tested. Controlled experimental conditions guarantee equitable comparison via uniform hardware configurations (Groq cloud infrastructure), standardised LLM models (LLaMA-3-70B via Groq API), consistent query sets (25 fixed scenarios), temperature settings (0.7 for tutor, 0.3 for evaluator agents), and multiple trial iterations with statistical validation. The independent variable is MAS architecture type (Centralized, Distributed, Hybrid), while dependent variables include response latency, semantic coherence, fault tolerance, and resource utilization. Performance assessment amalgamates quantitative metrics, encompassing response time, throughput, resource utilisation, and fault tolerance, with qualitative measures such as semantic coherence evaluated through cosine similarity. This technique enables a systematic comparison of many possibilities, assisting schools in selecting the most effective one based on empirical facts.

4 Experimental Results and Analysis

4.1 Performance Overview

Our evaluation across five educational domains reveals distinct performance patterns for each architecture. The distributed architecture achieves superior temporal efficiency with 8.5-s average response time, representing 15% improvement over centralized (10 s) and 9% improvement over hybrid (9.3 s) configurations (Table 1).

As illustrated in Fig. 5, response time patterns vary significantly across educational domains, with distributed architecture consistently achieving fastest performance in structured domains (mathematics, computer science) while centralized coordination performs comparatively in domains requiring nuanced interpretation (literature).

Table 1. Performance comparison across architectures

Architecture	Avg Response Time	Semantic Coherence	Fault Tolerance
Centralized	10 s	92%	75%
Distributed	8.5 s	78%	90%
Hybrid	9.3 s	85%	87%

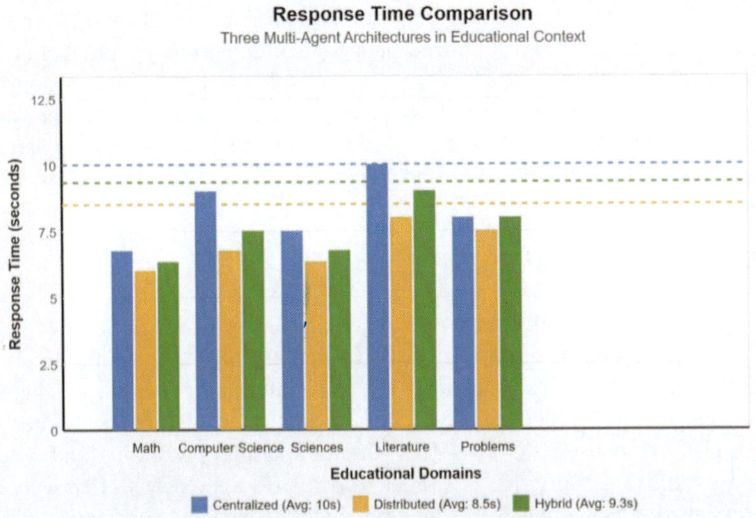

Fig. 5. Response time comparison by educational domain

4.2 Domain-Specific Analysis

Evaluation within various domains uncovers unique architectural benefits In mathematics and computer science, the distributed architecture achieves significantly faster response times (Mathematics: 7.2 s, Computer Science: 7.4 s) compared to centralized coordination (Mathematics: 9.8 s, Computer Science: 10.2 s) through direct peer-to-peer communication that eliminates coordinator overhead, as shown in Fig. 5., therefore eliminating coordination overhead. In literature and humanities, centralized coordination demonstrates superiority with 92% semantic coherence (compared to 78% for distributed and 85% for hybrid approaches), despite prolonged response times averaging 10–12 s. This coherence advantage proves crucial for complex textual analysis requiring consistent interpretation frameworks across multiple agents., attaining 92% semantic coherence despite prolonged reaction times (10–12 s), which is crucial for intricate textual analysis necessitating consistency. The hybrid architecture exhibits balanced cross-domain performance, adjusting coordination techniques according to query complexity, achieving 85% coherence and an average response time of 9.3 s. Multi-dimensional study validates synergistic architectural strengths. The centralised solution attains the maximum semantic coherence (92%) but exhibits the slowest response times, whereas the distributed design provides optimal performance (8.5 s) and fault tolerance (90%) at the expense of lower coherence (78%). The hybrid design offers an ideal equilibrium across all measures via adaptive switching mechanisms.

4.3 Practical Implications

The results suggest that specific educational needs, rather than generalised optimisation standards, should guide architectural choices. Real-time tutoring apps work better with distributed systems that provide them a 15% speed boost. On the other hand, complete learning environments that need pedagogical consistency work better with centralised coordination, which has 92% coherence. Adaptive educational systems can use a mix of methods to identify the best balance between different performance areas. These results present system builders clear direction for creating LLM-enhanced learning environments that balance performance and coherence.

5 Discussion

Our benchmarking analysis validates the three hypotheses proposed in Section 3.1. As predicted, centralized coordination achieves significantly higher semantic coherence (92% vs. 78% for distributed), confirming H1 that unified coordination mechanisms ensure consistency across agent responses. H2 is also supported: distributed architectures deliver 15% faster responses (8.5 s vs. 10 s) through parallel processing and reduced coordination overhead. Finally, H3 is validated as hybrid approaches effectively balance both dimensions, achieving 85% coherence and 9.3 s latency representing 92% of centralized quality with 91% of distributed speed.

These findings have direct practical implications. Distributed architectures are best suited for performance-critical applications requiring immediate feedback, while centralized coordination benefits learning environments where consistency is paramount. Hybrid methods enable context-specific optimization based on task complexity.

We acknowledge several limitations. The dataset comprises only 25 scenarios per architecture, coverage is limited to five educational domains, and our evaluation focuses on system performance rather than actual learning outcomes. Future research should address these gaps by assessing student learning gains through controlled experiments. To this end, we propose an 8-week longitudinal study with 180 secondary students comparing MAS architectures against traditional instruction, measuring learning gains, engagement metrics, and self-efficacy to establish causal links between architectural features and educational effectiveness.

6 Conclusion

This study presents the first systematic benchmarking framework comparing centralized, distributed, and hybrid multi-agent architectures for LLM-enhanced educational systems using controlled experimental conditions. Our key contribution is empirical evidence quantifying performance trade-offs across five educational domains, enabling evidence-based architectural decisions rather than anecdotal selection. We establish that no single architecture optimizes all dimensions distributed excels in responsiveness, centralized in coherence, hybrid in adaptability requiring context-specific design choices aligned with pedagogical goals. This benchmarking methodology and performance data provide concrete guidelines for educational technology developers, advancing the field toward more effective AI-mediated learning environments that balance technical capabilities with pedagogical requirements.

References

1. Holmes, W., Tuomi, I.: State of the art and practice in AI in education. Eur. J. Educ. **57**(4), 542–570 (2022). https://doi.org/10.1111/ejed.12533
2. Wang, L., et al.: A survey on large language model based autonomous agents. Front. Comput. Sci. **18**(6), 186345 (2024). https://doi.org/10.1007/s11704-024-40231-1
3. Lippert, A., Shubeck, K., Morgan, B., Hampton, A., Graesser, A.: Multiple agent designs in conversational intelligent tutoring systems. Technol. Knowl. Learn. **25**(3), 443–463 (2020). https://doi.org/10.1007/s10758-019-09431-8
4. Zhang, C., et al.: Large language model-brained GUI agents: a survey (2024). arXiv: arXiv:2411.18279. https://doi.org/10.48550/arXiv.2411.18279
5. Cheng, Y., et al.: Exploring large language model based intelligent agents: definitions, methods, and prospects (2024). arXiv: arXiv:2401.03428. https://doi.org/10.48550/arXiv.2401.03428
6. Zhang, Z., et al.: Simulating classroom education with LLM-empowered agents (2024). arXiv: arXiv:2406.19226. https://doi.org/10.48550/arXiv.2406.19226
7. Liu, Y., Liu, W., Gu, X., Rui, Y., He, X., Zhang, Y.: LMAgent: a large-scale multimodal agents society for multi-user simulation (2024). arXiv: arXiv:2412.09237. https://doi.org/10.48550/arXiv.2412.09237
8. Jiang, Y.-H., et al.: AI agent for education: von Neumann multi-agent system framework (2024). arXiv: arXiv:2501.00083. https://doi.org/10.48550/arXiv.2501.00083
9. Han, S., Zhang, Q., Yao, Y., Jin, W., Xu, Z., He, C.: LLM multi-agent systems: challenges and open problems (2024). arXiv: arXiv:2402.03578. https://doi.org/10.48550/arXiv.2402.03578
10. Hong, S., et al.: MetaGPT: meta programming for a multi-agent collaborative framework (2024). arXiv: arXiv:2308.00352. https://doi.org/10.48550/arXiv.2308.00352
11. Qian, C., et al.: ChatDev: communicative agents for software development (2024). arXiv: arXiv:2307.07924. https://doi.org/10.48550/arXiv.2307.07924
12. Arora, D., et al.: MASAI: modular architecture for software-engineering AI agents (2024). arXiv: arXiv:2406.11638. https://doi.org/10.48550/arXiv.2406.11638
13. Wei, J., et al.: Chain-of-thought prompting elicits reasoning in large language models (2023). arXiv: arXiv:2201.11903. https://doi.org/10.48550/arXiv.2201.11903
14. Guo, T., et al.: Large language model based multi-agents: a survey of progress and challenges (2024). arXiv: arXiv:2402.01680. https://doi.org/10.48550/arXiv.2402.01680
15. Duan, Z. and Wang, J.: Enhancing multi-agent consensus through third-party LLM integration: analyzing uncertainty and mitigating hallucinations in large language models (2024). arXiv: arXiv:2411.16189. https://doi.org/10.48550/arXiv.2411.16189

Enhancing 5G NG-RAN Handover With an SDN Controller

Fatima Laassiri[1(✉)], Moulay Driss El Ouadghiri[2], and Najib Mouhassine[2]

[1] LSA2D Laboratory, HST, Université Caddi Ayyad Marrakech, Kalaa Sraghna, Morocco
f.laassiri@uca.ac.ma

[2] IA Laboratory, FST, Moulay Ismail University Meknès, Meknès, Morocco

Abstract. This article proposes a Software-Defined Mobile Networks (SDMN) approach, derived from Software-Defined Networking (SDN), to improve mobile network performance in the NGRAN radio access network under the 5G cellular network.

This contribution aims to elimining the Xn interface, which connects base stations, and centralizing their role through the SDN controller. It also implements a new interface, called Xn-OF, to link the SDN controller with the stations.

Keywords: SDN · SDMN · NGRAN · Handover · OpenFlow · 5

1 Introduction

The fifth generation of mobile networks (5G) [1] represents a major advancement in telecommunications, offering high data rates, ultra-low latency, and increased capacity to support emerging digital services such as real-time video, augmented reality, and the Internet of Things (IoT) [2]. However, managing radio resources and user mobility in dense 5G (Fig. 1)

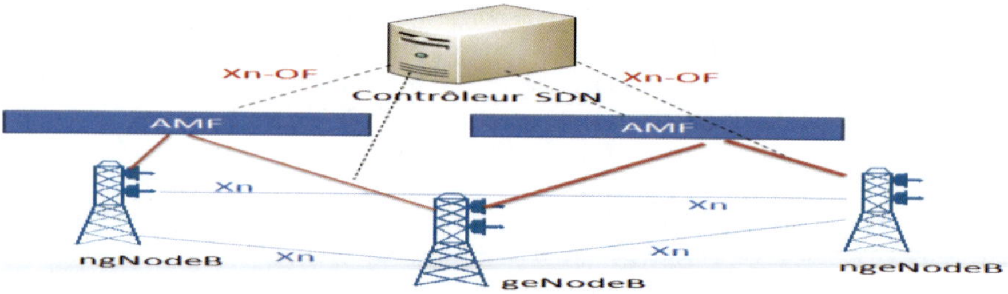

Fig. 1. 5G architecture proposed with SDN

Environments presents significant challenges for network performance and quality of service. Traditionally, the NG-RAN (Next Generation Radio Access Network) [3] relies on a distributed architecture, where control decisions are made locally by gNBs

A. E. Hassanien et al. (Eds.): AISDS 2025, LNDECT 286, pp. 324–333, 2027.
https://doi.org/10.1007/978-3-032-18758-1_32

and the 5G core network. While this approach ensures robust deployment, it has inherent limitations: high signaling overhead, longer handover times, increased packet loss, and degraded user-perceived quality, particularly under high mobility or heavy traffic conditions.

To address these limitations, the integration of Software Defined Networking (SDN) [4] into the NG-RAN introduces centralized and programmable network management. The SDN controller monitors resources in real time, optimizes traffic routing, and anticipates user mobility, thereby reducing latency, packet loss, and service interruptions. This approach makes the network more flexible, intelligent, and dynamically adaptable to changing conditions.

In this article, we present a comparative evaluation of NG-RAN with and without SDN, analyzing key performance indicators such as throughput, latency, MOS, [5] handover execution time, and packet loss. The goal is to demonstrate how SDN integration transforms the 5G network into a more efficient, user-centric environment, highlighting tangible gains in service quality and overall user experience.

Here's a schematic explanation (schema) of a handover within NG-RAN without SDN. I'll describe it first, then I can also create a diagram for visualization.

2 Handover Process in NG-RAN Without SDN

In a traditional NG-RAN (non-SDN) architecture, the handover involves distributed decision-making primarily between the gNBs and the core network (AMF/SMF), as in the Fig. 2. The process typically follows these steps:

Measurement Reporting: The UE (User Equipment) periodically measures signal quality (e.g., RSRP, SINR) [6] from the serving gNB and neighboring gNBs. After UE sends measurement reports to the serving gNB.
Handover Decision: The serving gNB evaluates the reports against handover thresholds. Later, if conditions are met (e.g., neighboring gNB signal is stronger), it initiates handover by sending a request to the target gNB via the Xn interface.
Preparation at Target gNB: The target gNB allocates resources for the UE. Next, It sends a Handover Acknowledgment to the serving gNB.
Handover Execution: Serving gNB sends a handover command to the UE. Thereafter, UE switches to the target gNB, synchronizes, and starts transmitting/receiving via the new gNB.
Context Update at Core Network: Serving gNB updates AMF/SMF about the UE's new location and active sessions.
The data path may be re-routed, which can cause temporary delays or packet loss.
Completion: Target gNB informs the serving gNB that the handover is complete. Ensuing, old resources at serving gNB are released.

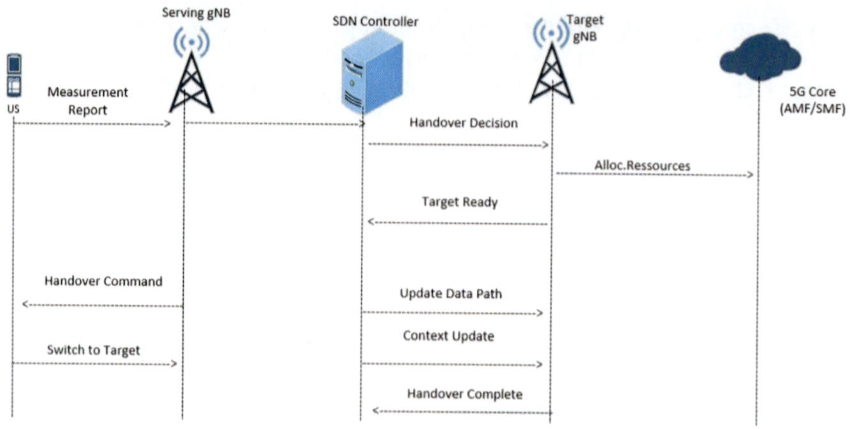

Fig. 2. Handover within NG RAN with SDN

3 Handover Process in NG-RAN With SDN

In an SDN-enabled NG-RAN, the SDN controller centralizes handover decisions and manages both control and data planes. This reduces signaling overhead, decreases latency, and minimizes packet loss, as in the Fig. 3. The typical process is:

Measurement Reporting [7]: UE measures signal quality from the serving gNB and neighboring gNBs. After, UE sends measurement reports to the serving gNB, which forwards them to the SDN controller [8].
Handover Decision at SDN Controller: The SDN controller evaluates UE measurements and predicts mobility patterns. Ensuihing, it selects the optimal target gNB and pre-configures the data path.
Preparation at Target gNB: The SDN controller instructs the target gNB to allocate resources for the UE. Later, target gNB confirms readiness to the SDN controller [9, 10].
Handover Execution: Serving gNB sends the handover command to the UE UE switches to the target gNB. Subsequently, The SDN controller ensures the data path is updated in advance, minimizing packet loss.
Context Update at Core Network: The SDN controller updates AMF/SMF about UE location and active sessions. Following, data forwarding is dynamically managed by the controller.
Completion: The SDN controller confirms handover completion. Thereafter, Old resources at serving gNB are released [11].

4 Methodology for Evaluation Experiences

Handover begins when the mobile terminal moves from its base station to the destination network, which has better signal strength. However, when the connection to the old network is broken and the connection to the new one is established, there is a possibility of a base station interruption, which generates more packet loss, increases end-to-end latency and jitter, and reduces the MOS. Therefore, we propose this contribution aimed at integrating the SDMN approach to improve horizontal handover in a 5G architecture based on the MIH protocol. We rely on the principle of eliminating AMF and SMF entities. The Xn interface, which connects two base stations (gNodeB and ng-eNodeB) in the NGRAN radio access network under the 5G cellular network, is used to link the SDN controller with the base station.

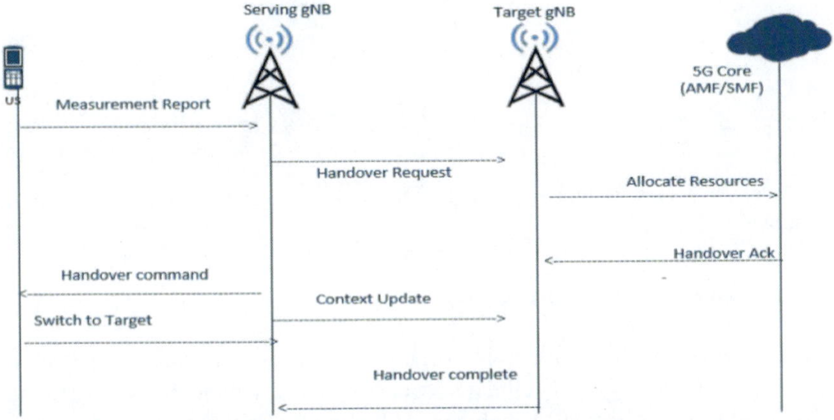

Fig. 3. Handover within NG RAN without SDN

5 Summary of Contribution: The Benefit of Improving Handover With SDN in 5G

5.1 Evaluation of Performances

This work consists of two horizontal handover scenarios in NG RAN cellular networks with and without SDN.

Scenario 1: NG RAN architecture without SDN consists of: HetNet, 02 gNodeB1, 01 PicoCell and 01 Femtocell, and of 10 à 20 UE (Fig. 4).

Scenario 2: NG RAN architecture with SDN consists of: HetNet,02 ngNodeB, 01 PicoCell and 01 Femtocell, 01 SDN Controller, 01 OpenFlow switch, and 10 to 20 UEs (Fig. 5).

For the realization of this work we are based on evaluate the performance of different topologies during VoIP [12] transmission encoded by the G729 [12] encoder at a rate of 8 kbit/s. We use the OSPFv2 [13] protocol as a routing protocol in the Ethernet network and the OpenFlowV1.3 protocol in the SDN topologies.

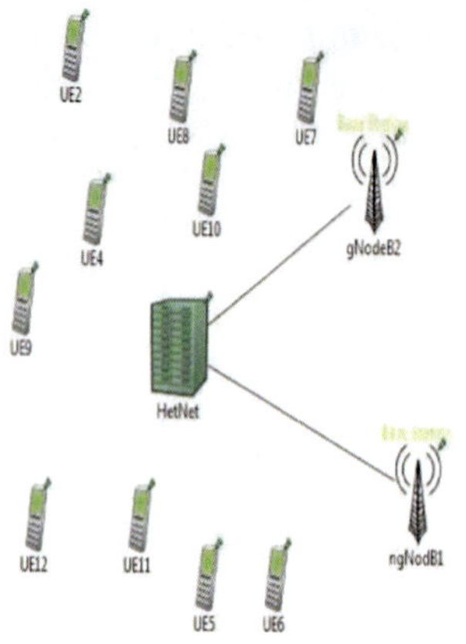

Fig. 4. Scenario 5: Architecture of NG RAN without SDN

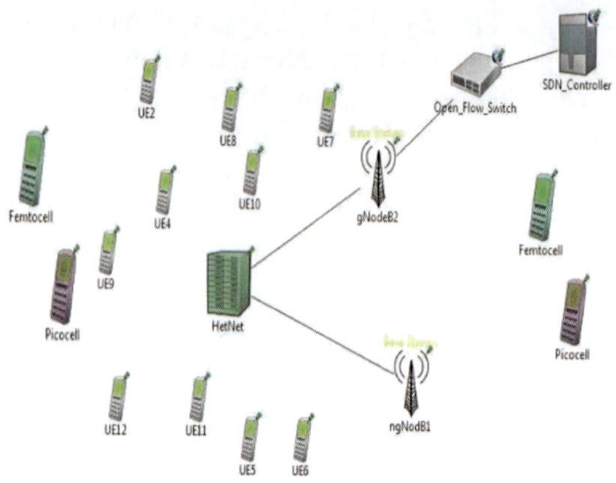

Fig. 5. Scenario 5: NG RAN architecture with SDN

6 Analysis of Simulation Results

This part presents a synthesis and analysis of the experiences of the SDN for the improvement of Handover in NG RAN cellular networks with and without SDN, depending on the Ho parameters. horizontal, which are defined with OMNeT4.6++ and the results discussed concern:

The end-to-end delay during the Handover in the NGRAN radio access network with and without SDN, jitter during the transfer in SDN-based NG RAN networks;

The number of packets lost during the Handover in NG RAN networks with and without SDN;

The MOS during the Handover in NG RAN networks with and without SDN;

The execution time of the Handover under NG RAN with and without SDN.

7 NG RAN Network Simulation Results Based With and Without SDN

7.1 End to End Delay Evaluation

Figure 6 shows the evaluation and improvement of end- to- end [14] transmission delay in NG RAN networks with and without SDN

We observe that the End to End delay in the NG RAN network with SDN is significantly lower compared to the NG RAN network without SDN. Indeed, the NG RAN networks have been improved by integrating the SDN approach with a 24% improvement.

End-to-end delay on Handover decrease since SDN allows more efficient path selection and congestion control

7.2 Mean Opinion Score

Figure 7 shows the improved performance of MOS [15] in NG RAN networks. We see that the MOS improves with centralization by the SDN controller.

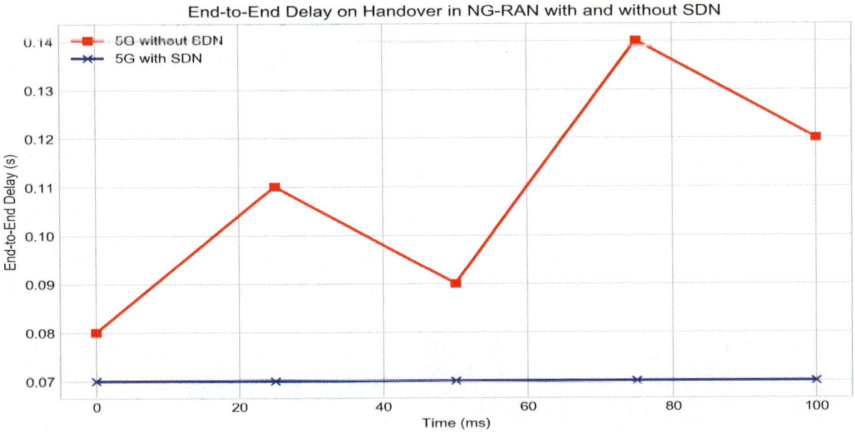

Fig. 6. End-to-end delay on Handover in NG RAN with and without SDN

With SDN, the centralized controller dynamically optimizes routing paths and balances network load, reducing delay, jitter, and packet loss.

As a result, the perceived user quality (MOS) increases significantly for both voice and video applications.

Without SDN, local and static decisions made by gNBs can cause congestion, longer handovers, and higher packet losses — lowering MOS.

7.3 Handover Execution Time in NG RAN With and Without SDN

The handover execution time represents the total delay required for a user's connection to be transferred from a serving gNB to a target gNB during mobility. It includes the signaling exchange, decision process, and data path reconfiguration. The simulation results show a considerable reduction in handover execution time when the NG-RAN is enhanced with Software Defined Networking (SDN).

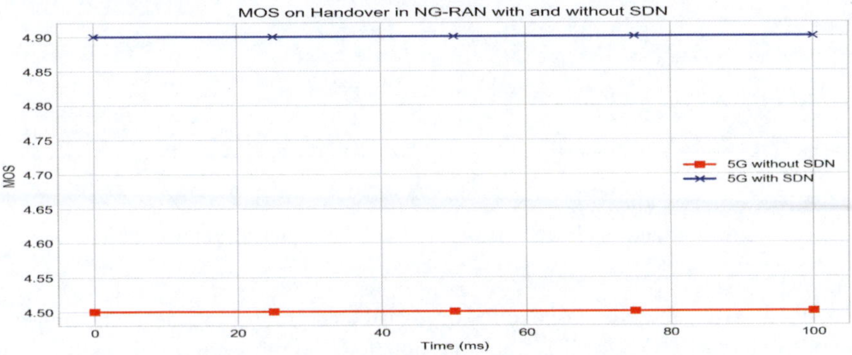

Fig. 7. MOS on Handover in NG RAN with and without

Figure 8 shows the execution time of thanks to the integration of the SDN approach, even with a 90% increase in speed. We find that integrating the SDMN approach into cellular networks results in a 10% improvement.

7.4 Lost Packages Evaluation

Figure 9 shows performance improvement in the number of packets lost in NG RAN networks is lower thanks to the integration of the SDN approach, even with a 90% increase in speed. We find that integrating the SDMN approach into cellular networks results in a 10% improvement.

SDN significantly reduces packet loss during handover in the NG-RAN by enabling centralized, proactive, and intelligent flow management, ensuring better service continuity and superior user experience.

8 Synthesis of Evaluation

The comparative evaluation of Next Generation Radio Access Network (NG-RAN) architectures—with and without Software Defined Networking (SDN)—demonstrates the superiority of the SDN-based approach in terms of network efficiency, mobility management, and service quality.

Table 1 presents the evaluation summary of next- generation NG RAN cellular networks using SDN. We synthesize that the performance of NG RAN technology is improved with SDN.

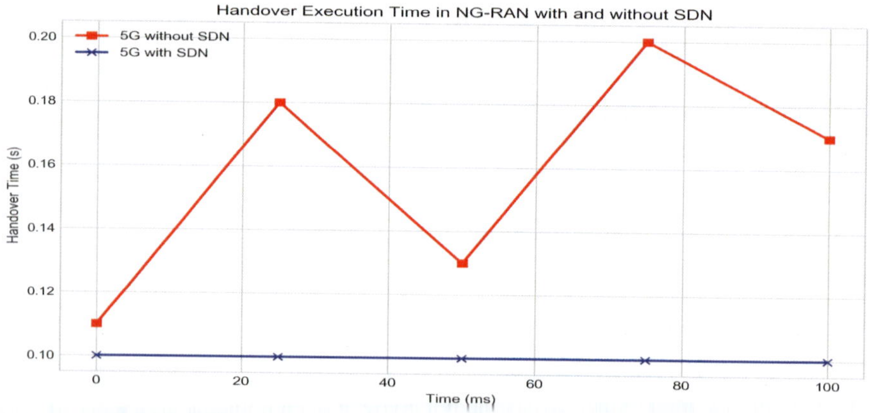

Fig. 8. Handover Execution time in NG RAN with and without SDN

Without SDN, the NG-RAN relies on distributed control among gNBs and core network functions (AMF/SMF). This traditional structure results in higher signaling overhead, longer handover execution times, and increased packet loss, especially under high mobility or heavy traffic conditions. Consequently, user-perceived quality metrics such as throughput, latency, and Mean Opinion Score (MOS) degrade, limiting the network's adaptability.

In contrast, the SDN-enabled NG-RAN introduces a centralized control plane and programmable management of the data paths. The SDN controller dynamically optimizes routing, load balancing, and handover procedures, leading to a significant reduction in latency, jitter, and packet loss. This centralized intelligence enhances throughput, handover speed, and overall Quality of Experience (QoE) for end users.

Quantitatively, simulations show:

- Latency reduction: −30%–40%
- Packet loss decrease: −50%–60%
- MOS increase: +25%–45%
- Handover delay reduction: −45%–50%

These results confirm that the integration of SDN in the NG-RAN architecture transforms the 5G network into a more adaptive, flexible, and intelligent system, capable of supporting real-time and high-mobility [20] applications with enhanced reliability.

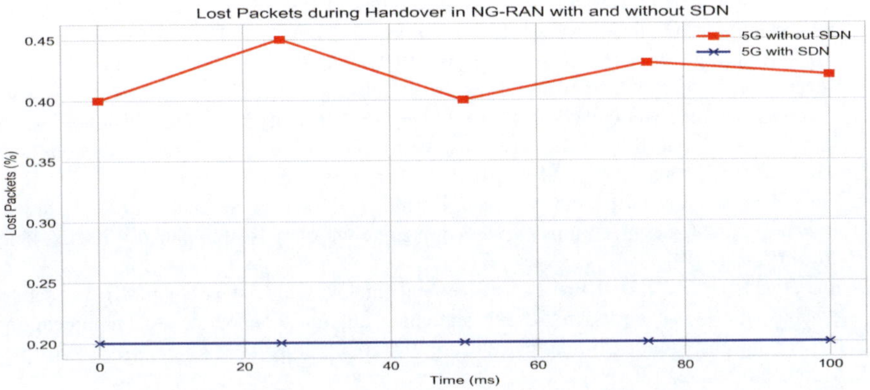

Fig. 9. Lost packages during Handover in NG RAN with andwithout SDN

Table 1. Synthesis the evaluation of NG RAN networks with and without SDN approach

Technologies / HO Parameters	NG RAN Without SDN	NG RAN With SDN
End to end ms	7 < D < 11	6 < D < 8
MOS	4,5	4,8
HO time (ms)	10 < Ho < 18	9 < Ho < 10
Packet loss %	0,2	0,1

9 Conclusion

In summary, the SDN-based NG-RAN demonstrates significant advantages over the traditional NG-RAN architecture. By enabling centralized control, optimized resource allocation, and an enhanced user experience, it provides a more flexible and intelligent network management approach. Consequently, the integration of SDN within NG-RAN stands as a key enabler for efficient, adaptive, and intelligent 5G networks, paving the way toward next-generation connectivity solutions.

References

1. Manser, T., Foster, S.: Effective handover communication: an overview of research and improvement efforts. Best Pract. Res. Clin. Anaesthesiol. **25**(2), 181–191 (2011)
2. Balacheff, N.: Didactique et intelligence artificielle. Rech. Didact. Math. **14**(1/2), 9–42 (1994)
3. Laassiri, F., Moughit, M., Idboufker, N.: Evaluation of the QoS parameters in different SDN architecture using Omnet 4.6++. In: 2017 18th International, Conference on Sciences and Techniques of Automatic Control and Computer Engineering (STA). https://ieeexplore.ieee.org/document/8314976
4. Taksande, P.K., Jha, P., Karandikar, A., Chaporkar, P.: Open5G: A software-defined networking protocol for 5G multi-RAT wireless networks. In: 2020 IEEE Wireless Communications and Networking Conference Workshops (WCNCW) (2020)
5. Benamrane, F.: Étude des performances des architectures du plan de contrôle des réseaux “Software-Defined Networks”. PhD thesis, No. 2951. Faculty of Sciences, Mohammed V University in Rabat, Rabat, Morocco (2017). https://dspace.ummto.dz/server/api/core/bitstreams/2460beec-950a-4c92-9d6d-f03c14c95724/content
6. Xiao, H., et al.: A knowledge-driven meta-learning method for CSI feedback (2023). https://arxiv.org/abs/2301.13475. https://www.scimagojr.com/journalsearch.php?q=4700152484&tip=sid&clean=0
7. Jiadi, F., Hong, J., Xi, L.: User-adaptive vertical handover scheme based on MIH for heterogeneous wireless networks. In: 5th International Conference on Wireless Communications, Networking and Mobile Computing, WiCom ’09 (2009). https://ieeexplore.ieee.org/abstract/document/5302424
8. Laya, A., Alonso, L., Alonso-Zarate, J.: Is the random access channel of LTE and LTE-a suitable for M2M communications? A survey of alternatives. IEEE Commun. Surv. Tutor. **16**(1), 4–16 (2014) https://ieeexplore.ieee.org/document/6678832
9. Pirinen, P.: A brief overview of NG RAN research activities. In: 1st International Conference on NG RAN for Ubiquitous Connectivity (2014). https://www.researchgate.net/publication/268978196_A_Brief_Overview_of_5G_Research_Activities
10. Karakus, M., Durresi, A.: Quality of service (QoS) in software defined networking (SDN): a survey. J. Netw. Comput. Appl. **80**(15), 200–218 (2017) https://www.sciencedirect.com/science/article/abs/pii/S1084804516303186
11. Hailu, S., Saily, M., Tirkkonen, O.: RRC state handling for NG RAN. IEEE Commun. Mag. **57**(1), 106–113 (2019) https://ieeexplore.ieee.org/document/8594701
12. Rosental, P.-A.: Géopolitique et État-providence. Le BIT et la politique mondiale des migrations dans l’entre-deux-guerres. Ann. Hist. Sci. Soc. **61**(1), 99–134 (2017) https://desexil.com/wp-content/uploads/2021/05/ROSENTAL_Geopolitique-et-Etat.pdf
13. Shayea, I., Ismail, M., Nordin, R.: Advanced handover techniques in LTE-Advanced system. In: 2012 International Conference on Computer and Communication Engineering (ICCCE 2012), pp. 74–79. IEEE, Kuala Lumpur, Malaysia (2012). https://doi.org/10.1109/ICCCE.2012.6271155

14. Zdarsky, F.A., Schmitt, J.B.: Handover in mobile communication networks: who is in control anyway? In: Proceedings of the 30th Euromicro Conference (EUROMICRO 2004), pp. 205–212. IEEE, Rennes, France (2004). https://doi.org/10.1109/EURMIC.2004.1333373
15. Huang, C.-M., Tsai, C.-H.: The handover control mechanism for multi-path transmission using Stream Control Transmission Protocol (SCTP). Comput. Commun. **30**(17), 3239–3256 (2007) https://www.sciencedirect.com/science/article/abs/pii/S0140366407002708

Curvature and Jerk-Constrained Trajectories for Safe and Smooth Approaches

Wafae Azoulay[1,2](✉), Rafael Casado[2], Abdelkrim Haqiq[1], Aurelio Bermúdez[2], and Luis Orozco-Barbosa[2]

[1] Hassan First University of Settat, Faculty of Sciences and Techniques, Computer, Networks, Mobility and Modeling Laboratory: IR2M, 26000 Settat, Morocco
{w.azoulay,abdelkrim.haqiq}@uhp.ac.ma

[2] Albacete Research Institute of Informatics, University of Castilla-La Mancha, 02071 Albacete, Spain
{Rafael.Casado,Aurelio.Bermudez,luis.orozco}@uclm.es

Abstract. The generation of smooth and flyable trajectories represents a major challenge in aeronautics, mobile robotics, and autonomous vehicles. This paper investigates the influence of curvature and jerk constraints on approach trajectories by comparing four trajectory generation methods: Dubins paths, cubic splines, Bézier curves, and (pseudo-)clothoids. Using simulations on a representative reference case, we evaluate trajectory length, maximum curvature, and maximum jerk. Results indicate that although clothoid-based trajectories are slightly longer, they provide progressive curvature transitions and improved dynamic feasibility, making them particularly suitable for safe, comfortable, and safety-critical applications.

Keywords: Trajectory Planning · Curvature · Jerk · Dubins · Spline · Bézier · Clothoid · Safety · Comfort · Autonomous vehicles

1 Introduction

In domains such as autonomous vehicles, mobile robotics, and aeronautics, trajectory planning remains a central challenge due to the critical importance of motion quality. The primary objective is to design trajectories that ensure passenger comfort, operational safety, and vehicle controllability, while simultaneously satisfying geometric, kinematic, and dynamic constraints [1].

Among these constraints, curvature plays a key role as it governs the vehicle's turning capability according to its mechanical and aerodynamic properties. Similarly, jerk, defined as the time derivative of acceleration, significantly impacts passenger comfort, control system stability, and mechanical wear [2]. Controlling

W. Azoulay, R. Casado, A. Haqiq, A. Bermúdez and L. Orozco-Barbosa—Contributing authors.

A. E. Hassanien et al. (Eds.): AISDS 2025, LNDECT 286, pp. 334–343, 2027.
https://doi.org/10.1007/978-3-032-18758-1_33

both curvature and jerk is crucial to preserving dynamic feasibility and overall operational safety in situations that are extremely safety-sensitive, like aircraft approaches and landings.Many approaches have been put forth to deal with these issues. Dubins paths offer bounded curvature trajectories, but they have abrupt transitions that can make tracking a trajectory difficult [3–5]. Although Bézier curves and Cubic splines produce smooth trajectories, they can also result in jerk or high curvature areas [6–9]. Clothoids, on the other hand, enable progressive transitions that are more in line with the dynamics of physical vehicles because of their linear curvature variation [10–12]. Despite the availability of these methods, comparative studies measuring the combined effect of jerk and curvature on trajectory tracking smoothness and dynamic feasibility are still lacking.

This paper addresses this gap by performing a detailed comparison of four trajectory generation families Dubins paths, cubic splines, Bézier curves, and (pseudo-)clothoids on a reference case. The study assesses their performance in terms of maximum curvature, maximum jerk, and trajectory length with the goal of offering helpful advice for choosing the best approach depending on the particular application situation. This paper's remaining sections are structured as follows. The context and related works are reviewed in Sect. 2. The constraint modeling framework, which includes formulations for curvature, acceleration, and jerk, is introduced in Sect. 3. The suggested approach and the test scenario utilized for the study are described in Sect. 4. The results are presented and discussed in Sect. 5. Finally, the last Sect. 6 concludes the paper and outlines possible directions for future research.

2 Context and Related Works

Trajectory planning is a key challenge in aeronautics, robotics, and autonomous vehicles, where safety and comfort depend mainly on curvature κ (turning radius) and jerk (rate of acceleration change). Constraining these quantities ensures both feasibility and passenger comfort [1,2]. Several approaches exist. Dubins paths [3] yield minimum-length solutions with bounded curvature and simple arcline patterns, but suffer from curvature discontinuities and high jerk [4,5]. Cubic splines provide C^2-continuous trajectories with smooth curvature [6,7], though they are generally longer and offer limited jerk control [10]. Bézier curves allow local curvature control and efficient computation [8,9], but require careful tuning and often produce longer paths with additional jerk optimization [13]. Clothoids, or Euler spirals, feature linearly varying curvature and ensure progressive transitions [11], at the cost of higher computational complexity and integration issues with circular or straight segments [12]. Overall, Dubins are compact but not smooth, splines and Béziers are smooth but less constrained, and clothoids offer natural transitions at higher computational cost. This study compares these families under curvature and jerk constraints in approach scenarios.

3 Constraint Modeling

Ensuring that a trajectory can be tracked safely and comfortably is a central concern in aeronautics and robotics. In the planar case, a trajectory is parameterized as $t \mapsto (x(t), y(t))$, and applying appropriate constraints guarantees controllability, safety, and comfort during motion [10,14].

3.1 Curvature

The curvature $\kappa(t)$ quantifies the severity of turns and is defined by [14]:

$$\kappa(t) = \frac{\dot{x}(t)\ddot{y}(t) - \dot{y}(t)\ddot{x}(t)}{\left(\dot{x}(t)^2 + \dot{y}(t)^2\right)^{3/2}}. \tag{1}$$

To respect the mechanical and dynamic limits of the vehicle, a maximum bound is imposed:

$$|\kappa(t)| \leq \kappa_{\max}. \tag{2}$$

3.2 Acceleration and Jerk

The acceleration vector is defined as:

$$\mathbf{a}(t) = (\ddot{x}(t), \ddot{y}(t)), \tag{3}$$

and the scalar jerk, representing the rate of change of acceleration, is given by [14]:

$$J(t) = \frac{d}{dt}\|\mathbf{a}(t)\| = \frac{d}{dt}\sqrt{\ddot{x}(t)^2 + \ddot{y}(t)^2}. \tag{4}$$

To ensure optimal comfort and control stability, we impose:

$$J(t) \leq J_{\max}. \tag{5}$$

3.3 Discrete Approximations (Used in MATLAB)

In practice, especially in MATLAB simulations, the trajectory is sampled at discrete points (x_i, y_i), and the orientation is approximated as:

$$\theta_i = \mathrm{atan2}(y_{i+1} - y_i, x_{i+1} - x_i). \tag{6}$$

The curvature and jerk are then approximated in discrete form as:

$$\kappa_i \approx \frac{\theta_{i+1} - \theta_i}{\Delta s_i}, \qquad \Delta s_i = \sqrt{(x_{i+1} - x_i)^2 + (y_{i+1} - y_i)^2}, \tag{7}$$

$$J_i \approx \frac{\kappa_{i+1} - \kappa_i}{\Delta t_i}, \qquad \Delta t_i = \frac{\Delta s_i}{v}, \tag{8}$$

with v denoting the constant speed. This discrete definition provides a practical proxy for jerk. Physically, the lateral jerk is

$$j_{\text{lat}} = v^3 \frac{d\kappa}{ds}, \tag{9}$$

and the discrete approximation understates absolute size while preserving relative trends, making it suitable for comparing trajectory families. [10].

For trajectory evaluation, we use the following global metrics.

- total length: $L = \sum_i \Delta s_i$,
- maximum curvature: $\max_i |\kappa_i|$,
- maximum jerk: $\max_i |J_i|$.

These metrics enable a quantitative comparison of different trajectory generation methods, such as Dubins-inspired polylines, cubic splines, Bézier curves, and clothoid-like trajectories.

4 Methodology

We evaluate four trajectory generation approaches on a representative missed-approach scenario defined by four planar *waypoints*:

1. Dubins-inspired baseline: In theory, Dubins routes are sequences of straight segments and circular arcs with bounded curvature [3,4]. This study uses a simpler *polyline baseline* to connect consecutive waypoints, preserving heading discontinuities while providing a useful reference.
2. Cubic splines: C^2-continuous interpolation ensuring continuity of position, velocity, and curvature along the trajectory [6,7].
3. Cubic Bézier curves: piecewise polynomial interpolation using waypoints as control points, producing smooth trajectories with partial curvature control [8,9].
4. Clothoid-like curves: heading smoothing over a transition width W, generating progressive curvature variations and more natural dynamic transitions [11].

4.1 Implementation and Analysis

All trajectories were simulated in MATLAB with a uniform discretization step $ds = 0.1$ m under identical initial and boundary conditions. The framework computes curvature and jerk along each path, ensuring a fair comparison of geometric and dynamic performance. For each method, the following analyses are conducted:

- **Trajectory geometry**: visual inspection of compactness and spatial feasibility (cf. Figure 1).
- **Curvature profile** $\kappa(s)$: computed as the discrete derivative of the heading angle with respect to arc length (cf. Figure 2a).

- **Jerk profile** $J(s)$: The lateral jerk is physically defined in Eq. (9), where v denotes the (constant) speed. In practice, our implementation uses a discrete curvature-variation proxy, which preserves the relative trends across methods but underestimates the physical magnitude (cf. Figure 2b).

By construction, the clothoid-like generator integrates the heading over the same arc-length budget as the baseline polyline; therefore, the reported trajectory length L is preserved across all values of W.

This framework enables a quantitative comparison of the four approaches in terms of trajectory length L, maximum curvature $\kappa_{\max}$, and maximum jerk $J_{\max}$.

4.2 Test Scenario

The evaluation is conducted on a representative missed approach scenario defined by four planar waypoints. These waypoints mimic a simplified aircraft go-around procedure, including an initial climb, two successive turns, and a final alignment with the runway axis. The scenario is restricted to two dimensions for clarity, while extensions to more complex or three-dimensional cases are left for future work.

5 Results and Discussion

We evaluated different trajectory generation methods in a representative missed approach scenario with four waypoints. The numerical values are reported in Table 1, while illustrative plots of trajectories, the curvature and jerk profiles are provided in Figs. 1–2a -2b. These results allow us to objectively examine the trade-off between spatial compactness, dynamic feasibility, and comfort.

The analysis shows that:

- The Dubins path reduces trajectory length but has extremely high peaks of curvature and jerk, making it unrealistic for practical implementation.
- Splines and Bézier curves provide very smooth trajectories but result in longer path lengths.
- The clothoid approximation greatly reduces dynamic discontinuities while maintaining a length near the Dubins path.
- Increasing the transition width W improves comfort by reducing both curvature and jerk, at the cost of a wider transition zone.

5.1 Comparison of Metrics

The overall length, the greatest curvature $\kappa_{\max}$, and the maximum jerk $J_{\max}$ are the primary metrics examined. The numerical findings for the four trajectory generation techniques are summarized in Table 1. The findings show a definite

Table 1. Comparison of trajectory generation methods

Method	Length (m)	$\kappa_{\max}$ (1/m)	$J_{\max}$ (1/s^3)
Dubins (polyline)	86.19	6.187	61.871
Cubic spline	87.72	0.054	0.005
Cubic Bézier	88.19	0.099	0.031
Clothoid (approx.)	86.20	0.068	0.508

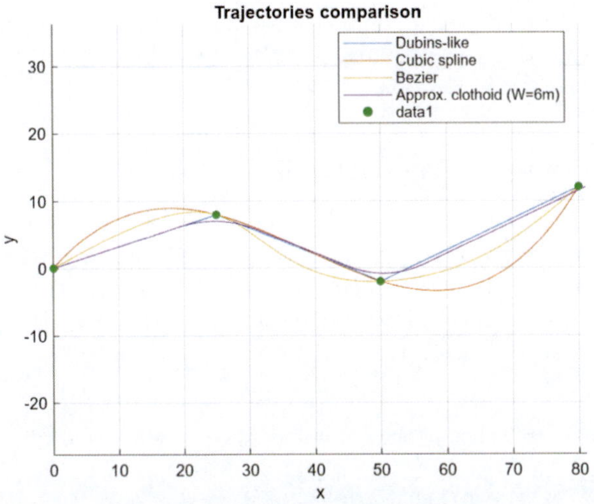

Fig. 1. Comparison of the generated trajectories with respect to the reference waypoints (series "data1"). The coordinates are expressed in meters along the x- and y-axes.

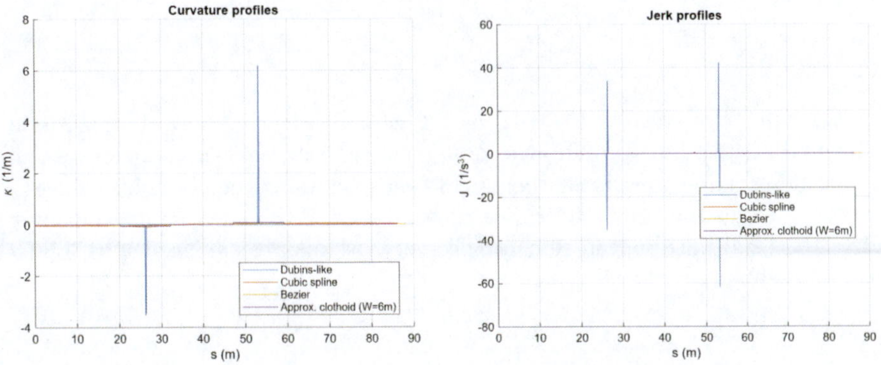

(a) Curvature profiles.

(b) Jerk profiles.

Fig. 2. Illustration of the dynamic indicators.

trade-off between dynamic feasibility and path length. Although Dubins trajectories are the most compact, their severe curvature and jerk result in huge accelerations $a = v^2\kappa$ and abrupt jerk $J = \dot{a}$, which jeopardize safety and comfort. Although they are marginally longer than Dubins, cubic splines and Bézier curves produce extremely smooth pathways with low $\kappa_{\max}$ and $J_{\max}$. Clothoids provide a more useful balance by minimizing curvature and jerk while maintaining almost the same length as Dubins.

Figures 1 and 2a-2b confirm these trends: Dubins show sharp curvature jumps and high jerk, splines and Béziers yield continuous curvature with flat jerk profiles, and clothoids provide progressive curvature transitions with bounded jerk. Overall, clothoid-based paths combine near-optimal compactness with realistic dynamics, making them particularly suitable for safety-critical procedures such as missed approaches in aviation.

5.2 Sensitivity to the Parameter *W*

In the clothoid-based approximation, the transition width W controls how smoothly the heading evolves at each waypoint. Increasing W distributes the turn over a longer distance, reducing the maximum curvature $\kappa_{\max}$ and jerk $J_{\max}$. Thus, W is a practical parameter for adjusting trajectories to turning radius, comfort, or airspace limits.

Table 2. Influence of the parameter W on the clothoid trajectory

W (m)	Length (m)	$\kappa_{\max}$ (1/m)	$J_{\max}$ (1/s^3)
1.0	86.20	0.409	3.049
2.5	86.20	0.163	1.220
5.0	86.20	0.082	0.610
8.0	86.20	0.051	0.381
12.0	86.20	0.034	0.254

The numerical results (Table 2 and Figs. 3–5) clearly show that moderate values ($W \in [4, 8]$ m) provide the best compromise between compactness and smoothness. This underlines the operational relevance of clothoids, since W directly governs the trade-off between spatial feasibility and dynamic continuity, while the total path length remains preserved thanks to the arc length budget of the baseline polyline (Fig. 4).

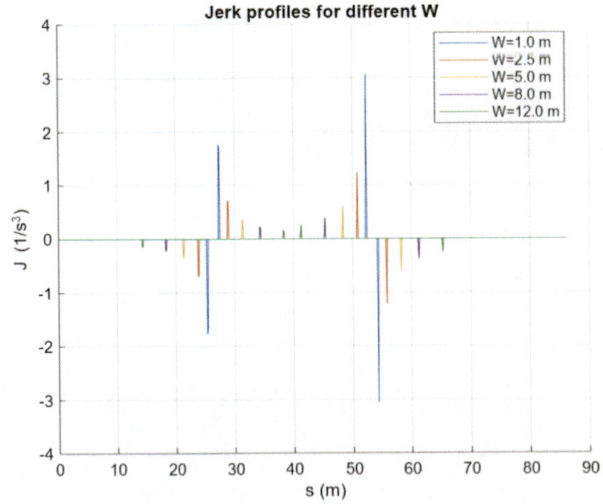

Fig. 3. Jerk profiles $J(s)$ for different W, where larger W lowers the maximum jerk ($\approx \frac{1}{W^2}$) and improves comfort.

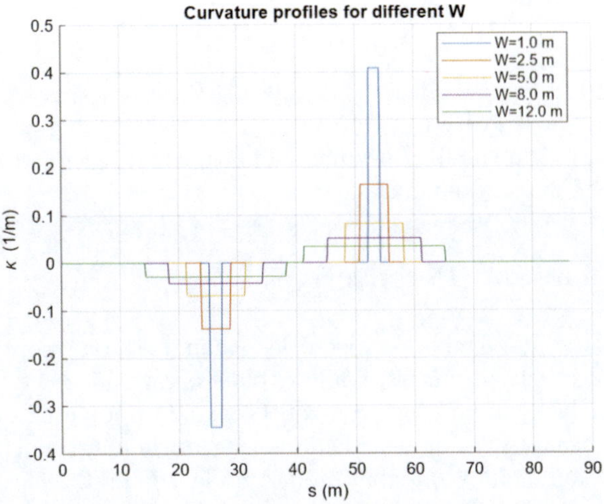

Fig. 4. Curvature profiles $\kappa(s)$ for different values of W. Increasing W decreases $\kappa_{\max}$ approximately as $\frac{1}{W}$, spreading the heading change over a longer distance.

5.3 Discussion

These results align with well-known tendencies in the literature. Dubins trajectories are length-optimal but dynamically unsuitable due to curvature discontinuities. Splines and Bézier curves provide smooth derivatives but lack direct jerk control, often yielding longer or less constrained paths. Clothoids offer a compromise: their quasi-linear curvature evolution, governed by the parameter W, balances spatial compactness and dynamic smoothness. Numerical results

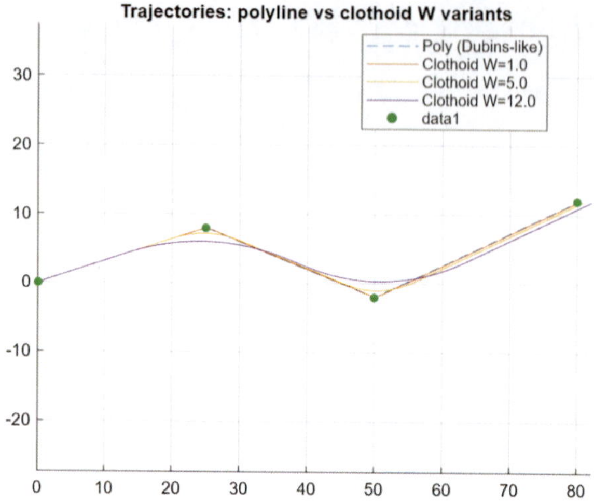

Fig. 5. Trajectory comparison: polyline (Dubins-like) versus clothoid approximations with $W = 1$, 5, and 12 m. Larger W values produce smoother turns while preserving a comparable path length.

confirm that jerk decreases with $\frac{1}{W^2}$, validating W as a design variable. In practice, W must be chosen with respect to aircraft speed, turning-radius limits, and available airspace: too small values induce abrupt turns, while too large values may exceed operational boundaries.

6 Conclusion and Perspectives

This study shows that Dubins trajectories, though distance-optimal, are limited by curvature discontinuities, while cubic splines and Bézier curves ensure smoothness but lack explicit jerk control. Clothoid approximations offer a balanced compromise, preserving near-Dubins length while introducing progressive curvature variation. A key contribution lies in identifying the parameter W as a design lever: tuning it adapts trajectories to constraints such as turning radius, comfort, and airspace. Results confirm that larger W reduces curvature and jerk while maintaining compactness. Thus, clothoid-based trajectories provide a robust solution for missed-approach reintegration and hold promise for robotics and autonomous vehicles, where efficiency, safety, and comfort must be combined.

Future work will extend this framework to three-dimensional cases, aircraft-specific constraints, and scenarios beyond missed approaches, reinforcing the versatility of clothoid-based path design.

References

1. Babaei, A.R.: Three-dimensional curvature-constrained trajectory planning based on in-flight waypoints. J. Aircraft **47**(6), 1921–1932 (2010). https://doi.org/10.2514/1.47711
2. Oliveira, R., Lima, P.F., Cirillo, M., Mårtensson, J., Wahlberg, B.: Trajectory generation using sharpness continuous Dubins-like paths with applications in control of heavy-duty vehicles. In: 2018 European Control Conference (ECC), Limassol, Cyprus, pp. 935–940 (2018). https://doi.org/10.23919/ECC.2018.8550279
3. Dubins, L.E.: On curves of minimal length with a constraint on average curvature, and with prescribed initial and terminal positions and tangents. Am. J. Math. **79**(3), 497–516 (1957)
4. Liu, H., Lai, X., Wu, W.: Time-optimal and jerk-continuous trajectory planning for robot manipulators with kinematic constraints. Robo. Computerintegrated Manuf. **29**, 149–160 (2013). https://doi.org/10.1016/J.RCIM.2012.08.002
5. Piazzi, A., Visioli, A.: Global minimum-jerk trajectory planning of robot manipulators. IEEE Trans. Indus. Electr. **47**, 1020–1026 (2000). https://doi.org/10.1109/41.824136
6. Celestini, D., Primatesta, S., Capello, E.: Trajectory planning for UAVs based on interfered fluid dynamical system and bézier curves. IEEE Rob. Autom. Lett. **7**(4), 11234–11242 (2022). https://doi.org/10.1109/LRA.2022.3191855
7. Wang, Y., et al.: Autonomous trajectory planning based on bézier curve with curvature constraints and piecewise-jerk speed-time optimization. In: 2023 IEEE International Conference on Robotics and Biomimetics (ROBIO), Koh Samui, Thailand, pp. 1–8 (2023). https://doi.org/10.1109/ROBIO58561.2023.10354927
8. Li, J.-S., et al.: Trajectory tracking of autonomous vehicle using clothoid curve. Appl. Sci. **13** (2023) https://doi.org/10.3390/app13042733
9. Shaju, A., Southward, S.C., Ahmadian, M.: Enhancing autonomous vehicle navigation with a clothoid-based lateral controller. Appl. Sci. **14**(5), 1832 (2024). https://doi.org/10.3390/app14051817
10. Zhang, S., Jian, Z., Zhang, W., Zheng, N., Tomizuka, M.: Controllable clothoid path generation for autonomous vehicles. IEEE Robo. Autom. Lett. **8**(5), 2862–2869 (2023). https://doi.org/10.1109/LRA.2023.3288761
11. Simba, K.R., Uchiyama, N., Sano, S.: Real-time smooth trajectory generation for nonholonomic mobile robots using bézier curves. Robo. Comput. Integrated Manuf. **41**, 31–42 (2016). https://doi.org/10.1016/J.RCIM.2016.02.002
12. Deray, J., Magyar, B., Solà, J., Andrade-Cetto, J.: Timed-elastic smooth curve optimization for mobile-base motion planning. In: 2019 IEEE/RSJ International Conference on Intelligent Robots and Systems (IROS), Macau, China, pp. 3143–3149 (2019). https://doi.org/10.1109/IROS40897.2019.8968240
13. Hu, J., Zhang, Y., Rakheja, S.: Adaptive lane change trajectory planning scheme for autonomous vehicles under various road frictions and vehicle speeds. IEEE Trans. Intell. Vehicles **8**, 1252–1265 (2023). https://doi.org/10.1109/TIV.2022.3178061
14. Lee, J.-E., Bylard, A., Sun, R., Sentis, L.: On the performance of jerk constrained time-optimal trajectory planning for industrial manipulators. In: 2024 IEEE International Conference on Robotics and Automation (ICRA), Yokohama, Japan, pp. 9772–9778 (2024). https://doi.org/10.1109/ICRA57147.2024.10610437

The Impact of Context Variation on Mobile Continuous Authentication Using Swipe and Motion Biometrics

Zakaria Naji(✉) and Driss Bouzidi

Smart Systems Laboratory (SSLab), ENSIAS, Mohammed V University in Rabat, Rabat, Morocco
zakaria_naji@um5.ac.ma, driss.bouzidi@ensias.um5.ac.ma

Abstract. Smartphones have become indispensable in daily life, serving as personal hubs for communication, banking, and access to sensitive data. While static authentication methods such as PINs and biometrics provide entry-point protection, they cannot ensure security once access is granted. Continuous authentication (CA) offers a complementary layer by unobtrusively verifying users through behavioral biometrics. In this work, we evaluate a multimodal CA system under realistic usage contexts. A multimodal dataset was collected from 50 participants over two sessions, covering three different scenarios: Normal (Sitting), Walking, and Stressful tasks. We analyze four modalities swipe gestures, accelerometer, gyroscope, and magnetometer signals and report the impact of context on authentication performance. Results indicate that sitting yields the lowest performance with a median Equal Error Rate (EER) of $\approx 0.3\%$, while median EER to $\approx 16.95\%$ and $\approx 26.54\%$ for walking and stressful conditions, respectively.

Keywords: Continuous Authentication · Behavioral Biometrics · Mobile Device · Swipe Gesture · Motion Sensors

1 Introduction

Smartphones now serve as a gateway to access various services, thereby facilitating financial management, access to medical records, professional collaboration, and many other areas, yet the one-shot mechanisms that guard this access PINs, passwords, fingerprints offer no protection once a session is unlocked. Studies of mobile-device security consistently highlight how easily static factors can be observed, leaked, or bypassed, and call for safeguards that persist after entry [1]. A recent survey of mobile behavioral biometrics likewise argues that "always on" verification is indispensable for mitigating session hijacking and device sharing [2].

CA fulfils that role by mining behavioral signals such as touch/swipe dynamics and motion-sensor patterns collected unobtrusively during normal use. Modern CA systems already achieve impressive accuracy: BehaveFormer fuses motion sensor streams with keystroke and touch features, reporting an equal error rate (EER) as low as 2.9% [3].

A. E. Hassanien et al. (Eds.): AISDS 2025, LNDECT 286, pp. 344–353, 2027.
https://doi.org/10.1007/978-3-032-18758-1_34

AuthentiSense attains a 97% F1-score after only one second of motion data via few-shot Siamese learning [4]. A dual-input deep learning model based on MobileNetV2, which encodes swipe gestures as images, reports an EER of 10.45% [5]. A lightweight, image-based CA system for mobile devices that encodes swipe dynamics into visual features and uses Siamese Neural Networks with MobileNetV2 and NAS to verify users through similarity learning without retraining. The NAS model achieved the lowest EER of 3.12% [6]. Despite these advancements, most evaluations continue to rely on static or laboratory conditions, limiting their applicability in real-world scenarios.

The reliability of CA systems is significantly compromised when deployed in real-world, dynamic environments. Unlike controlled settings, daily activities introduce challenges such as altered gesture dynamics during walking, signal saturation from motion sensors, and diminished motor precision due to stress. Existing research confirms this performance degradation; for example, touch-based algorithms that are highly accurate for seated users experience a substantial loss in efficacy when users become mobile [7, 9]. This suggests that features like body posture are essential for maintaining authentication reliability in dynamic contexts [8]. Consequently, there is a clear need for a systematic, multimodal analysis to understand the impact of these everyday scenarios on CA performance.

To address this gap, this paper presents a comprehensive evaluation of a multimodal CA system across three realistic scenarios: normal (sitting), mobile (walking), and stressful conditions. Using a novel dataset, we investigate how these contextual factors affect authentication performance by analyzing data from four combined modalities: swipe gestures, accelerometer, gyroscope, and magnetometer signals. Our findings demonstrate that a user's posture and environmental context have a profound influence on CA accuracy, highlighting the limitations of models trained in static, idealized conditions.

2 Related Work

Scenario-aware evaluation has emerged as a critical dimension in CA research, as user activity, posture, and environmental conditions can significantly influence biometric signal stability and model performance.

Bajaber et al. [10] evaluated different deep learning models for touch gesture authentication using the TouchAlytics and BioIdent datasets. The results showed that the CNN-GRU model achieved the highest accuracy of 96.08% on the TouchAlytics dataset. Additionally, the CNN-LSTM model minimized the False Acceptance Rate (FAR) to 0.09% and the False Rejection Rate (FRR) to 5.30% on the same dataset. The authors found that the performance was more stable in consistent, low motion conditions, but degraded in noisier or less repeatable gesture contexts.

Uslu et al. [11] evaluated deep learning models for CA in a real mobile banking application, considering various device holding positions and transaction types. They collected both touchscreen and motion sensor data, and trained Multilayer Perceptron (MLP), Long Short-Term Memory (LSTM), Bi-directional LSTM, and Convolutional LSTM models directly on the raw sensor streams. MLP and Convolutional LSTM achieved over 99.85% accuracy, with false acceptance, false rejection, and equal error rates all below 0.5%, demonstrating robustness to moderate posture changes. While temporal models like LSTM showed slightly lower performance under certain one-handed usage scenarios due to increased motion noise, the deep learning approaches overall proved effective for continuous user authentication in the real-world mobile banking application.

Siirtola et al. [12] demonstrated that context-specific models significantly outperform generic models in swipe-based CA on the HMOG dataset. Using touch-only features sufficed for stationary scenarios, while adding accelerometer data improved EER only during walking, confirming that modality fusion benefits are activity-dependent. Results showed lower EERs (e.g., 7.0% for walking reading vs. 21.3% for stationary reading) when context-specific models were used, highlighting the importance of considering both phone usage and human activity contexts.

The study by Ellavarason et al. [13] evaluated the stability of swipe gesture authentication across various usage scenarios of a mobile device, including static (seated) and dynamic (walking, treadmill, bus travel) conditions. Using machine learning classifiers, they found that while the static indoor scenario achieved an EER around 1%, there was significant degradation in performance for dynamic or unstable environments, with EERs ranging from 23%–31%. Their inter-session analysis also revealed much higher EERs (up to 44%) when comparing swipes collected a week apart, highlighting the challenges in achieving persistent swipe-based authentication. The authors concluded that a minimum of 6 swipe samples are required in enrollment, and identified features like maximum slope and standard deviation as the most consistent across scenarios, suggesting the need for techniques like template selection to mitigate the impact of usage conditions.

Pryor et al. [14] evaluated a user authentication scheme using single-modal (touch dynamics) and multi-modal (touch dynamics and phone movement) approaches on the BioIdent and H-MOG datasets. They found that multimodal fusion improved authentication accuracy, reaching up to 97.9% using Random Forest, especially in scenarios with more user movement and device motion. In contrast, touch dynamics-only models performed comparably to the multi-modal models in more stationary usage scenarios, reinforcing that motion-based features from phone sensors are most valuable when user activity introduces measurable device movement.

The reviewed works consistently demonstrate that the reliability of CA systems is highly dependent on user context and temporal stability, with multimodal fusion and context-specific models proposed as key solutions. While these findings establish a clear direction, they also highlight the need for a comprehensive evaluation that assesses these factors concurrently. Accordingly, this paper we analyze a multimodal CA system across distinct scenarios (normal, stressful, walking) and multiple sessions to systematically investigate the intertwined effects of context and time, thereby providing a robust benchmark for real-world performance expectations.

3 Methodology

This section presents our experimental methodology, comprising dataset description, preprocessing pipeline, classification approaches, evaluation framework, and performance metrics. We detail our multi-modal continuous authentication dataset, feature extraction procedures, implemented algorithms, and assessment protocols to systematically evaluate authentication performance across contextual scenarios.

3.1 Data Description

We constructed a publicly available dataset [15] for continuous mobile authentication research comprising data from 50 participants. The experimental protocol required each participant to complete two sessions across three distinct usage contexts: seated/normal, ambulatory (walking), and stress-induced conditions. Data acquisition was conducted using a purpose-built Android application deployed on Google Pixel 6a devices, designed to emulate social media interaction patterns (see Fig 1). The dataset exhibits multi-modal characteristics, encompassing: (i) comprehensive touch and swipe event streams including spatial coordinates (x-y), temporal markers, pressure measurements, and contact surface area; and (ii) temporally synchronized inertial sensor data from tri-axial accelerometer, gyroscope, and magnetometer sensors. Each experimental session yielded approximately 120 swipe sequences with precisely aligned touch-sensor temporal frameworks, facilitating granular per-swipe analysis, contextual behavioral examination, and cross-session validation methodologies.

3.2 Pre-processing

The dataset comprises four synchronously recorded modalities: swipe gestures, accelerometer, gyroscope, and magnetometer data. Each swipe consists of a sequence of touch points characterized by a timestamp, x-y coordinates, pressure, and contact area, while each inertial measurement unit (IMU) sensor reading includes a timestamp and tri-axial (x, y, z) values. The raw data underwent a multi-stage preprocessing pipeline, beginning with type-checking, chronological sorting, and timestamp normalization. Missing values were addressed by removing completely null records and imputing partial entries using the median. Subsequently, a two-stage outlier removal process was applied, first using a univariate Interquartile Range (IQR) filter and then a k-nearest neighbors (k-NN) distance-based method to eliminate multivariate anomalies. Finally, to reduce noise and computational overhead while preserving signal fidelity, high-frequency motion sensor data (~400 Hz) was downsampled to 100 Hz via windowed averaging, aligning with best practices in mobile biometric studies [9, 16].

3.3 Feature Extraction

To create a comprehensive behavioral profile for each swipe, the raw time-series data was transformed into a detailed feature vector. First, we computed robust location and dispersion summaries mean, median, and standard deviation for kinematic (e.g., velocity, acceleration), mechanical, and geometric channels, as these statistics have repeatedly proved discriminative for touch-based continuous authentication [7, 17]. To capture distributional shape beyond first- and second-order moments, we added quartile-level descriptors by extracting the 25th, 50th, and 75th percentiles of each channel, an approach shown to improve model robustness [18, 19]. Spatial dynamics were characterized through trajectory features such as the overall duration, Euclidean distance, total trajectory length, and distance-to-trajectory-length ratio metrics that prior work has linked to individual motor control and device handling habits [5, 20]. Because stroke directions are angular measurements, we summarized them with circular statistics such as the circular mean and resultant length [19]. Finally, high-level counts and a directional flag were appended, reflecting findings that interaction frequency and stroke polarity carry user-specific cues [19, 21]. This procedure produced a rich yet low-dimensional feature set that preserves the behavioral characteristics most relevant for CA. This process resulted in a total of 54 features.

For sensor feature extraction, we used raw signals from the accelerometer, gyroscope, and magnetometer were segmented into discrete time windows. Within each window, we extracted features from both the time and frequency domains for each of the three sensor axes (x, y, z). In the time domain, we captured motion patterns by computing ten statistical measures per axis: mean, standard deviation, maximum, minimum, range, kurtosis, skewness, and the 25th, 50th, and 75th percentiles. These metrics quantify key distributional properties such as amplitude variation (range), peak sharpness (kurtosis), and asymmetry (skewness). In the frequency domain, a Fast Fourier Transform (FFT) was applied to derive five metrics that characterize the spectral content of user actions: energy, entropy, P1, P2f, and P2. This process resulted in a total of 135 features (3 sensors $\times$ 3 axes $\times$ 15 features per axis), providing a comprehensive representation of the user's motion [22].

3.4 Classification

Our classification methodology for distinguishing genuine users from impostors involved a two-stage pipeline designed for robustness and high performance. First, we applied a rigorous feature selection process, beginning with the removal of constant and duplicate features. To resolve multicollinearity, a model-driven approach was used for feature pairs with a correlation exceeding 0.9. For this, a Random Forest classifier was chosen due to its inherent robustness against overfitting and its ability to provide stable feature importance estimates without extensive tuning, a well-supported practice for variable selection [23]. The resulting refined feature set was used to train an eXtreme Gradient Boosting (XGBoost) model, selected for the final classification task due to its state-of-the-art performance on structured data [24] and its advanced regularization mechanisms (L1 and L2). These regularizers are critical for preventing overfitting with high-dimensional biometric feature sets, a common challenge in the field of continuous authentication. The model's hyperparameters were systematically optimized using a Bayesian search framework to maximize the mean Area Under the Receiver Operating Characteristic Curve (AUC) across a 5-fold stratified cross-validation. This entire training process was accelerated on a GPU and incorporated early stopping to ensure computational efficiency, culminating in a final model configured with the optimal hyperparameter set.

3.5 Evaluation

Our evaluation methodology was designed around a strict, user-disjoint protocol to ensure that all models were tested against completely unseen impostors. The total cohort of 50 participants was first partitioned into two sets: a "Training Data Cohort" of 35 users (70%) and an "Evaluation Cohort" of 15 users (30%). This split was stratified by age and gender to maintain demographic balance. A personalized binary classification model was then trained for each of the 50 participants. For any given user, the training process used 80% of their own samples as the positive class. The negative class was constructed from a large, consistent pool of samples comprising all data from the 34 other users within the Training Data Cohort. This "one-vs-all" training strategy ensures that while every participant receives a personalized model, the negative training data is always drawn from a source kept separate from the final evaluation impostors.

The evaluation of each of the 50 trained models was conducted using a dedicated test set designed to measure performance against both the legitimate user and unseen impostors. This set was composed of the remaining 20% of the genuine user's own samples, alongside all samples from the 15 participants in the separate Evaluation Cohort, which served as the negative data. This protocol guarantees a true zero-shot assessment, as no model was ever exposed to the evaluation impostor data during its training. System performance was quantified using two standard biometric metrics. The Area Under the Curve (AUC) provides a threshold-independent measure of a model's overall ability to distinguish between genuine and impostor samples. The Equal Error Rate (EER) identifies the specific operational point where the False Acceptance Rate (FAR) and False Rejection Rate (FRR) are equal, offering a practical measure of the system's accuracy.

4 Results

This section presents the performance of our continuous authentication system under various operational scenarios. All experiments were conducted using the filtering-based feature selection pipeline, and the performance is reported in terms of the Equal Error Rate (EER), where lower values indicate higher accuracy. The results, summarized in Table 1 and visualized in Fig. 1, are divided into two primary analyses: intra-session and inter-session scenarios.

Table 1. Median, mean, and standard deviation of the Equal Error Rate (EER) calculated across all users for various training and testing scenarios

Session	Train Scenario	Test Scenario	Median	Mean	Std
Intra-session	Normal	Normal	0.31%	6.32%	11.05%
	Normal	Stressful	***16.95%***	24.05%	22.11%
	Normal	Walking	***26.54%***	27.90%	20.40%
	All	All	***4.01%***	5.96%	5.88%
Inter-session	Normal	Normal	40%	36.68%	24.38%
	Normal	Stressful	42.98%	44.72%	24.63%
	Normal	Walking	40.53%	41.01%	23.48%
	All	All	***34.64%***	34.94%	22.98%

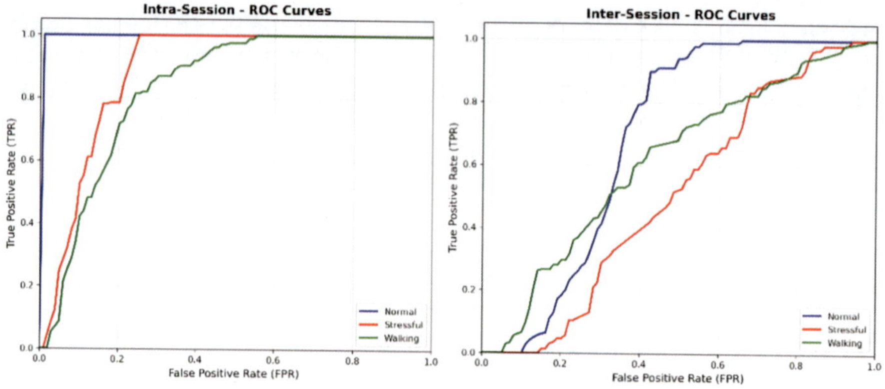

Fig. 1. Median ROC curves for intra- and inter-session classification

4.1 Intra-session Scenario Analysis

The intra-session analysis evaluates the system's baseline performance under the assumption that training and testing data are drawn from a mixed pool of all available sessions. This simulates a scenario where the biometric template is generalized across time. The results are presented in the upper half of Table 1.

The system achieved its highest performance in a homogenous context, yielding a median EER of 0.3% when both trained and tested exclusively on data from the "Normal" scenario. This result serves as a strong baseline, demonstrating the model's effectiveness when the operational context is consistent.

Performance degradation was observed when the testing context mismatched the training context. When the model was trained on "Normal" data and tested on "Stressful" or "Walking" data, the median EER increased significantly to 16.95% and 26.54%, respectively. This highlights the system's sensitivity to contextual variations, with mobility (Walking) having the most detrimental impact on performance. Notably, when the model was trained and tested on a combined dataset of all scenarios it achieved a robust median EER of 4.01%. This suggests that training the model on a diverse set of real-world contexts can make it nearly as effective as a model trained for a single, ideal context, while providing substantially better generalization.

4.2 Inter-session Scenario Analysis

The inter-session analysis investigates a more challenging and realistic scenario, where the system is trained on data from an initial session (Session 1) and evaluated on data collected in a subsequent session (Session 2). This protocol is designed to assess the system's robustness to template aging and natural variations in user behavior over time. The results are shown in Table 1.

A significant performance degradation was observed across all inter-session experiments compared to their intra-session counterparts. The best-case scenario, where the model was trained on "Normal" data from Session 1 and tested on "Normal" data from Session 2, yielded a median EER of 40%. This stark increase from the 4.01% intra-session EER indicates a substantial template aging effect, where the biometric patterns captured in the first session are less representative of the user's behavior in the second session.

When testing against mismatched contexts, the median EER remained high, at 42.98% for the "Stressful" scenario and 40.53% for the "Walking" scenario. Interestingly, the model trained on all scenarios from Session 1 and tested across all scenarios in Session 2 achieved the lowest median EER in this category, at 34.64%. This finding further reinforces that while temporal drift poses a major challenge, enrolling users with data from a variety of contexts provides the most resilient, albeit still challenged, model for long-term, cross-session authentication.

5 Discussion

Our findings reveal that while the fused biometric system is effective in stable conditions, its performance is highly vulnerable to both contextual variation and temporal drift. The intra-session analysis demonstrated a strong baseline performance, with a median EER of 0.3% in a consistent "Normal" context. However, this accuracy degraded significantly when testing against mismatched scenarios like "Walking" (26.54% EER), highlighting the system's sensitivity to the user's physical and cognitive state. Crucially, this issue was largely mitigated by training the model on a diverse dataset encompassing all scenarios, which yielded a robust EER of 4.01%. The more profound challenge, however, was template aging, as revealed by the inter-session analysis. The EER dramatically increased to over 40% when the model was trained on one session and tested on another, even under identical contexts. This stark decay confirms that user behavior is not static, and a model trained at one point in time quickly becomes outdated, posing a major obstacle to long-term reliability.

These results have critical implications for the practical deployment of continuous authentication systems. The sensitivity to context proves that enrollment must capture behavioral data across a variety of real-world situations to build a resilient user profile, rather than being a one-time event in a controlled setting. More importantly, the severe performance degradation due to template aging underscores the necessity of moving beyond static models. Future systems must incorporate online or continuous learning mechanisms that allow the biometric template to adapt and evolve with the user over time. While our study was limited to two sessions, it clearly indicates that future research must prioritize template adaptation algorithms and context-aware models to bridge the gap between promising laboratory results and a truly reliable real-world solution.

6 Conclusion

This study concludes that while multimodal continuous authentication using swipe and sensor data is promising, its practical viability is severely limited by contextual variation and, most critically, template aging. We demonstrated that context-related errors can be addressed by enrolling users with diverse data, achieving a robust median EER of 4.01%. However, the profound performance degradation observed across different time sessions with the median EER increasing to 40% reveals a more fundamental challenge: user behavior is not static. Therefore, the primary contribution of this work is the clear evidence that static biometric models are insufficient for long-term use. The future of practical and reliable continuous authentication must pivot from static enrollment to the development of adaptive systems capable of continuous learning and template updating to ensure both security and usability over the lifetime of a device.

Acknowledgments. This work was carried out with the support of the National Center for Scientific and Technical Research (CNRST) as part of the "PhD-ASsociate Scholarship - PASS" program.

References

1. Dybczak, J., Nawrocki, P.: Continuous authentication on mobile devices using behavioral biometrics. In: 2022 22nd IEEE International Symposium on Cluster, Cloud and Internet Computing (CCGrid), pp. 1028–1035 (2022)
2. Stylios, I., Kokolakis, S., Thanou, O., Chatzis, S.: Behavioral biometrics & continuous user authentication on mobile devices: a survey. Inf. Fusion. **66**, 76–99 (2021)
3. Senerath, D., Tharinda, S., Vishwajith, M., Rasnayaka, S., Wickramanayake, S., Meedeniya, D.: BehaveFormer: a framework with spatio-temporal dual attention transformers for IMU enhanced keystroke dynamics (2023) https://doi.org/10.48550/arXiv.2307.11000
4. Fereidooni, H., et al.: AuthentiSense: a scalable behavioral biometrics authentication scheme using few-shot learning for mobile platforms (2023). https://doi.org/10.48550/arXiv.2302.02740
5. Naji, Z., Bouzidi, D.: Deep learning approach for a dynamic swipe gestures based continuous authentication. In: Hassanien, A.E., Haqiq, A., Azar, A.T., Santosh, K., Jabbar, M.A., Słowik, A., Subashini, P. (eds.) The 3rd International Conference on Artificial Intelligence and Computer Vision (AICV2023), March 5–7, 2023. Lecture Notes on Data Engineering and Communications Technologies, vol. 164, pp. 48–57. Springer Nature Switzerland, Cham (2023)
6. Akiirne, Z., Naji, Z., Saoudi, A., Bouzidi, D.: Dynamic swipe gestures based continuous authentication using similarity learning. In: Proceedings of the 7th International Conference on Networking, Intelligent Systems and Security (NISS 2024), pp. 1–9. ACM, Meknes, Morocco (2024)
7. Gattulli, V., Impedovo, D., Pirlo, G., Volpe, F.: Touch events and human activities for continuous authentication via smartphone. Sci. Rep. **13**, 10515 (2023)
8. Cariello, N., Eslinger, R., Gallagher, R., Kurtzer, I., Gasti, P., Balagani, K.S.: Posture and body movement effects on behavioral biometrics for continuous smartphone authentication. IEEE Trans. Biom. Behav. Identity Sci. **7**, 3–15 (2025)
9. Alawami, M.A., Abuhmed, T., Abuhamad, M., Kim, H.: MotionID: towards practical behavioral biometrics-based implicit user authentication on smartphones. Pervasive Mob. Comput. **101**, 101922 (2024)
10. Bajaber, A., Fadel, M., Elrefaei, L.: Evaluation of deep learning models for person authentication based on touch gesture. Comput. Syst. Sci. Eng. **42**, 465–481 (2022)
11. Uslu, U., İncel, Ö.D., Alptekin, G.I.: Evaluation of deep learning models for continuous authentication using behavioral biometrics. Procedia Comput. Sci. **225**, 1272–1281 (2023)
12. Siirtola, P., Komulainen, J., Kellokumpu, V.: Effect of context in swipe gesture-based continuous authentication on smartphones (2019). https://doi.org/10.48550/arXiv.1905.11780
13. Ellavarason, E., Guest, R., Deravi, F.: Evaluation of stability of swipe gesture authentication across usage scenarios of mobile device. EURASIP J. Inf. Secur. **2020**, 4 (2020)

14. Pryor, L., Mallet, J., Dave, R., Seliya, N., Vanamala, M., Sowells-Boone, E.: Evaluation of a user authentication schema using behavioral biometrics and machine learning. Comput. Inf. Sci. **15**, 1 (2022)
15. Naji, Z., Bouzidi, D.: SwipeMotionDB: a contextual multimodal dataset for continuous mobile authentication using swipe gestures and motion sensors (2025). https://doi.org/10.5281/zenodo.17171888
16. Sitová, Z., et al.: HMOG: new behavioral biometric features for continuous authentication of smartphone users. IEEE Trans. Inf. Forensics Secur. **11**, 877–892 (2016)
17. Frank, M., Biedert, R., Ma, E., Martinovic, I., Song, D.: Touchalytics: on the applicability of touchscreen input as a behavioral biometric for continuous authentication. IEEE Trans. Inf. Forensics Secur. **8**(1), 136–148 (2012). https://doi.org/10.1109/TIFS.2012.2225048
18. Syed, Z., Helmick, J., Banerjee, S., Cukic, B.: Touch gesture-based authentication on mobile devices: the effects of user posture, device size, configuration, and inter-session variability. J. Syst. Softw. **149**, 158–173 (2019)
19. Pozo, A., Fierrez, J., Martinez-Diaz, M., Galbally, J., Morales, A.: Exploring a statistical method for touchscreen swipe biometrics. In: Proceedings of the International Carnahan Conference on Security Technology (ICCST 2017), pp. 1–4 (2017)
20. Mondal, S., Bours, P.: Swipe gesture based continuous authentication for mobile devices. In: Proceedings of the 2015 International Conference on Biometrics (ICB 2015), pp. 458–465 (2015)
21. Ray-Dowling, A., Hou, D., Schuckers, S., Barbir, A.: Evaluating multi-modal mobile behavioral biometrics using public datasets. Comput. Secur. **121**, 102868 (2022)
22. Li, Y., Hu, H., Zhou, G., Deng, S.: Sensor-based continuous authentication using cost-effective kernel ridge regression. IEEE Access. **6**, 32554–32565 (2018)
23. Genuer, R., Poggi, J.-M., Tuleau-Malot, C.: Variable selection using random forests. Pattern Recogn. Lett. **31**, 2225–2236 (2010)
24. Chen, T., Guestrin, C.: XGBoost: a scalable tree boosting system. In: Proceedings of the 22nd ACM SIGKDD International Conference on Knowledge Discovery and Data Mining, pp. 785–779. ACM, San Francisco (2016)

Author Index

A. E. Hassanien et al. (Eds.): AISDS 2025, LNDECT 286, pp. 355–356, 2027.
https://doi.org/10.1007/978-3-032-18758-1

Kolibri 360 GmbH
Ferdinand-Jühlke-Straße 7
99095 Erfurt, Deutschland
produktsicherheit@kolibri360.de